Handbook of Mobile Radio Networks

For a listing of recent titles in the *Artech House Mobile Communications Library*, turn to the back of this book.

Handbook of Mobile Radio Networks

Sami Tabbane

Artech House
Boston • London

Library of Congress Cataloging-in-Publication Data
Tabbane, Sami.
    Handbook of mobile radio networks / Sami Tabbane.
       p.   cm.—(Artech House mobile communications library)
    Includes bibliographical references and index.
    ISBN 1-58053-009-5 (alk. paper)
    1. Mobile communication systems—Handbooks, manuals, etc.   I. Title.
    II. Series.
  TK6570.M6 T33   2000
  621.3845—dc21
                                                99-054639
                                              CIP

British Library Cataloguing in Publication Data
Tabbane, Sami
    Handbook of mobile radio networks.—(Artech House mobile communications
    library)
    1. Mobile communication systems   2. Mobile radio stations
    I. Title
    621.3'845

    ISBN 1-58053-009-5

Cover design by Igor Valdman

© 2000 ARTECH HOUSE, INC.
685 Canton Street
Norwood, MA 02062

International Standard Book Number: 1-58053-009-5
Library of Congress Catalog Card Number: 99-054639

10 9 8 7 6 5 4 3 2 1

CONTENTS

CHAPTER 1

GENERAL INTRODUCTION

Over the past two decades, telecommunications has undergone significant and revolutionary changes. Since the end of the 1980s, the importance of mobile systems has grown prolifically—whether that growth is measured in terms of research, investment, income, users, or traffic.

There is no doubt that the mobile communications revolution has been one of the most important items in the field of telecommunications since the beginning of the 1990s. The key aspect of this revolution has been the transition from communication between two fixed and cable-linked access points to a wireless communication mode between mobile users. Although it is too early to predict the socioeconomic changes caused by mobile services, many observers believe that the mobility of telecommunications services will have a fundamental impact on the way of life and work habits of an increasing number of people.

Since the beginning of the 1990s, the demand for mobile services has experienced an exponential growth in many countries. It is likely that this demand will continue to increase, at least to the point where mobile radio systems are available to more than half the population of the developed countries. Mobile radio communications systems (see Figure 1.1), mainly voice, have a peculiarity, when compared with other technologies, that the demand always exceeds availability. Yet, unlike the market for microcomputers, for example, the usefulness of a mobile terminal to almost anyone irrespective of age or social class is evident.

The object of this introductory chapter is to give the reader a few points of reference within the general context in which mobile systems are developing. The historical background of these systems is presented in Section 1.1, together with some of the reasons for their rapid development. In Section 1.2, the major differences between fixed and mobile networks are identified. Section 1.3 focuses on the role of standardization and regulatory bodies. Finally, in Section 1.4, the content of the different chapters of this book is introduced.

1.1 HISTORY AND DEVELOPMENT

The history of mobile radio communications can be divided into three main phases. The first concerns the theoretical prediction and demonstration of the existence of radio waves. The second phase saw the development and evolution of both equipment and techniques, although their use was limited to certain sections of the population. The third and final phase is the provision of mobile radio communication services on a large scale for the general public.

1.1.1 Phase One: Theoretical Foundations

During this first phase, a kind of prehistory in the field of mobile radio services, the fundamental principles of radio phenomena were established, but systems were still experimental and limited to laboratory use.

This phase began in 1678 with the work of Huygens on the phenomena of light reflection and refraction. It was Fresnel who, in 1819, demonstrated the wave-like nature of light. In 1865, Maxwell established his famous equations unifying electrical, magnetic, and light phenomena, but it was not until 1887 that Hertz first demonstrated the possibility of transmitting and detecting an electromagnetic wave between two points a few meters apart. At the end of the century, in 1897, Ducretet succeeded in establishing radio contact over a distance of a few kilometers. But it was Marconi who, by building on his predecessors' work, brought radio out of the laboratory and gave it practical applications. Marconi recognized that longer waves propagate over a larger distance and used them to good effect, so that by 1896 his apparatus could provide communications of up to 3 km.

1.1.2 Second Phase: Development and Applications

Three years after demonstrating mobile radio communication between a ship and an island in 1898, Marconi set up the first transatlantic radio transmission between Europe and the United States. In 1901, he successfully demonstrated the transmission and reception of the Morse code for the letter *S* across the Atlantic Ocean between England and Newfoundland, over 3,000 km at an operating frequency of less than 1 MHz. The first commercial service started six years later, at frequencies less than 100 kHz, when Marconi established the first radio communications system to be used in maritime services.

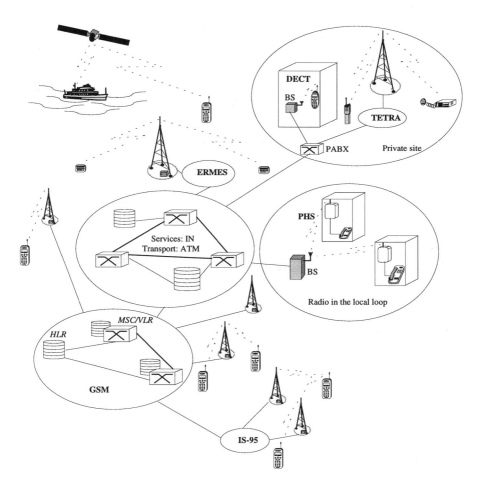

Figure 1.1 A few current mobile networks.

At the beginning of the 20th century, the size of transmitter/receiver radio stations was considerable. For example, in 1902, a military radio station used for telegraphy required a petrol-driven engine for the towing of the device, which itself consisted of a 1-kW generator mounted on a trailer and followed by another trailer for the transmitter and the receiver. The equipment was very far from the levels of integration and miniaturization of present-day systems.

It was by virtue of technological advances and improvements in equipment performance that radio systems would become truly mobile. The development of the diode by Fleming and de Forest, and Lieben's invention of the triode, among other developments, made it possible to reduce the size of the equipment and develop radio-telephonic services in place of previous applications that allowed only telegraphy services.

From the beginning of the 20th century, mobile radio services have enjoyed growing success in almost all areas, professional (e.g., administration, security, trade, health) and private. Their usefulness was recognized first by the maritime services, then by public security services such as police forces and fire brigades. Later, it caught on with the private sector, particularly in the energy and oil industries and by public transport and taxi companies. Take-up was most marked in the United States with an annual growth rate of over 20% in the second half of the 1950s. At the beginning of that decade there were only a few thousand users.

The extension of radio systems and the ability to define the bandwidth using filters led to spectrum control, first by local administrations, then by the World Administration Radio Conference (WARC, now the WRC). In 1906, the first WARC conference was held and recommendations were issued regarding the assignment of radio frequency bands.

The radio frequencies regulated at the beginning of the 20th century were those at the low end of the band. In 1912, for example, the WARC only regulated frequencies below 3 MHz, deeming higher frequencies commercially useless and leaving them to nonprofessional (e.g., amateur) radio systems.

From 1930 to 1960, the range of frequencies brought into service for radio communications gradually increased to include the metric, decimetric, and centimetric wavelengths (see Appendix 1A). Frequency modulation began to be used.

The Second World War accelerated the development of radio systems, and their use was extended into civilian life (e.g., taxi companies and ambulances) in the 1950s. However, the equipment remained heavy and occupied a large space in the vehicles in which they were installed.

1.1.3 Third Phase: Mobile Services for the General Public

In this phase, technical progress and the development of communications systems allowed mobile radio communications systems to reach the general public.

The first large-scale public communications networks were cellular systems (see Chapters 6, 7, and 12). Analog cellular systems were developed in the 1970s. In 1979, the first cellular system, called Advanced Mobile Phone Service (AMPS) was installed in Chicago, followed in 1980 by the High Capacity Mobile Telephone System (HCMTS) in Tokyo. During the 1980s, analog cellular systems became widespread in many countries. In 1981, the Nordic Mobile Telephone (NMT) was introduced in Scandinavia. In 1985,

France's Radiocom 2000 became operational, as did the United Kingdom's TACS and Germany's C 450 system.

At the same time, mass-produced, low-cost cordless systems reached impressive growth rates. Cordless telephones (Chapter 10) outnumbered fixed telephones in some countries, including the United States.

In the 1990s, paging systems (Chapter 11) achieved high growth rates, with penetration indices as high as 20% or 30% in places like Hong Kong and Singapore. However, it is GSM (Chapter 12), with its ISDN style of services and the availability of international roaming, that most significantly illustrates the mobile revolution of the 1990s.

1.1.4 Evolution of Mobile Systems

Since the beginning of the 1990s, the mobile communications market has developed beyond all expectations. In 1983, it was predicted that by the year 2000, fewer than ten million Americans would be using cellular phones [1]. In 1995, there were already more than 20 million.

Among all existing mobile systems (see the second part of this book), cellular systems constitute the main mobile radio communications market. There was approximately a 70% increase in the number of users in Western Europe in 1997, 50% in Northern America, 55% in Australia and Asia, and more than 200% in the largest Latin American markets. This rapid growth is due to a combination of the following factors:

- The deregulation of telecommunications with the advent of new operators. In many countries, the first telecommunications services to be affected by deregulation were the mobile services, as private operators were granted licenses. For them, radio systems were the quickest way of setting up a communications infrastructure at a reasonable price (with returns on investments within a few months in the best cases).
- The general improvement in the standard of living, noticeable in the demand for convenience in the area of communications, both in professional and in private activities.
- Economic activity has shifted towards third-world markets, whose activities are based on trade and communication. At the same time, the growing size of urban centers has made journey times in and out of these areas increasingly longer and more variable. The use of radio communications has therefore improved the productivity of professionals who spend time traveling.
- Technological progress has resulted in miniaturization and, above all, in the lower costs that have put radio communications within reach of a large number of people. Indeed, technology has generally been the limiting factor in the ability to offer mobile services.
- In developed countries, mobile systems enable companies to keep in touch with their employees which, at the same time, helps reduce operating costs. This, in turn,

allows operators to increase the number of telephone terminals and the volume of traffic.

- In developing countries, mobile systems offer the possibility of installing a telecommunications infrastructure necessary to the economic development of the country in only a few months, without having to install a fully wired infrastructure (see WLL systems for an example in Chapter 10).

Current (second half of 1998, Table 1.1) and predicted figures (see Figure 1.2) for the number of subscribers to cellular telephone networks around the world are presented below.

Table 1.1

Cellular services penetration rates in different countries [2]

Country	Penetration rate
Japan	24 %
Sweden	48 %
U.K.	19 %
Italy	31 %
Finland	55 %
Denmark	32 %
Hong Kong	35 %
Australia	28 %
Germany	16 %
Portugal	26 %

The most striking figures are those for the penetration indices of cellular services in Scandinavia.

By the turn of this century, the total number of wireless users is expected to be in excess of 200 million, compared with a number for fixed line systems of 700 million. According to some forecasts, the number of mobile phone subscribers may reach one billion by 2010, surpassing fixed phone lines [3].

Figure 1.2 shows demand estimates in cellular services up to the year 2001 [4].

1.1.5 Equipment Development

The first mobile radio systems used devices containing vacuum tubes requiring powerful batteries. Before 1930, only receivers could be mobile, allowing one-way communication.

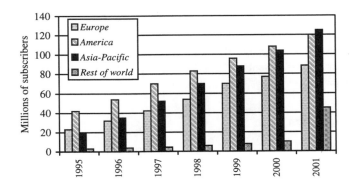

Figure 1.2 Estimated progression of the demand in cellular services.

After World War II (during which radio communications technology made rapid advances), it became possible to produce transmitters and receivers fitted in vehicles that operated in the VHF band (see Appendix 1A). Later, in the early 1950s, radio equipment became small enough to be carried around. The main applications of the first portable equipment were military communications. From 1957 onward, vacuum tubes were progressively replaced by transistors. Reduction in device size by over 50%, lower power consumption, and better reliability resulted. The first portables appeared in the 1960s. As a result, military applications gave way to civil applications.

The technology of digital integrated circuits that makes it possible to manufacture frequency synthesizers did not lead immediately to the design of new systems. It was not until a few years later, with the development of the microprocessor, that significant progress in the field of radio equipment was achieved; the new terminals were composed of a transmitter, a receiver, a frequency synthesizer, a modem, and a microprocessor.

The next breakthrough occurred with the inception of integrated circuits in the mid-1970s. Today, Large Scale Integration (LSI) and Very Large Scale Integration (VLSI) circuits are part of all kinds of mobile radio equipment, making them lighter, smaller, and cheaper.

The most important changes occurred at the level of component integration. From about 1,000 components per terminal in 1992, the number fell to half that figure five years later. In the year 2000, this figure will probably not exceed 100. Moreover, the present trends show a yearly 25% price reduction combined with a reduction in size and weight in similar proportions. At the same time, the battery life between charges has gone up by 50%. To give just one example, the size of the first GSM terminal was 350 cm^3 in 1992. Five years later, its volume was brought down to 100 cm^3, and it is predicted that it will reach 50 cm^3 before the end of the century [5].

1.2 DIFFERENCES BETWEEN FIXED NETWORKS AND MOBILE NETWORKS

The use of a radio interface in communications introduces a certain number of differences as compared with cable communication. In this section, some of the problems linked to the use of radio channels are presented. They will be further developed in the course of the first part of this book.

1.2.1 Limited Spectrum

The radio spectrum, and consequently the available capacity for radio access, is generally limited by national and international regulations. Unlike in fixed communications, where a growing population and demand can easily be satisfied by the installation of extra cables to connect subscribers to the network, the capacity of the radio spectrum cannot be extended arbitrarily. This problem is partly solved by use of the cellular technique, in which the service zone is divided into cells, each of which is served by a base station. The radio spectrum is used again and again as many times as possible in other geographic locations. The cellular technique is not simple to implement, however, as shall be seen in Chapters 6 and 7.

The limitation on spectrum has also led to the use of spectrally efficient modulation schemes and source compression coding, which eliminates irrelevant data so as to reduce the throughput to be transmitted. The objective of designers and operators of mobile networks is to transmit as much information as possible through fixed bandwidth channels.

1.2.2 Fluctuating Quality of Radio Links

Whereas in a wired network, transmission links have a constantly high quality, the radio link is subject to many problems due to the movement of its users, as well as changes in the environment (obstacles, moving reflectors, and various types of interferers). In a digital link, these problems will cause a fluctuation in Bit Error Rate (BER) on the radio link (see Chapter 2).

1.2.3 Unknown and Variable Access Points

Unlike wired connections where users communicate via fixed access points on the network, access to a mobile radio network allows users to change their access point dependent on their location, even during a single connection. This is because mobile access operates in two ways. First, the network allows the system to locate a subscriber wherever he or she is within the network (location management), and, second, the network allows communications to go on uninterrupted when the subscriber moves from one access point to another in the course of a call (handover function). Cellular systems are the major users of these functions, which are discussed in Chapter 8.

One last aspect related to the use of mobile radio channels is that as a broadcast medium. In other words, potentially anyone can tune in and listen to a communication. It is therefore necessary to implement security functions that ensure the confidentiality of communication, and the authentication of terminals attempting to gain access to the system. These aspects are examined in Chapter 5.

1.3 MANAGEMENT OF THE SPECTRUM AND STANDARDIZATION

Radio spectrum is a limited resource shared by many users (e.g., television and radio broadcasters, telecommunications operators, the armed forces, government, companies, general public, amateurs). The coordination and administration of the allocation of spectrum among these different users is performed at national and international levels. Therefore, national administrations (e.g., the FCC in the U.S.) must collaborate with users and industry to establish rules and procedures for the planning and use of frequency allocations. These allocations are in turn based on coordinating agreements reached at an international level.

1.3.1 International Organizations

The setting up of the first international telecommunications links within Europe required the conclusion of bilateral and later multilateral agreements. The International Telegraphic Union was set up as early as 1865. It soon acquired worldwide recognition and later became the International Telecommunications Union (ITU). ITU is a specialized United Nations agency whose headquarters are in Geneva, Switzerland. It has more than 150 members of government in charge of all the policies related to radio, telegraphy, and telephony. At ITU, the world is divided into three regions: region 1: Europe, Africa, and the Middle East; region 2: America and Greenland; region 3: Asia and Australasia. Historically, these three regions have developed more or less independent frequency plans that best address their local needs.

The ITU comprises three branches: ITU-R is in charge of radio communications, ITU-T is in charge of nonradio standardization, and ITU-D is in charge of telecommunications development. In terms of radio systems, the basic tasks of this organization are the management of spectrum, and the regulation of the use of radio communications systems to minimize interference among radio stations of different countries.

The radio communications branch (ITU-R) supervises the activities of the World Radiocommunications Conference (WRC) and the International Frequency Registration Board (IFRB) and its specialized offices. It examines the technical and operational issues concerning radio communications systems, a task previously performed by the International Radiocommunications Consultative Committee (CCIR). The role of ITU-R is to organize the WRC with the help of other international organizations. ITU-R deals with technical, operational, and regulatory problems regarding radio communications, with the exception of interconnection protocols between radio communications systems and public networks.

Another task is the standardization of the characteristics of radio communications systems. The results of its activities are summarized in the form of reports and recommendations published every four years.

The IFRB acts as a controller of the radio resource. It registers radio frequencies and provides technical information to members. Drawing on the work done by ITU-R, in the course of WRC meetings, the IFRB and national administrations work on agreements for the use of frequencies and their planning, and provide documentation accordingly. These agreements are followed by actions at a national level.

ITU-T organizes the World Telecommunications Standardization Conferences (WTSC), in collaboration with the Telecommunications Standardization Bureau (TSB). It controls studies of the standardization of signals, systems, networks, and telecommunications services, and has taken over the activities of the former International Telegraph and Telephone Consultative Committee (CCITT).

1.3.2 European Organizations

The European Conference of Postal and Telecommunications Administrations (CEPT) coordinates the regulatory bodies in Europe. Radio matters are dealt with by the European Radio Committee (ERC), which comprises three permanent study groups. The ERC issues three types of documents: reports that are studies concerning a particular subject, recommendations whose application is not compulsory, and enforced recommendations that signatory countries are obliged to implement. It works more and more with the European standardization body, The European Telecommunications Standards Institute (ETSI). For example, in 1982, it was CEPT that decided to initiate the study of a pan-European mobile communications system (GSM; see Chapter 12).

Created in 1987, ETSI includes radio operators, manufacturers, and administrations. This organization is in charge of standardization, and its work involves defining technical specifications for radio systems operated in Europe. To manage its standardization program, ETSI has created technical committees (TC) that have in turn created technical subcommittees (TSC), where the work of development, compilation, and preservation of European Telecommunications Standards (ETS) as well as ETSI technical reports is effectively carried out. Seven categories of members participate in ETSI activities. These are administrations, manufacturers, network operators, users, providers of services in member countries of CEPT as well as associate members from non-European countries, observers without the right of vote, and advisors. ETSI has already published a large number of equipment specifications, among which the most widely known are GSM, DECT, ERMES, TETRA, and HIPERLAN.

1.3.3 Some U.S. Organizations

The American National Standards Institute (ANSI) Committee T1 on telecommunications was created in response to the breaking up of the national Bell Telephone Company system in 1984, which meant that de facto standards would no longer be available. T1 membership comprises four types of interest groups: users and general interest groups, manufacturers, interexchange carriers, and exchange carriers [6].

Committee T1 structure has a number of functionally oriented technical subcommittees: performance and signal processing; network interfaces and environmental considerations; interwork operations, administration, maintenance, and provisioning; systems engineering, standards planning, and program management; services, architecture, and signaling; digital hierarchy and synchronization. In addition to serving as the forum that establishes ANSI telecommunication network standards, Committee T1 technical subcommittee has the job of drafting U.S. candidate technical contributions to the ITU.

The other important standardization body in the United States is the Telecommunications Industry Association (TIA). The TIA is a full-service trade organization that provides its members with numerous services, including government relations, market support activities, educational programs, and standard-setting activities. Its technical committee was, among a total of six committees, accredited by ANSI in the United States to standardize telecommunications products. Technology standardization activities are reflected by TIA's four product-oriented divisions: user premises equipment, network equipment, mobile and personal communication equipment, and fiber optics.

Finally the Institute of Electrical and Electronics Engineers (IEEE) is a scientific organization that has been particularly active in developing standards. IEEE standards boards are responsible for standards adoption within the IEEE. Proposed standards are developed in technical committees of the IEEE societies.

1.4 CONTENTS OF THE BOOK

This book is made up of two main parts. The first part introduces the basic techniques and concepts of terrestrial mobile radio systems, and the second part contains a description of the most important present-day systems.

1.4.1 First Part (Chapters 2-8)

In this first part, a view of the essential features characterizing terrestrial mobile radio systems is presented along with problems encountered.

In Chapter 2, a simple description of the different characteristics of the mobile radio environment and their effect on mobile radio links is given. The most important models used to predict signal propagation are introduced at the end of this chapter.

In Chapter 3, multiple access techniques, how bandwidth is shared among several users, are defined. The main random access protocols that allow a mobile to acquire a radio link accessible by several mobiles at a time are also explained.

Chapter 4 recalls the different elements of a radio modem and indicates how they are used to eliminate transmission problems. Space-Division Multiple Access (SDMA) is introduced. The main advantage of this technique, primarily designed for cellular systems, is to overcome interference problems.

Chapter 5 deals with security, one of the most important problems linked to the use of the radio resource. This concerns the protection of a system against intruders or fraudulent users and the protection of the confidentiality of user communications. Steps taken to address these different problems are specified.

The cellular concept, defined in Chapter 6, makes it possible to implement frequency reuse so as to serve an almost limitless number of subscribers over an almost limitless area. The basic principles underlying cellular systems are explained in detail in this chapter. Techniques that allow the improvement of intrinsic capacity of cellular systems, such as power control and frequency hopping, are discussed.

The design of cellular systems in terms of engineering and cell planning, along with the steps in this process, are the themes of Chapter 7.

Relative to the fixed network, cellular radio generates a vast amount of signaling and complex management overhead. Two important functions must be provided using all this information if true mobility is to be achieved. The first one is *handover*, and the second is location management, or *roaming*. These two processes are explained in detail in Chapter 8.

1.4.2 Second Part (Chapters 9-14)

The second part is devoted to terrestrial mobile radio systems. Five types of systems are identified. These are professional radio communications systems, cordless systems, paging systems, cellular networks, and data transmission systems. Several digital system standards are covered, particularly TETRA, DECT, PHS, ERMES, GSM, IS-95, PDC, IEEE 802.11, HIPERLAN, MOBITEX, and CDPD.

Chapter 9 is devoted to the oldest of the mobile telecommunications systems, namely private mobile radio (PMR). The two principal PMR families, conventional systems and trunked systems, are described. A description is given of the very particular context in which these systems, intended for professional applications, are designed and what sets them apart from other types of systems.

Cordless systems are undoubtedly the most popular bidirectional radio systems, because they are accessible to the largest number of people. Such systems are covered in Chapter 10 with fixed radio distribution systems, or wireless local loop (WLL), which is gaining more interest in developed countries, as well as in countries with a poor telecommunications infrastructure.

The simplest mobile radio service consists of broadcasting messages in one direction only. This service is offered by paging systems. Being the simplest systems in terms of management and operation, their appeal continues to grow, especially among young people. The most important paging systems are described in Chapter 11.

Chapter 12 gives a brief description of the most widely available cellular systems. Because the problems linked to the management of these systems are dealt with in the first part of this book, this chapter limits its scope to the description of the main characteristics of these systems, namely their architecture and radio interface.

Data transmission imposes constraints that are different from those imposed by voice transmission systems. In Chapter 13, data transmission systems operating over large areas (a region or even a country), as well as local systems (wireless local area networks), are examined.

Chapter 14 concludes with a look at the future of mobile radio services and systems such as wireless ATM and "software radio."

APPENDIX 1A
FREQUENCY BANDS AND SOME SERVICES AND APPLICATIONS

Frequency

Band	Applications
Infrared optical domain	
3,000 GHz	
Micrometric waves	Remote sensing, laser communications, optical space communications
300 GHz	
Extremely high frequency (EHF) millimetric band	Radar, experimental, radio astronomy, simple radio communications
30 GHz	
Super high frequency (SHF) centimetric band	Microwave links, satellite communications
3,000 MHz	
Ultra high frequency (UHF) decimetric band	TV broadcasting, cellular systems, aircraft and weather radar, Cordless telephones, radio communications for personal use
300 MHz	
Very high frequency (VHF) metric band	TV and FM radio broadcasting; shipping and aircraft communications; radio paging, mobile terrestrial communications, PMR
30 MHz	
High frequency (HF)	Shortwave broadcasting; shipping and aircraft communications
	CB (citizen band), Amateur radio
3000 kHz	
Medium frequency (MF)	Medium wave broadcasting, marine and aviation guidance beacons Loran for navigation of ships and aircraft
300 kHz	
Low frequency (LF)	Navigation of ships and aircraft Marine and aviation guidance beacons
30 kHz	
Very low frequency (VLF)	Navigation, sonar, aircraft
3 kHz	
Extra low frequency (ELF)	Military and submarine communications

REFERENCES

[1]   Zysman, G., "Les réseaux radiotéléphoniques," *Pour la Science*, No. 217, Nov. 1995, pp. 44–48.

[2]   *Mobile Communications International*, June 1998.

[3]   Shafi, M., et al., "Wireless Communications in the Twenty-First Century: A Perspective," *Proceedings of the IEEE*, Vol. 85, No. 10, pp. 1,622–1,638, Oct. 1997.

[4]   *Mobile Communications International*, Feb. 1997.

[5]   Kuisma, E., "Technology Options for Multimode Terminals," *Mobile Communications International*, p. 71–74, Feb. 1997.

[6]   Dimolitsas, J., "Standards Setting Bodies," *Mobile Communications Handbook*, Editor-in-Chief Jerry D. Gibson, CRC Press, 1996.

SELECTED BIBLIOGRAPHY

[1]   Goodman, D. J., "Trends in Cellular and Cordless Communications," *IEEE Communications Magazine*, pp. 31–40, June 1991.

[2]   Kucar, A., "Mobile Radio: An Overview," *IEEE Communications Magazine*, Nov. 1991, pp. 72–85.

[3]   Lee, W., "*Mobile Communications Design Fundamentals*," J. Wiley, New York, 1993.

CHAPTER 2

PROPAGATION IN A MOBILE RADIO ENVIRONMENT

The mobile radio transmission channel is one of the most variable of all communications media. Radio waves, as they propagate through space, experience multiple corruption due to the morphology, temperature, and humidity of the environment through which they are traveling. As a consequence, unlike transmission on a fixed link (e.g., copper or fiber-optic cable) where the characteristics of the medium are very well controlled and easily predicted, mobile radio transmissions suffer large fluctuations in both time and space. For these reasons, the mobile radio channel is extremely difficult both to control and to model.

The aim of this chapter is to introduce the basic parameters that characterize the behavior of the mobile radio channel. The intention is to introduce the terminology used when describing the behavior of this medium. These basic definitions enable the presentation, in later chapters, of the impact of channel properties on system implementation (e.g., multiple access, protection against propagation impairments, handover management, planning process, power control). A few simple examples have been chosen to illustrate these channel impairments and their most significant effects. Several books referred to at the end of this chapter deal with these topics in a more detailed and theoretical manner. In these references, the reader can find more complete descriptions and explanations dealing with the characteristics of the mobile radio channel.

The first part of this chapter shows some of the basic concepts of antennas. The second part describes principal effects experienced by the mobile radio channel. Finally, the third part introduces the most widely used radio coverage prediction methods.

2.1 ANTENNA BASIC ELEMENTS

Radio waves are a component of the electromagnetic spectrum. A radio signal can be represented as a progressive electromagnetic wave (Figure 2.1). It induces an electric field component that allows the measurement of the power density of the radiated wave.

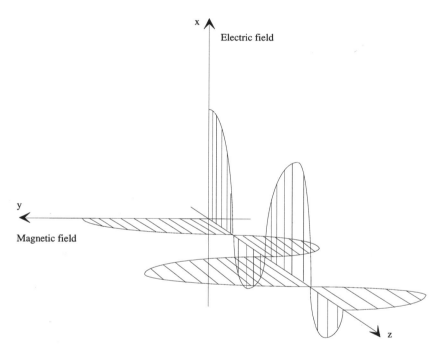

Figure 2.1 Electromagnetic wave (electric and magnetic components).

An electromagnetic wave propagates along a given direction. The concept of antenna *directivity* is based on this property. Generally speaking, a *radio antenna* is a structure, usually made from a good conductor, associated with the interface between the transmitted (or received) wave connected to the transmitter (or receiver), and the wave propagating through the air. The shape and size of an antenna is designed such that it efficiently radiates or receives electromagnetic energy. Its basic function consists in linking this energy between the *ether* and a device such as a transmission line, a coaxial cable, or a wave-guide.

2.1.1 Principal Antenna Characteristics

An antenna consists of one or several radiating elements through which an electric current circulates. In a wire antenna, for example, the current oscillates vertically or horizontally along the structure (vertical or horizontal) and transmits radiation concentrated in the horizontal (vertical) plane of the antenna. The antenna has a characteristic called *reciprocity,* which enables it to receive and transmit with similar characteristics at a given frequency.   Generally, antennas are classified as *omnidirectional, directional, phased arrays, adaptive,* and *optimal.* The first three of these are described in the following section. The principal characteristics used to characterize an antenna are: *radiation pattern, directivity, gain,* and *efficiency.*

2.1.1.1 Radiation Pattern

The *radiation pattern* of an antenna is defined as the relative distribution of electromagnetic power in space. More precisely, the IEEE organization defines the antenna radiation pattern as "the graphic representation of the radiation characteristics of the antenna according to space coordinates. These characteristics include the radiation intensity, the field power, the phase, and the polarization."

The radiation pattern identifies the field or power variations according to two spherical coordinates, θ and ϕ. The corresponding patterns represented by a three-dimensional shape are obtained after measuring and recording several two-dimensional plots. Note that in most applications, measurements are achieved for only a small number of points on the diagram as a function of ϕ, and for particular values of θ (Figure 2.2). Many shape types such as *isotropic, directional,* and *omnidirectional* can be identified in accordance with the shape of the radiation pattern. To radiate efficiently, the antenna size must be comparable to or greater than the wavelength of the signal to be transmitted or received [1].

2.1.1.1.1 Isotropic Antenna

"Isotropic" refers to a theoretical point source antenna, and it uniformly radiates power in all directions. While ideal, it is impossible to realize in practice, but it is often taken as a reference for real antennas.

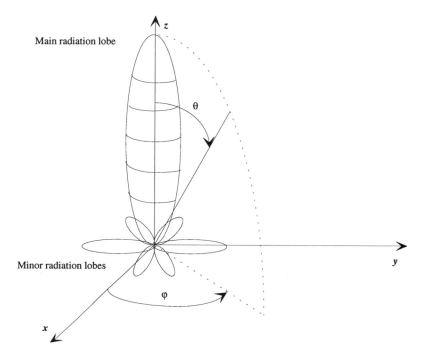

Figure 2.2 Example of an antenna radiation pattern.

2.1.1.1.2 Directional antennas

A *directional* antenna transmits or receives waves more efficiently in certain directions than in others. This antenna has a horizontal and vertical heterogeneous radiation pattern, and is used, for example, in sectored cells (see Chapter 6). Directional antennas are used for the extension of radio system coverage and for frequency reuse purposes. The gains that can be obtained from these types of antenna are typically between 9 and 16 dB (see Appendix 2A).

2.1.1.1.3 Omnidirectional antennas

In case of an *omnidirectional* antenna, the radiation pattern is essentially nondirectional in azimuth and directional in height. The horizontal radiated power can be increased by implementing several dipoles over the same vertical axis, such that power and phase addition can be obtained from each dipole. The typical gain values of such antennas are between 6 and 9 dB. Each doubling of the number of dipoles produces an increase in gain of 3 dB. The principal limit to this increase is the size of the antenna array. Omnidirectional antennas are used in mobile stations to receive and transmit so that users are not obliged to orient their antenna in specific directions.

2.1.1.2 Directivity

"Directivity" is used to describe an antenna that does not radiate uniformly in all directions, and the variation of its' signal intensity. The concept of directivity indicates that an antenna can concentrate the energy in specific directions. The value D (measured in dBi, i.e., relative to an isotropic antenna) of antenna directivity is defined as the ratio between the power of the transmitted radiation in a particular direction to that of the reference antenna [2].

2.1.1.3 Effective Area

The effective area or *aperture* (also called *antenna cross-section* or *capture area*), represented as A_e, is a notional surface that is slightly different from the geometrical surface which, in the case of a receiving antenna, collects the electromagnetic radiation. In the case of a directional antenna such as a microwave dish, the effective area is approximately the area of the dish perpendicular to the direction of maximum radiation.

2.1.1.4 Polarization

Consider a linear antenna such as the one represented in Figure 2.3. In this case, it is the orientation of the electric field $\vec{E}$ that determines the polarization of radiation from the antenna. If the antenna is placed parallel to the ground, $\vec{E}$ is horizontal and the antenna is said to have a *horizontal polarization*. If the antenna is vertical, it is said to have *vertical polarization*. When the polarization is vertical, the radiated energy along the xx' axis is almost nil, and the radiation in the plane P perpendicular to xx' is at a maximum.

2.1.1.5 Gain

The *power gain* or *antenna gain*, expressed in dB, is defined as 4π times the ratio of the radiation intensity in a given direction to the total power delivered to the antenna. The gain is related to the *directional gain D* by the relation $G = \eta.D$, where η is the efficiency factor of the antenna.

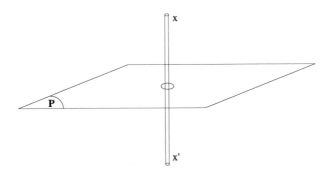

Figure 2.3 Linear antenna with vertical polarization.

The power P_S of the signal collected at the receiver antenna is an important parameter, as it must be greater than the power of the noise P_N (in dBm). The signal-to-noise ratio, expressed in dB, is $(S/N) = P_S - P_N$, where P_S and P_N are normally expressed in dBm. An improvement of this ratio can be obtained, for example, through the use of a high-gain antenna.

Table 2.1

Some typical antenna characteristics

Antenna	A_e	D
Short dipole	$3\,{}^2/8$	1.5
/2 dipole	0.1306. 2	1.64
Isotropic	${}^2/4$	1

The antenna gain and the effective area are related by the following expression:

$$G = \frac{4\pi}{\lambda^2} A_e \qquad (2.1)$$

For example, for a short dipole (see 2.1.2.1), $G = 1.5$ and $A_e = 3\lambda^2/8\pi$ (see also Table 2.1).

2.1.2 Common Antennas

The commonly used antennas (e.g., cars, TV, satellite) represent only a small number of available antenna systems. For instance, specialized and high-performance communication links, radar systems, and scientific experiments require the design of highly complex antenna systems [1]. Actually, there is an almost endless variety of structural shapes that can be used for an antenna. In practice, however, only simple and economical structures are designed. Among the most commonly used antennas, we introduce four main groups: *wire* antennas, *aperture* antennas, *array* antennas, and *reflector* antennas.

2.1.2.1 Wire Antennas

Wire antennas are the most widely used, as they are implemented for different types of use in vehicles, buildings, ships, planes, and so on. Wire antennas exist in different forms, but the most usual are rectilinear antennas (dipole antennas), loop antennas, and helical antennas (the latter is widely used on cellular handsets to minimize the physical size). A combination of these different shapes is also possible (Figure 2.4).

The dipole antenna is often considered as a reference antenna in mobile radio systems because it is the least complex. Loop antennas are not always circular and can take any other shape: square, rectangle, ellipse. Folded (looped) dipoles are commonly used at base station sites, as their characteristic impedance can be easily matched to feeder cables. Omnidirectional base station antennas often consist of several dipoles assembled one above the other on the same mounting. Mobile stations are often equipped with quarter wave ($\lambda/4$) monopole antennas.

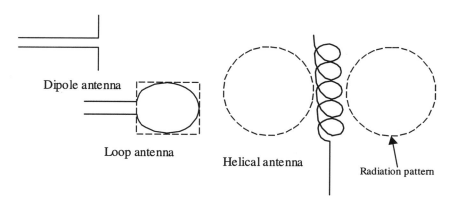

Figure 2.4 Examples of linear antennas.

2.1.2.2 Aperture Antennas

"Aperture" antennas consist of a group in which radiation is considered to emanate from an aperture. The commonly used antennas are of two types: the *parabolic* reflector antenna and the *horn* antenna (Figure 2.5).

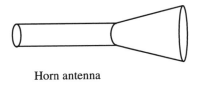

Horn antenna

Figure 2.5 Aperture antenna example.

An aperture-type antenna must have an aperture length and width of at least several wavelengths to obtain a high gain. Aperture antennas are used for transmissions at high frequencies. This type of antenna is often used for satellite communication and radar.

2.1.2.3 Array Antennas

Combining several dipoles (or other elementary radiators) into an array may be used to produce a directional radiation beam. Array antennas are designed for special cases that require radiating characteristics that cannot be obtained from a single element.

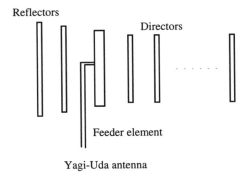

Yagi-Uda antenna

Figure 2.6 Array antenna example.

By using a set of radiating elements arranged in a given geometrical and electrical way (array), it is possible to obtain particular and adjustable radiation characteristics. The arrangement of the array can be made such that the radiation patterns of each element are added to one another, to produce a maximum radiation pattern pointing in one or several given directions, and a minimum pattern in the others.

Yagi antennas (Figure 2.6) are, for example, very commonly used for terrestrial television reception, and sometimes used in PMR systems to supply intersite links. In cellular systems, it is very common to have sectored cells where each antenna covers an arc of 120 degrees. These antennas are usually phased arrays of dipoles mounted in front of a panel reflector. By careful spacing of the dipoles, varying degrees of electrical downtilt can be built into the radiation pattern in order to assist in cellular planning by restricting the maximum propagation distance, and also increase the ground-level field strength at median distances.

2.1.2.4 Reflector Antennas

Reflector antennas consist of a source radiating over a reflecting area. These antennas are widely used in space communications (thousands to millions of kilometers communication range). The parabolic reflector is the most common shape for these antennas (Figure 2.7).

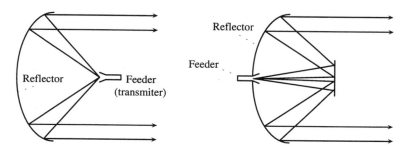

Figure 2.7 Reflector antenna examples.

2.1.3 Coupling Loss Between Antennas

When multiple antennas are mounted on a tower at a base station site, coupling between antennas supplying different services or coupling between transmit and receive antennas for the same service can present a problem.

The coupling loss (expressed in dB) between two close antennas is defined as the ratio between the power fed at the input of one of the antennas, and the power received at the output of the other. Being close, the coupling loss between the two fixed antennas is a fixed parameter. Transmit to receive coupling degrades the quality of the received signal by introducing wideband transmitter noise and/or receiver blocking, which effectively reduces the receiver sensitivity. Coupling between channels that transmit different signals can cause intermodulation products (see definition in 2.3.2.3), which may inject spurious signals into the system and again desensitize the receiver.

Generally speaking, antenna coupling loss can be classified as one of the following three types.

2.1.3.1 Vertical Coupling Loss

To obtain a vertical coupling loss with both antennas vertically polarized, the two antennas are installed on a vertical axis, one above the other. The value of the coupling loss depends on the distance between the two antennas. For example, at 900 MHz, a 30-dB coupling loss can be obtained with a distance of about 30 cm (say, about 2 wavelengths) between the antennas.

2.1.3.2 Horizontal Coupling Loss

In this case, if the same two antennas are placed in the same horizontal plane, a greater distance is needed for the same coupling loss than in the previous case. At 900 MHz, a 30-dB coupling loss requires a separation between the two antennas of about 1.5m.

2.1.3.3 Oblique Coupling Loss

To generate an oblique coupling loss, the antennas are placed at an angle one to the other. The coupling loss is at its greatest as the angle between the antennas approaches 90 degrees.

2.1.4 Parameters to Be Specified When Designing or Selecting Antennas

Electrical parameters:

- Gain in dB relative to an isotropic radiator (dBi) or half-wave dipole (dBd),
- Voltage standing wave ratio (VSWR),
- Radiation pattern, specify the beamwidth in the azimuth and elevation planes,
- Front-to-back ratio on directional antennas,
- Input power,
- Intermodulation performance,
- Bandwidth.

Mechanical:

- Structural design of antennas and supports,
- Electrolytic contact potentials,
- Conformity to severe environmental specifications.

2.1.5 Power (Link) Budget

Figure 2.8 represents the gains and losses budget between the transmitter and the receiver [3].

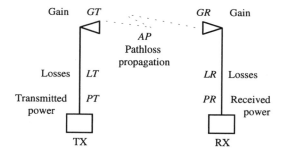

Figure 2.8 Losses and gain budget between a transmitter and a receiver.

The received power *PR* can be expressed as follows:

$$PR = PT - LT + GT - AP + GR - LR \tag{2.2}$$

where *LT* (dB) represents the losses between the transmitter and the transmitting antenna, and *LR* (dB) represents the losses between the receiver and the receiving antenna.

The losses *LT* and *LR* are mainly due to the *feeders* and *connectors* (transmission lines linking the antenna to the radio transmitter) and to the transmission/reception equipment (e.g., filters, duplexers, attenuators). In order to reduce the losses *LT* and *LR*, feeders can be shortened, if possible. The different kinds of losses, represented by the parameter *AP*, are dealt with in the Sections 2.2. and 2.3.

In Chapter 6, the principles of power budget calculation are detailed.

2.2 MOBILE RADIO PROPAGATION

Radio wave propagation is affected by the following mechanisms:

- Reflection (Section 2.2.4),
- Diffraction (Section 2.2.5),
- Scattering that occurs when the medium through which the waves travel contains objects with dimensions that are small compared with the wavelength, and the number of objects per unit volume is quite large.

During its propagation between transmitting and receiving antennas, the radio signal experiences deleterious effects due to a number of phenomena. This degradation affects the signal quality, and induces errors in received messages that leads to a loss of information for either the user or the system. The signal degradation resulting from propagation in the mobile radio environment can be classified by type. The principal ones are gathered under the so-called three-level model. They are

- *Pathloss* due to the distance covered by the radio signal;
- Signal attenuation resulting from *shadowing* effects introduced by obstacles between the transmitter and receiver;
- *Fading* of the signal caused by numerous effects, all of which are related to the multipath propagation phenomenon.

The three-level model does not take into account external jamming caused by other radio transmissions. Therefore, to these three basic effects, the following have to be included which also have an impact on signal behavior:

- Jamming caused by interference (*co-channel* or *adjacent* channel interference) resulting from other transmissions. Avoiding this kind of impairment is of great importance when designing frequency reuse based systems such as cellular,
- Jamming resulting from ambient noise as opposed to direct interference; for example, induced from other radio transmitters or motor vehicle ignition systems.

Propagation characteristics differ with the environment through and over which the radio waves travel. In general, several types of environment can be identified and are classified according to the following parameters:

- Terrain morphology,
- Vegetation density,
- Buildings, density and heights,
- Open areas,
- Water surfaces.

There are four classes of environment that are generally accepted and characterized (see Table 2.2).

Table 2.2

Different types of radio propagation environments

Environment type	Description
Dense urban	A central business area that consists of many high, close buildings, made of concrete, glass, or iron. They are usually more than 12 floors high and composed of structures such as financial institution offices, public administrations, and private accommodation.
Urban	Business and residential area consisting of several very close, concrete buildings. They are about 10 to 15 floors high.
Suburban	Decentralized business area with residential housing and buildings of 2 to 5 floors made of brick, iron, and concrete.
Rural	Business and residential population spread over open areas with significant vegetation and few man-made structures.

These environments can also be classified based on the *ground occupation rate* (or GOR), which defines the ratio between the area covered by the buildings and the total area. According to GOR, urban environments are characterized by a GOR > 1; suburban environments are characterized by a GOR around O.4, and rural environments are characterized by a GOR < 0.1.

2.2.1 Pathloss

2.2.1.1 Free-Space Pathloss

To define free-space propagation, consider an isotropic point source consisting of a transmitter with a power P_t W. At distance d from this source, the power transmitted is spread uniformly on the surface of a sphere of radius d.

The power density at the distance d is then as follows:

$$S_r = \frac{P_t}{4\pi.d^2} \tag{2.3}$$

If the receiving antenna is placed at a distance d from the source, it will receive a power proportional to its effective area A_e. The power received by the receiving antenna is then equal to:

$$P_r = \frac{P_t A_e}{4\pi . d^2}$$ (2.4)

Using the formula (2.1), we get:

$$P_r = \frac{P_t G_r}{[4\pi(d/\lambda)]^2}$$ (2.5)

Finally, replacing the isotropic source by a transmitting antenna with a gain G_t, the power received at a distance d of the transmitter by a receiving antenna of gain G_r becomes:

$$P_r = \frac{P_t G_r G_t}{[4\pi(d/\lambda)]^2}$$ (2.6)

In decibels the propagation pathloss (*PL*) is given by:

$$PL(dB) = -10.\log_{10}\left(\frac{P_r}{P_t}\right) = -10.\log_{10}\left[\frac{G_t.G_r.\lambda^2}{(4\pi)^2.d^2}\right]$$ (2.7)

Note: Relationship (2.6) can be applied only when $d \geq 2d_a^2/\lambda$, where d_a is the size of the antenna. The free space propagation model is ideal and can only sensibly be applied to satellite systems and to short-range line-of-sight propagation.

Pathloss attenuation is dependent on frequency. In mobile radio communications, the propagation loss (measured by the difference between the signal power at the transmitter and the signal power at the receiver) is least at low frequencies, becoming larger as the frequency increases. The losses also vary with the polarization of the signal.

Note: Pathloss attenuation is a function of frequency and affects channel allocation in radio communication systems.

Radio communication systems that consist of mobile terminals and base stations (as in cellular systems) usually define uplink channel frequencies (mobile to base station links) that are different from those reserved for the downlink channels (base station to mobile links). The frequencies allocated to this type of system are situated in two bands:

- A frequency sub-band for the uplink channel; it is reserved at the low end of the spectrum allocated to the system;
- A second frequency sub-band for the downlink channel; it is reserved at the high frequency end of the spectrum allocated to the system.

The reason for this choice is justified by the variation of attenuation with respect to frequency. It is preferable to allocate to the mobiles (whose power is lower than that of the base stations) the frequencies that are less sensitive to attenuation. This is why the GSM system has reserved the 890-915 MHz band for the uplink channels and the 935-960 MHz band for the downlink channels.

Example 1: Free-space pathloss calculation
   Consider that the transmitting and receiving antennas have the following characteristics: $P_t = 10W$, $G_t = 2$ dB, and $G_r = 2$ dB.
   At 900 MHz, the power received at

- 1 m from the transmitter is equal to $P_r \cong 28$mW
- 2 m from the transmitter is equal to $P_r \cong 7$mW

At 450-MHz frequency and at a distance of 1m from the transmitter, the received power is equal to: $P_r \cong 113$ mW.

2.2.1.2 Mask or Shadowing Effects

The most important losses experienced by the signal are due to natural or man-made obstacles. This effect is called *mask effect* or *shadowing effect*. The higher the number of obstacles between the transmitter and the receiver, the greater the signal is attenuated.
   Important differences in the received signal are experienced that depend on whether the transmitter and the receiver are

- *In line-of-sight of sight*, in which case no obstacle is encountered on the direct path between the transmitter and the receiver;
- *Nonline-of-sight of sight*, in which case a direct path does not exist, so one or several obstacles are encountered by the radio waves between the transmitter and the receiver.

This masking effect gives rise to *slow fading* because of time and space and environmental variations, contrary to *fast fading* (see Section 2.3).
   The mask effect can be modeled by a log-normal law, which in dB becomes a normal law. If A_s is the attenuation in dB and σ the standard deviation (in urban environments, usually $\sigma = 6$ dB), the probability P (for A_s to be greater or equal to x dB) is given by the following formula:

$$P(A_s \geq x) = \frac{1}{\sigma\sqrt{2\pi}} \int_x^\infty e^{-\frac{u^2}{2\sigma^2}} \qquad (2.8)$$

2.2.2 Attenuation Caused by Vegetation

In rural areas, trees contribute an important proportion of the radio signal attenuation. In urban areas, where the number of trees is generally not very significant, their effect is negligible. The attenuation caused by trees depends on parameters such as their height, shape and mass, the time of year, and on the ambient humidity.

Example 2: Calculation of attenuation due to vegetation.
The propagation losses caused by vegetation have been studied and addressed by many authors and researchers. Here we present the formula introduced by Weissberger. It can be applied to frequencies between 230 MHz and 95 GHz. This model consists of the following formulas:

$$L = 1.33 F^{0.284} d_f^{0.588} \quad \text{for } 14 \le d_f \le 400\text{m} \tag{2.9}$$

$$L = 10.45 F^{0.284} d_f \quad \text{for } 0 \le d_f < 14\text{m} \tag{2.10}$$

Where L is the loss in dB, F is the frequency in GHz, and d_f, in meters, is the distance traveled by the signal through the trees. The difference in attenuation between trees with leaves and those without can vary by several dBs. Radio planners would normally build in a margin of 10 dB to allow for this.
Using the above formula at 1 GHz frequency and for a row of trees 5m high, located between the transmitter and the receiver, the propagation loss is $L_{1\ GHz} = 52.25$ dB. At 900 MHz, it is $L_{900\ MHz} = 50.71$ dB.

2.2.3 Attenuation Due to Atmosphere

Signal attenuation through the atmosphere is mainly due to molecular absorption by oxygen for frequencies ranging between 60 and 118 GHz, and from water vapor in the 22 , 183, and 325-GHz bands.
Rain is the most significant of these atmospheric effects. It results in energy absorption by the water drops themselves and as a secondary effect, energy is scattered by the drops. Unlike absorption by the gases, which has a constant effect, the losses resulting from rain are only noticeable for less than 1% of the time. It depends on the intensity of the rain and on the transmission frequency used. This reduction is perceptible only when frequencies are greater than 1.5 GHz (e.g., 0.01 dB/km in the case of heavy rain, which represents a significant loss for a satellite link).

2.2.4 Diffraction and Fresnel Region

In a multipath environment, diffraction of the radio signal takes place when the electromagnetic wave front meets a surface that has sharp irregularities (edges). The secondary waves resulting from the obstruction give rise to bending of the waves about the irregularity (Figure 2.9).

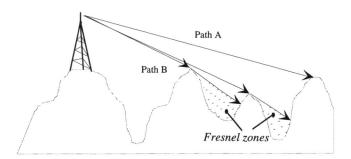

Figure 2.9 Knife-edge diffraction and Fresnel regions.

Energy is diffracted into the shaded areas, which is called the "Fresnel region."

2.2.5 Multipath

A radio signal radiates in all directions, and depending on the environment type, will be reflected or absorbed by the obstacles it encounters. As a consequence, in an urban environment, the reflected waves are more numerous than in a rural area because the number of reflectors is higher. The radio wave could be reflected by any kind of obstacle such as a mountain, a building, a truck, a plane, or even a discontinuity in the atmosphere (ducting).

Reflection from a building depends on the height, size, material and orientation of the building, and on the direction of travel of the radio waves. A reflection can produce a major signal in areas masked by other buildings or obstacles if the surfaces are arranged in such a way that they form a sort of waveguide. In some cases, the reflected signal is highly attenuated, whereas in others almost all the radio energy is reflected and only a small amount is absorbed (the case of an almost perfect reflector).

Multiple reflections thus produce many paths between the receiver and the transmitter (*multipath propagation*). The multipath propagation phenomenon has two effects: one is positive and the other negative.

2.2.5.1 Multipath Propagation Positive Effects

The main advantage of the multipath propagation is that it enables communication even when the transmitter and the receiver are not in line-of-sight conditions (Figure 2.10). Multipaths allow radio waves effectively to "go through" obstacles (e.g., mountains, buildings, tunnels, underground parking lots) by getting around them, and help ensure more or less continuous radio coverage.

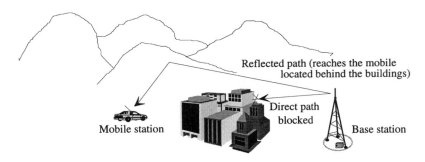

Figure 2.10 Multipath propagation—bypassing obstacles.

2.2.5.2 Multipath Propagation Negative Effects

Multipaths also cause many signal impairments. The three main ones are delay-spread; interference between paths coming from the transmitter, which create fast fluctuations of the received signal (*Rayleigh fading*); and random frequency modulation due to *Doppler shifts* on the different paths.

2.2.6 Delay-spread

The reflected paths are usually longer than the direct path, which means that their signals reach the receiver later than those from the direct path (Figure 2.11). The signals that originate from the same transmitter can arrive at the receiver with different delays and at different times. The delay-spread or multipath spread can be calculated according to the following simplified formula:

$$\text{Multi - path spread} = \frac{\text{longest path - shortest path}}{c} \tag{2.11}$$

where c represents the speed of light.

The channel impulse response delay-spread depends on physical factors such as the direction, reflectivity, and distance between the reflecting objects (e.g., buildings, mountains, walls, vehicles). The delay-spread values range from a few tens of nanoseconds (indoor environment) to several microseconds (outdoor environment).

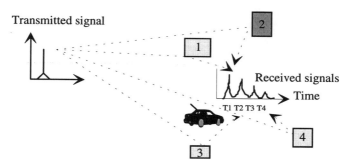

Figure 2.11 Multipath propagation—Delay-spread.

Example 3: Delay-spread calculation for indoor and outdoor environments

Consider a signal propagating outdoors as shown in Figure 2.12. Path *1* has a length of 1,000m, path *2* is 1,600m, and path *3* is 2,500m. Using (2.11), the signal delay spreads at the mobile antenna are

- Between path *1* and path *2*: $\Delta_{12} = \dfrac{t_2 - t_1}{c} = \dfrac{600}{300.10^6} = 2\ \mu s$

- Between path *1* and path *3*: $\Delta_{13} = \dfrac{t_3 - t_1}{c} = \dfrac{1{,}500}{300.10^6} = 5\ \mu s$

For the case of indoor propagation as shown in Figure 2.13, the same calculations are performed.

Path *1* has a length of 3m, path *2* has a length of 5m, and path 3 has a length of 8m. In this case, the signal reaches the mobile antenna with the following delays:

- Between path *1* and path *2*: $\Delta_{12} = \dfrac{t_2 - t_1}{c} = \dfrac{2}{300.10^6} \cong 7\ ns$

- Between path *1* and path *3*: $\Delta_{13} = \dfrac{t_3 - t_1}{c} = \dfrac{5}{300.10^6} \cong 17\ ns$

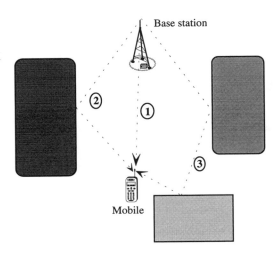

Figure 2.12 Example of multipath in an outdoor environment.

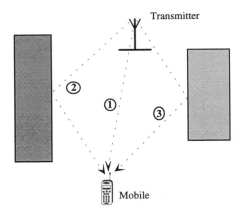

Figure 2.13 Example of multipath propagation in an indoor environment.

In a digital system, especially if it operates at a high bit rate, the delay-spread causes every symbol (or information unit) to overlap with its previous and following one. It results in the intersymbol interference (ISI) phenomenon as shown below.

In the case of propagation through a multipath channel, symbols arriving at the receiver are as drawn in Figure 2.14. At time *0*, only the sample that corresponds to the first symbol (denoted by *1*) has a power not equal to zero. At times *T*, *2T*, and *3T*, for instance, samples corresponding to the first symbol have no power, and the sample corresponding to the following symbol (denoted by *2*) is received.

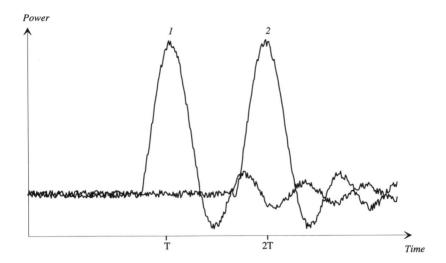

Figure 2.14 Symbol transmission without ISI.

Assuming the same modulation and coding schemes (see Chapter 4), when the information rate and thus the symbol rate increases, the samples become more frequent. But the ISI phenomenon (Figure 2.15) causes some information to be lost because the symbols overlap. In this way, ISI limits the system information rate.

Example 4: Intersymbol interference effects

Considering four-level modulation (i.e., 2 bits per symbol) and a 200 ksymbol/sec rate (that is 400 Kbps), each symbol has a duration of $1/200.10^3 = 5$ µs. A delay-spread of 1 µs leads to an intersymbol overlap of about 20%. In Table 2.3, we have reproduced the delay-spread values measured in different environments.

Definition: channel bandwidth

In CEPT countries, radio regulating agencies have a very strict definition for the *channel bandwidth*. In narrowband systems with 12.5 kHz channel separation, power in the adjacent channels (upper and lower) must be at least 60 dB below the carrier. In a 25 kHz system, this value must be at least 70 dB. In all digital cellular networks such as GSM, these figures are significantly relaxed as channel allocations are coordinated and adjacent channels are not used within the same cell. The digital modulation technique also permits the use of a lower C/I ratio.

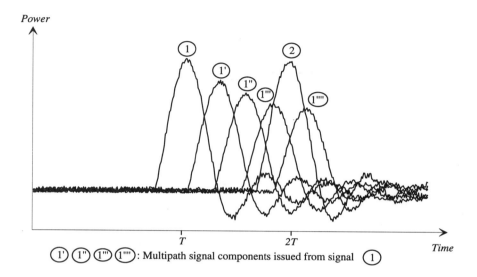

(1') (1") (1''') (1'''') : Multipath signal components issued from signal (1)

Figure 2.15 Intersymbol interference.

Table 2.3

Delay-spread comparison in different environments

Environment type	Delay-spread in s
Free space (open area)	<0.2
Rural area	1
Mountainous area	30 50
Suburban area	0.5
Urban area	3
Indoor area	0.1

Definition: wideband and narrowband channels
The delay-spread is used as an indicator to differentiate between *wideband channels* and *narrowband channels*. If the maximum channel delay-spread is equal or greater to the bit duration T, the channel is said to be *wideband*. If the maximum channel delay-spread is lower than T, the channel is said to be *narrowband*.

In narrowband systems, the propagation attenuation and fading statistics are the main causes of signal degradation. In wideband systems, the delay-spread is dominant because of the intersymbol interference that it creates. Thus, a signal with a bandwidth w larger than B_c (where B_c is the channel coherence bandwidth, see next definition) is highly affected by the channel. In this case, the channel is said to be *frequency selective*. On the other hand, if $B_c > w$, the channel is said to be *frequency nonselective* or *flat faded*. In the time domain, the channel can be considered as being frequency nonselective if the symbol duration T_S is greater than the delay-spread T_m.

Definition: coherence bandwidth
The channel *coherence bandwidth* is a statistical measure of the range of the frequencies over which the radio channel can be considered as "flat" (i.e., a channel in which all spectral components have approximately equal gain and linear phase). The channel coherence bandwidth can be experimentally determined. It corresponds to the minimum frequency separation between two frequencies when the correlation factor is equal to 0.5. In [3] for example, the coherence bandwidth is defined from the autocorrelation factor connecting the field amplitudes received at two separate frequencies with a quantity *df*. The coherence bandwidth is, by definition, the value of *df* for which the autocorrelation coefficient is equal to 0.4.

Note: The delay-spread relates to the channel coherence bandwidth *(Δf)* by the Fourier transform $(\Delta f)_c \approx 1/T_m$ [5]. Using this formula, systems using channels that are substantially narrower than the coherence bandwidth are referred to as *narrowband systems*. *Wideband systems* are those using channels wider than the coherence bandwidth.

2.2.7 Rayleigh Fading

The different paths taken by the same signal have different propagation times and cause interference at the receiver antenna. If two paths have the same propagation attenuation and if their propagation delays differ by exactly an odd number of half-wavelengths, the two radio signals can cancel each other completely at the antenna. If the period is an even multiple of half-wavelengths, the two waves can add positively and create a signal with an amplitude double that compared with one over a single path.

In all other cases, the resulting signal finds a value ranged between these limits. This channel fluctuation is called *small-scale fading,* and is seen as random time variations of phase caused by reflections from an obstacle that has altered the signal in both phase and amplitude. The amplitude of this fading has been described in the form of random variable distribution functions according to Rayleigh, Rice, or Nakagami [6].

Propagation conditions vary considerably from one environment to another and yield different types of small-scale fading that depend on the following factors:

- Transmitted signal bandwidth,

- Received signal delay-spread,

- Random phase and amplitude of the multipath components,

- The mobility of the transmitter, the receiver, or of surrounding objects that cause temporal fluctuations to the channel impulse response.

Rayleigh fading is usually called *fast fading* because it occurs over very short $\lambda/2$ intervals of space.

Example 5: Multipath phenomenon modeling in an indoor environment

Let the radio wave described by the function $Ae^{j2\pi f}$, $e^{j\Phi}$ with A the signal amplitude and $\Phi = 2\pi.f.d/c$ the phase at a distance d from the transmitter [7]. With L paths between the transmitter and the receiver, the signal reaches the receiver with phase and amplitude Φ_r and A_r such that

$$A_r e^{j\Phi_r} = A_0 \sum_{i=1}^{L} \frac{a_i e^{j\Phi_i}}{d_i}$$

where

$a_i = \prod_{j=1}^{K_i} a_{ij}$ is the reflection factor (a_{ij}: reflection factor of the jth reflection of the ith path) and

$\Phi_i = -2\pi.f.d/c$. For example, in the case of a direct path, $a = +1$. The received power is then equal to

$$P_r = P_0 \left| \sum_{i=1}^{L} \frac{a_i}{d_i} e^{j\Phi} \right|^2 \tag{2.12}$$

with $P_0 = P_t G_t G_r (\lambda/4\pi)^2$.

In Figure 2.16, the distances d_2 and d_3 are as follows: $d_2 = 2.\sqrt{\frac{d_1^2}{4} + 20^2}$ and $d_3 = 2.\sqrt{\frac{d_1^2}{4} + 100^2}$.

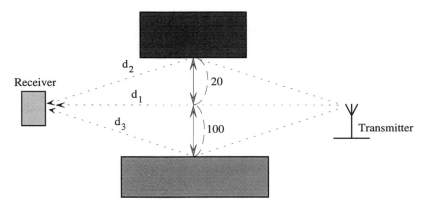

Figure 2.16 Indoor multipath propagation scenario.

The transmitting and receiving equipment characteristics are $P_t = 10\text{W}$, $G_t = G_r = 2$ dB, and $f = 900$ MHz. If we suppose $a_1 = +1$, $a_2 = a_3 = -0.5$, the distribution of the signal depending on d_i is as follows:

$$P_r = P_0 \left| \sum_{i=1}^{3} \frac{a_i}{d_i} e^{j\Phi} \right|^2 .$$

The signal behavior at the receiver antenna and for different frequencies is shown in Figure 2.17. In mobile radio, the most common distribution is according to a statistical law known as the *Rayleigh distribution*. For this reason, this phenomenon is very often called *Rayleigh fading*, and the mobile radio environment is called a *"Rayleigh environment."*

Note: The Rayleigh distribution function

$$f(r) = \frac{2r}{\alpha} \exp\left[-\frac{r^2}{\alpha} \right] \tag{2.13}$$

where $\alpha/2$ is the mean signal power and r the signal envelope (≥ 0).

The most significant signal fades are very deep and lead to signal attenuation of between 20 and 50 dB.

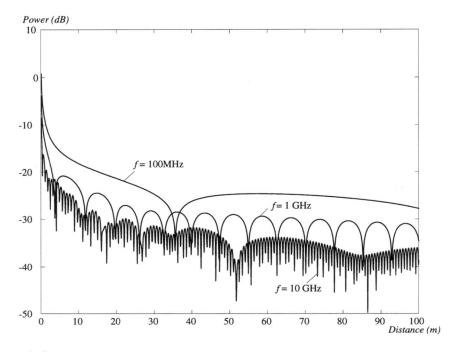

Figure 2.17 Signal power variations in a multipath propagation environment

Example 6: Calculation of time length between fades.

At a frequency $f = 900$ MHz ($\lambda = c/900 = 1/3$m), a vehicle moving at a speed of 100 km/h (i.e., 28 m/s) experiences on average $28/(\lambda/2) = 168$ fades per second of variable depth. The duration between two fades is about $\lambda/2v \cong 6$ ms.

The mean duration of a deep fade depends on the speed v of the receiver and on the frequency. It can be expressed by the following formula [8]:

$$\frac{\sqrt{2\pi}\left[e^{R^2}-1\right]}{\beta v R}$$

where $\beta = 2\pi/\lambda$ and R (dB) is such that $\log_{20}(R)$ gives the mean depth of the observed fade.

Example 7: Calculation of the mean fading duration

At 900 MHz and in an outdoor environment with a mean depth of -20 dB and $R = 0.1$, the mean fading duration is 35.57 ms for a mobile moving with a velocity of 50 km/h, and of 0.988 sec for a mobile moving at 1.8 km/h (Figures 2.18 and 2.19).

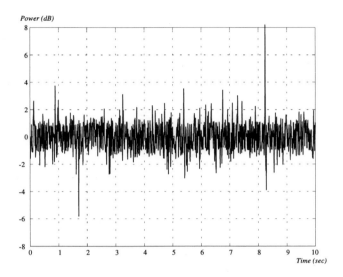

Figure 2.18 Channel behavior for a mobile of velocity 50 km/h (f = 900 MHz).

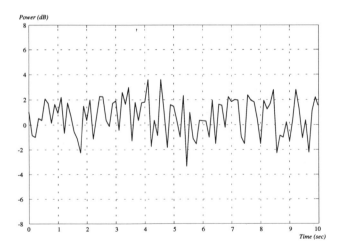

Figure 2.19 Channel behavior for a mobile with the velocity 1.8 km/h (f = 900 MHz).

Error bursts experienced by slow-moving vehicles are therefore longer than those experienced by fast-moving vehicles.

To summarize, multipath fading has two main consequences: first, it imposes a limit on the symbol rate, and second, it introduces bursts of errors in transmissions. Rayleigh fading (Figure 2.20 for its 3D representation) is the main source of errors that occur in radio transmissions and appear in the form of error bursts that put a limit on the intelligibility of the transmitted information. For instance, a bit error rate (BER) of 10^{-3} is frequently considered adequate for speech applications, but for data communications, a BER of 10^{-6} is usually necessary, although often very difficult to accomplish.

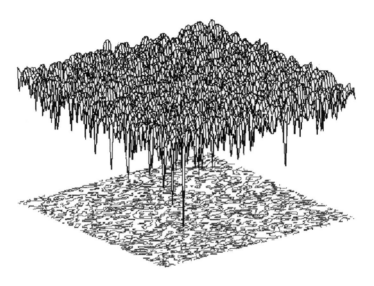

Figure 2.20 Three-dimensional representation of a Rayleigh environment (signal power level and its projection on the ground plan).

The Rayleigh fading model applies only when there are many reflected multipaths and they are dominant compared with the direct path. This model is mainly applied to outdoor environments. In some cases, when line-of-sight propagation exists, a direct path is dominant. This is the typical case for satellite communications and some indoor applications. The resultant signal is then the sum of the signal from the direct path and those of the signals coming from the reflected paths. This type of distribution is called a *Ricean distribution* and is expressed as follows:

$$\Pr(r) = \frac{r}{\sigma^2}.\exp\left[\frac{-r^2+v^2}{2.\sigma^2}\right].I_0.\left(\frac{r.v}{\sigma^2}\right) \qquad (2.14)$$

with $r \geq 0$,

$$I_0\left(\frac{r.v}{\sigma_2}\right) = \frac{1}{2\pi} \int_0^{2\pi} \exp\left(\frac{v.r.\cos\theta}{\sigma^2}\right) d\theta$$

v is the envelope of the direct component and σ^2 is proportional to the reflected component. In the special case when $v = 0$, the expression becomes a Rayleigh function because the direct path no longer exists.

2.2.8 Doppler Shift

The *Doppler effect* is a phenomenon that is caused by the relative velocity of the receiver to the transmitter. It leads to a frequency variation of the received signal called *Doppler shift* or *Doppler spread* (Figure 2.21).

This frequency shift depends on two factors: the direction of the movement and the speed of the receiver with respect to the transmitter. If we consider λ as the wavelength and f the carrier frequency of the transmitter, the frequency of the signal received by a mobile having a speed v with respect to the transmitter is

$$f' \cong f - v/\lambda$$

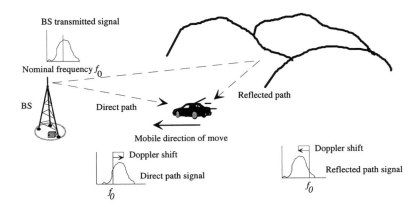

Figure 2.21 Doppler effect in the presence of multipaths.

The Doppler shift is expressed as follows [9]:

$$\text{Doppler Spread} = f_0 \frac{\Delta v}{c} \tag{2.15}$$

where Δv is the relative speed of the transmitter with respect to that of the receiver, f_0 is the carrier frequency, c represents the light speed, and Ψ_D is the angle between the received signal and the speed vector of the receiver.

Doppler shift introduces significant random frequency modulation in the signal and has an effect the multipath signals, some with a positive and others with a negative frequency shift. Doppler effect can be considered as causing temporal decorrelation of the multipaths and is often called the *time-selective fading effect*.

Example 8: Calculation of Doppler shift

Consider the situation represented in Figure 2.22. At frequency $f_0 = 900$ MHz, the signals transmitted by a mobile moving at 60 km/h speed will be received by the base station on path *1* with a shift of +50 Hz and on path *2* with a shift of −50 Hz.

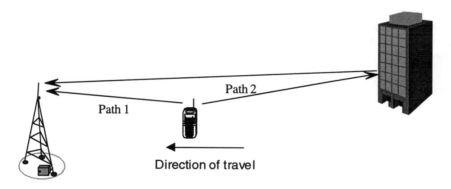

Figure 2.22 Doppler shift of the reflected and direct paths.

2.2.9 Indoor Environment Propagation Characteristics

Systems implemented in indoor environments, such as the cordless telephones (see Chapter 10) or wireless local area networks (see Chapter 13), are subject to particularly complex propagation conditions that are specific to this kind of environment.

Two principal factors characterize indoor radio propagation:

- *Masking effect* (e.g., due to people, furniture, or walls),
- *Multipath distortion* (reflection from obstacles such as walls, floors, or ceilings).

Movement around the antennas can lead to local deep fading (up to 30 dB). On the other hand, the multiplicity of many types of obstruction and obstacles increases the multipath phenomenon and thus the fades. In the indoor context, it is more difficult to predict signal propagation than outdoors. In these conditions, radio planning is very difficult to achieve in practice, and this is one of the reasons why dynamic allocation algorithms are used in systems specifically designed for indoor use. One way used to eliminate shadow fading and multipath distortion is to position base stations and terminals in line-of-sight or use diversity techniques (see Chapter 4).

Indoor and outdoor channels are identical in their basic function; that is, they must both accept multipath spreading caused by a large number of reflectors [10] and can both be described with the same mathematical models. However, several important differences exist between the two kinds of models:

- The conventional mobile radio channel (used for communication between a base station situated on a high site and slow moving mobiles) is stationary in time but not in space. This is caused by the presence of fixed big objects (buildings) which determine spreading of the signal. Compared with the latter, the effects of moving people or vehicles are insignificant. On the other hand, the indoor channel is neither stationary in time or space. Temporal variations of the indoor channel statistics are due to movement of people and equipment surrounding the mobile terminal antennas.
- Compared to the mobile outdoor channel, the indoor channel is characterized by deeper fades and higher variations in the average signal level.
- The fast movement and the high speeds typical of outdoor mobile users are absent in indoor environments. Doppler shift is consequently negligible in the case of indoor transmission.
- The delay-spread for a mobile channel is typically many dozens of nanoseconds if only the immediate surroundings of the mobile are considered and up to 100 µs if reflections from obstacles such as mountains or skyscrapers are considered. The indoor channel is characterized by delay-spreads of less than a microsecond. Consequently, the indoor bit-rate can be much higher.

2.2.10 Propagation in Dense Urban Environments

In urban environments buildings are high and set close to each other. Implementing high traffic capacity systems requires the installation of antennas situated on short masts (5 m) on the roofs of the buildings [11]. Under these conditions it is impossible to neglect the local topology when working out coverage prediction calculations. Even in the simplest cases (e.g., line-of-sight propagation along a street), propagation follows a law that has two slopes. At first the slope is gentle, then it steepens after the *break point* (see definition in Section 2.4.1.2.). At 900 MHz, the first slope has a gradient approximating a square law (this is the free-space path loss) and the break point is situated at a distance of between 200 and 300m from the transmitting antenna. After the break point, the slope changes significantly and the attenuation gradient can take between 4th and 8th power law values. The situation is more complicated if there are buildings in the radio path.

In nonline-of-sight cases, the *corner effect* can take place, where the field strength undergoes a very rapid degradation of about 20 to 25 dB within a few meters around the corner of a building.

2.2.11 Conclusions

In summary, the characteristics of fading differ very much depending on whether the object is fixed or mobile. In fixed communications, shadow fading introduces a small delay-spread that requires less channel compensation than for mobile communication. In the latter, fast fading is the dominant phenomenon to consider. In a rural environment, channel fluctuations are dominated by the direct path and the received signal usually follows a Ricean fading pattern. In an urban environment, where indirect signal paths prevail, multipath signals dominate and reception usually follows a Rayleigh fading distribution.

2.3 INTERFERENCE AND NOISE

In addition to the impairments experienced by the signal and caused by the propagation phenomena previously described, there are other signals that disturb the wanted signal. They can be classified under the general term of *noise*. Noise, in its broadest definition, consists of any undesired signal experienced by a receiver [12]. The subject of noise and its reduction is probably the most important phenomenon that a transmission engineer must face. Indeed, noise is a major limiting factor in a telecommunication system's performance.

In this section, a difference is drawn between what is referred to as *noise* (and which is due, for instance, to thermal noise) and *interference* (which can be cochannel or adjacent channel interference in FDMA or TDMA, or multiple access interference in CDMA).

2.3.1 Noise

There are two principal categories of noise. First, noise emanating from outside the system and second, noise that originates from within the system and which creates interference independent of external conditions. Internal sources of noise can be categorized into two types. First, there is the noise generated from logic devices and their resultant switching currents, and second, that which is produced within components due to electrical disturbances at the molecular level. The influence of the first group can be reduced, even eliminated; whereas the background noise cannot. The most important is the Brownian movement of the electrical particles in a thermodynamic equilibrium or under the influence of the magnetic fields. This type of noise is a time invariant process and is characterized by three principal components, which are (see also Appendix 2B):

- Thermal noise,
- Shot noise,
- Low-frequency additive noise (proportional to $1/f$).

2.3.2 Interference

In a mobile radio system, the radio links are affected by two types of interference:

- *Co-channel interference*, resulting from transmissions on the same frequency, and
- *Adjacent channel interference*, resulting from transmissions on adjacent frequencies.

Interference is undoubtedly one of the most important problems to consider when designing a radio communications system. As communications systems grow rapidly, it is almost impossible to set up a system free from interference. In frequency reuse systems (see Chapters 6, 7, and 12), for example, interference is permanent and has to be allowed for in dense traffic environments.

2.3.2.1 Adjacent Channel Interference

The main source of adjacent channel interference is the use of channels very close in frequency. The aim of doing this is to maximize the spectral efficiency of the system. Minimizing adjacent channel interference is important when adjacent channels are used on the same site or on neighboring sites. The performance limits of equipment makes it difficult to eliminate this type of interference. The performance limitations of transmitter amplifiers and receiver filters make the total suppression of adjacent channel interference virtually impossible. In practice, if a channel spacing greater than 5 channels is used, this tends to eliminate the risk of disturbance on an adjacent channel.

The adjacent channel interference phenomenon is illustrated in Figure 2.23. Two transmissions take place on adjacent channels and inevitably signal components spread beyond the channel limits and are received by the adjacent channel receiving equipment. The signal transmitted on a channel is always received with some level of power, never zero, by the adjacent channel receiving equipment. A signal transmitted on an adjacent channel is an interferer for a given channel when its power is large enough to disrupt the wanted signal.

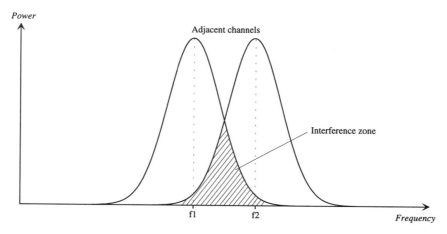

Figure 2.23 Adjacent channel interference.

The adjacent channel interference value can be estimated from the following expression [8]:

$$I_{ca} = K.log_{10}[\Delta f/(B/2)]/0.3 \tag{2.16}$$

where:

- K [in dB/oct or dB/dec] and is the filter characteristic slope,
- $\Delta f = |f_2 - f_1|$ with f_1 and f_2 as the two adjacent frequency channels,
- B is the channel bandwidth.

When planning a system, the channel allocation process (see Chapter 7) takes into account the adjacent channel interference protection ratio (ACIPR). It is the ratio between the power of the signals transmitted on adjacent channels at the point where the level of interference between them is capable of disturbing the communication [13].

2.3.2.2 Co-channel Interference

Co-channel interference occurs when signals transmitted on a frequency f_1 are jammed by other signals transmitted on the same frequency (Figure 2.24). This phenomenon occurs very often in frequency reuse systems such as cellular (see Chapter 6).

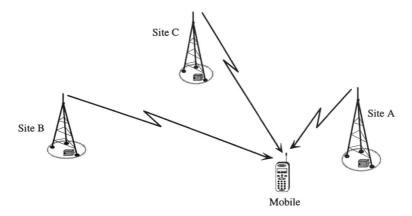

Figure 2.24 Co-channel interference (A, B, and C transmit on the same channel).

The performance indicator used to measure the quality of the received signal, depends on the useful signal (C) and the co-channel interference level (I). It is denoted by C/I (Figure 2.25). The C/I ratio is a random variable affected by random phenomena such as the mobile location, Rayleigh fading, the log-normal masking effect, antenna characteristics, and the location of the transmitters and receivers. The co-channel interference level is as follows:

$\sum_{j \in J} I_j$ where I_j is the signal power level issued from the transmitter j.

The C/I ratio can then be expressed as follows:

$$\frac{C}{I} = \frac{C}{\sum_{j \in J} I_j} \tag{2.17}$$

The impact of the C/I ratio on cellular architecture is described in Chapter 6.

2.3.2.3 Intermodulation

The intermodulation phenomenon is due to the mixing of out of band signals with a resultant occurring on the user signal frequency. It can happen at any nonlinear junction (rusty bolt effect) but usually at the transmitter or receiver.

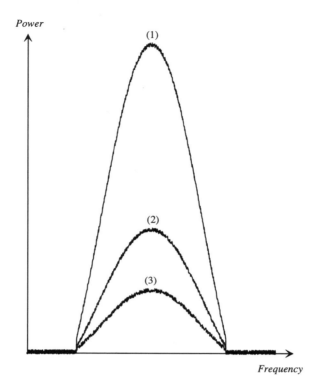

Figure 2.25 Co-channel interference.

Intermodulation at the receiver is due to the presence, at the input of the receiver, of several signals transmitted at a high power level on different frequencies. This phenomenon is produced by harmonics of the signals which have been generated in nonlinear components. If f is the frequency of the user signal, f_1 and f_2 the frequencies of the disturbing signals, an intermodulation product can appear if $2.f_1 - f_2 = f$ and more generally is $|m.f_1 - n.f_2| = f$, with m and n integers. The most significant intermodulation products correspond to the lowest values of m and n.

Intermodulation at the transmitter phenomena can also appear when many transmitters are placed on the same site and their antennas are very close or from different channels transmitting through the same amplifier. The intermodulation products are created in the nonlinear power stages of the transmitters where they generate supplementary spurious radiation. For example, the transmitted signals at transmitter *1* antenna (frequency f_1) reach transmitter *2* antenna (frequency f_2). The second harmonics of f_2 and f_1 generate a frequency $f_3 = 2.f_2 - f_1$ radiated by the antenna 2. In the same way, transmitter *1* generates a frequency $f_4 = 2.f_1 - f_2$. Furthermore, spurious emissions are generated at frequencies $f_5 = 3.f_2 - 2.f_1$ and $f_6 = 3.f_1 - 2.f_2$ (with lower amplitudes than f_3 and f_4).

2.4 PROPAGATION PREDICTION MODELS

One of the most important issues in the design, implementation and operation of a land mobile radio system is a knowledge of the received signal strength and its fluctuations at each point in the system coverage (see Chapter 7). This knowledge establishes the coverage area and the interference problems for each site. Knowledge of the signal level can be obtained by field measurements, but they are costly, especially if they must be made for each site in the system. It is very important for a radio planner to have a computer aided system to predict simultaneously both coverage area and interference problems. Such a system may be based on one or several propagation prediction methods. These methods use channel models, which are sets of mathematical expressions into which channel characteristics, obtained from previous field measurements can be inserted.

Traditionally, propagation predictions have considered the average power of a signal at a given distance from the transmitter as well as the variability of the signal near a particular point. The statistical representation of average signal power for an arbitrary transmitter/receiver distance is useful in estimating the transmitter radio coverage. Mobile propagation is usually studied in terms of large-scale or small-scale effects. More precisely, propagation models characterizing the signal strength at large transmitter/receiver distances (several hundreds to several thousands of wavelengths) are called *large-scale models* whereas models that describe rapid fluctuations received at short distances (of several wavelengths) or at short time periods (of about or less than a second) are called *small-scale models* [14].

In the following, only large-scale models, which are used to predict radio coverage, particularly of cellular systems (see Chapter 7), are discussed. In this field, there are two fundamental approaches dealing with the behavior of a transmission channel. The first consists of modeling the channel in a statistical way. The second method consists of using a direct analytical resolution of the propagation equations or predict the signal paths through the propagation medium (ray tracing).

Propagation models that are available take into account the topography, the type of environment, and the materials from which they are made. The kind of model chosen depends on the level of desired estimation: approximate or precise. Moreover, the available data for the coverage area can play a fundamental role. After a prediction has been estimated radio measurements must be taken to validate the model. Two types of model resulting from these approaches are *theoretical models*, based on theoretical modeling, and *empirical models. Semi-empirical models* combining both these approaches have been produced. Empirical models are based on mean values and establish simple relationships between the attenuation and the distance between the receiver and the transmitter. Their main advantage is that they can include all the factors that affect propagation. Their main drawback is that they must be calibrated and validated for each environment. Furthermore, the validity of empirical models is limited not only by the accuracy with which initial measurements are made, but also by the extent to which the environment for these measurements adequately represents the physical environment in which the model is ultimately applied. On the other hand, theoretical models cannot take into account all factors affecting propagation and require the use of complex topographic databases.

2.4.1 Statistical Methods

2.4.1.1 Plane Earth Propagation Model

The simplest propagation formula uses the *free-space propagation* formula. In this case, the decrease in the power is inversely proportional to the square of the distance. This means that at a distance *2d*, the power received is equal to a quarter of that received at a distance *d*. Expressing this in terms of decibels, it means that the signal experiences a decrease of 6 dB when the distance is doubled. However, in reality the signal does not spread out in all directions equally. Part reaches the receiver via reflections from the ground. The following formula was suggested by *Norton and Bullington* (see Appendix 3 for calculation details) to take into account Plane Earth reflections:

$$P_r \approx P_t G_t G_t \left(\frac{h_b h_m}{d^2} \right)^2 \qquad (2.18)$$

2.4.1.2 Break-Point Propagation Model

When observing line-of-sight propagation in a microcellular environment, it can be observed that the average signal level can be modeled by two factors, n_1 and n_2. The power decreases according to a curve with two slopes. The two regions are separated by a *break point* (Figure 2.26). At 900 MHz, this break point is located between 200 and 500m from the transmitter. For example, at the break point, the slope of the signal power level attenuation curve may change from a function of d^{-2} to a function of $(d^{-2} - d^{-8})$ [15].

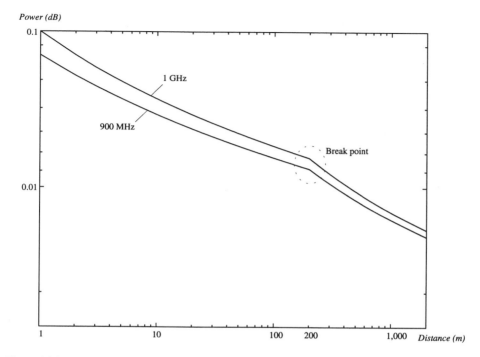

Figure 2.26 Break-point pathloss propagation.

Theoretically, the break point can be determined in association with Fresnel zone clearance [16]. The first Fresnel zone is an ellipsoid whose foci are the transmitting and receiving antennas. More precisely, if d_{TR} is the distance between the two antennas, d_1 the distance from any of the antennas to a point on the ellipsoid, and d_2 the distance between this point and the other antenna, then $d_1 + d_2 = d_{TR} + \lambda/2$. The break point is thus defined as the distance between the two antennas for which the ground just begins to obstruct the first Fresnel zone. This distance is given by the formula:

$$d = \frac{1}{\lambda}\sqrt{\left(A^2+B^2\right)^2 - 2\left(A^2+B^2\right)^2\left(\frac{\lambda}{2}\right)^2 + \left(\frac{\lambda}{2}\right)^4}$$

(2.19)

where $A = h_1 + h_2$ and $B = h_1 - h_2$, h_1 and h_2 are the transmitting and receiving antennas heights.

At high frequencies, this expression can be approximated by the following:
$d = 4h_1.h_2/\lambda$.

An empirical model based on measurements taken in the center of Stockholm were used to describe the break-point phenomenon [15]. The first part determines the *line-of-sight* (LOS) propagation, and the second the non-LOS propagation. In both cases, the signal is separated into three factors using a global loss component depending on the transmitter-receiver distance. These are: a component that depends on the transmitter-receiver distance, a component that follows a log-normal distribution for the shadow fading, and a component that models the fast fading. The received signal is expressed as follows:

$$P_r = P_t - 32.4 - 20.log(f) - L(d) + F \qquad (2.20)$$

where P_r (P_t) is the received (transmitted) power in dBm, f is the frequency in GHz, $L(d)$ is the attenuation in dB as a function of the distance d (in meters), and F is the shadow fading that follows a log-normal distribution (in dB).

$L(d)$ is given by the following formula:

$$L(d) = \begin{cases} k_1.log(d) & d \le D \\ (k_1 - k_2)log(D) + k_2.log(d) & d > D \end{cases} \qquad (2.21)$$

where d is the transmitter-receiver distance and D is the receiver break-point distance (when $k_1 = k_2 = 20$, it becomes a free-space propagation law). The typical values of these parameters are:

200 m $\le D \le 500$m, $15 \le k_1 \le 30$, and $40 \le k_2 \le 80$.

2.4.1.3 Okumura-Hata Empirical Formula

Okumura-Hata's formula is presently the most widely used propagation prediction formula in cellular planning software tools (see Chapter 7). This empirical formula has been produced by Hata based on the measurements made by Okumura in the Tokyo suburbs [17, 18]. This formula includes the following parameters:

f: frequency (in MHz) between 150 and 1,500 MHz,
h_b: height (in m) of the base station, between 30 and 300m,
h_m: height (in m) of the mobile station, between 1 and 20m,
d: base station-mobile station distance (in km), between 1 and 20 km.

The basic principle of Okumura-Hata's formula, and of its variants, first consists of calculating the free-space pathloss. An attenuation factor is then added to this component.

In an urban center, the pathloss L_u (defined as the difference between the transmitted power and the received power) is given by the following formula:

$$L_u(dB) = 69.55 + 26.16log_{10}f - 13.82log_{10}h_b - A(h_m)$$
$$+ (44.9 - 6.55log_{10}h_b)log_{10}d \ (dB) \qquad (2.22)$$

The correction factor $A(h_m)$ is defined as follows:

For medium or small sized cities:

$$A(h_m) = [1.1log_{10}(f) - 0.7]h_m - [1.56log_{10}(f) - 0.8] \text{ dB} \tag{2.23}$$

where $1 \text{ m} \le h_r \le 10 \text{ m}$.

For large sized cities:

$$A(h_m) = 8.29log_{10}(1.54h_m) - 1.1 \text{ dB} \quad \text{if } f \le 200 \text{ MHz,} \tag{2.24}$$

or

$$A(h_m) = 3.2log_{10}(11.75h_m) - 4.97 \text{ dB} \quad \text{if } f > 200 \text{ MHz} \tag{2.25}$$

To generalize this formula for suburban areas, the pathloss propagation formula is modified and becomes [19]:

$$L_{su}(dB) = L_u - 2\left[log_{10}\left(\frac{f}{28}\right)\right]^2 - 5.4 \tag{2.26}$$

with $A(h_m)$ determined by (2.23).

In the case of a rural area, that is a quasi-open space, the pathloss propagation L_{rqo} is given by the formula:

$$L_{rqo}(dB) = L_u - 4.78\left[log_{10}(f)\right]^2 + 18.33log_{10}(f) - 35.94 \tag{2.27}$$

with $A(h_m)$ determined by (2.23).

Finally, in the case of a rural area (i.e., an open space with very few obstacles), the pathloss propagation L_{ro} is given by the formula

$$L_{ro}(dB) = L_u - 4.78\left[log_{10}(f)\right]^2 + 18.33.log_{10}(f) - 40.94 \tag{2.28}$$

with $A(h_m)$ determined by (2.23).

In the curves of Figures 2.27 and 2.28, the values of the parameters used are $h_b = 50m$, $h_m = 1.5m$, and $f = 900 \text{ MHz}$.

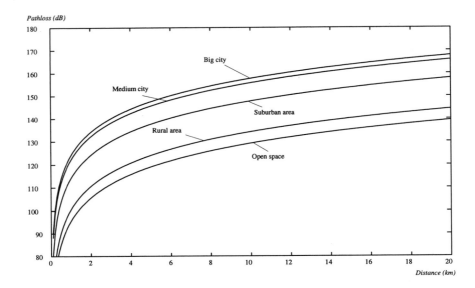

Figure 2.27 Pathloss propagation versus the environment.

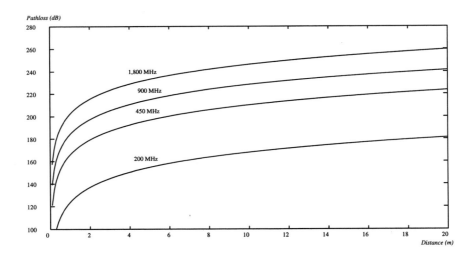

Figure 2.28 Pathloss propagation versus the frequency band.

The Okumura-Hata formula is widely used and enables the calculation of pathloss propagation for 50% of locations at cell boundaries, which corresponds to 70% of the locations inside the cell.

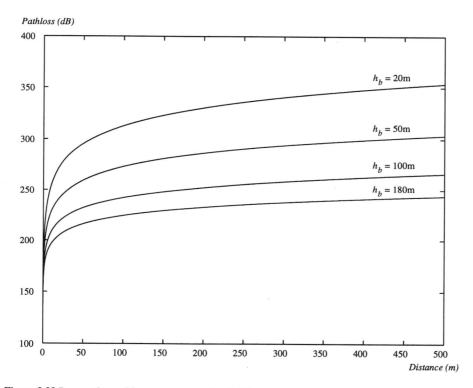

Figure 2.29 Propagation pathloss versus transmitter height.

To obtain good coverage for 90% of locations, a margin of several decibels (7 dB for 2 W terminals and 5 dB for 8 W terminals for example) should be added. As the Okumura-Hata model does not take into account the exact contours (diffraction points, for example) of the terrain, the use of topographic maps and field measurements are necessary to complete the model's prediction.

Figure 2.29 shows the signal pathloss for different base station antenna heights. The mobile station antenna height is 2 m and the transmission frequency is 900 MHz. It should be noted that there is greater attenuation when the antenna height is low. This is of particular interest when high traffic density systems are implemented. Under these circumstances, it is better to implement base stations with low antennas so that the system can benefit from a better frequency reuse scheme (see Chapter 6).

2.4.1.4 Walfish-Ikegami/COST 231 Hybrid Model

The European research group COST 231 have defined a model that combines both empirical and deterministic approaches to calculate the propagation pathloss values in an urban environment in the 900 and 1800-MHz frequency bands [20]. This model takes into account free-space pathloss, diffraction loss, and losses introduced by the roofs of surrounding building. It is based mainly on the models of Walfish and Bertoni [21] and of Ikegami [22].

The COST 231 model applies to urban environments for the following values:

$800 \text{ MHz} \leq f \leq 2 \text{ GHz}$,

$200\text{m} \leq d \text{ (km)} \leq 5{,}000\text{m}$,

$4\text{m} \leq h_b \leq 50\text{m}$, and

$1\text{m} \leq h_m \leq 3\text{m}$.

The model is composed of three terms:

$$L_u(\text{dB}) = \begin{cases} L_f(\text{dB}) + L_{rts}(\text{dB}) + L_{msd}(\text{dB}) & \text{if } L_{rts} + L_{msd} > 0 \\ L_f(\text{dB}) & \text{if } L_{rts} + L_{msd} \leq 0 \end{cases} \qquad (2.29)$$

where L_f is the free-space pathloss given by the following formula:

$$L_f = 10.log_{10}G_t + 10.log_{10}G_r - 20.log_{10}f - 20.log_{10}d_{km} - 32.44 \qquad (2.30)$$

$L_{rts}(\text{dB})$ is the diffraction loss and $L_{msd}(\text{dB})$ is a reflection term. $L_{rts}(\text{dB})$ is based on Ikegami's model.

$$L_{rts} = -16.9 - 10log_{10}w + 10log_{10}f + 20log\Delta h_{Mobile} + L_{ori} \qquad (2.31)$$

where L_0 is the free-space propagation, L_{rts} is the diffraction loss of the roof towards the street, L_{ori} is an empirical correction factor that takes into account the street orientation, Δh_m is the difference (in m) between the building height (h_r) and that of the mobile (h_m), w is the street width (in m), f is the frequency in MHz, d is the transmitter-receiver distance (in km),

$$\text{and } L_{ori} = \begin{cases} -10 + 0.354\psi & \text{for } 0° \leq \psi < 35° \\ 2.5 + 0.075(\psi - 35) & \text{for } 35° \leq \psi < 55° \\ 4.0 + 0.114(\psi - 55) & \text{for } 55° \leq \psi \leq 90° \end{cases} \qquad (2.32)$$

where ψ is the street orientation with respect to the direct radio path (in degrees). L_{msd} (multiscreen diffraction loss) is given by the following formula:

$$L_{msd} = L_{msd}(dB) + K_a + K_d.\log_{10}d + K_f\log_{10}f - 9\log_{10}b$$

$$\text{with } L_{bsh} = \begin{cases} -18\log_{10}(1+\Delta h_b) & for \ h_b > h_r \\ 0 & for \ h_b \le h_r \end{cases} \tag{2.33}$$

$\Delta h_b = h_b - h_r$ with h_r as the base station height in meters and b is the building separation in meters

$$K_a = \begin{cases} 54 & for \ h_b > h_r \\ 54 - 0.8\Delta h_b & for \ d \ge 0.5km \ \text{ and } \ h_b \le h_r \\ 54 - 0.8\Delta h_b.d/0.5 & for \ d < 0.5km \ \text{ and } \ h_b \le h_r \end{cases} \tag{2.34}$$

$$K_d = \begin{cases} 18 & for \ h_b > h_r \\ 18 - 15.\Delta h_b/h_r & for \ h_b \le h_r \end{cases} \tag{2.35}$$

$$K_f = \begin{cases} -4 + 0.7(f/925 - 1) & \text{(medium size cities and suburbs)} \\ -4 + 1.5(f/925 - 1) & \text{(urban centers)} \end{cases} \tag{2.36}$$

If detailed data of the building structure is not available, COST 231 recommends the following default values: 20m $\le b \le$ 50m, $w = b/2$, $h_r = 3$ (number of floors), roof height = 3m, and $\phi = 90°$.

2.4.1.5 Indoor Propagation: The Empirical Motley Model

The Motley model allows pathloss calculations (in dB) using the following formula:

$$L = L_0 + 10.n.\log_{10}(d) + \sum_{i=1}^{I} K_{fi}.L_{fi} + \sum_{j=1}^{J} K_{wj}.L_{wj} \tag{2.37}$$

where $L_0 = 37$ dB is the loss at the reference point (1m, recommended by Motley), n is the attenuation factor (recommendation: $n = 2$), d is the transmitter-receiver distance (in meters), I is the number of floors, J is the number of the different kinds of walls, L_{fi} is the loss factor for floor category i, L_{wj} is the loss factor for wall category j (see example of values in Table 2.4), K_{fi} is the number of crossed floors of category i, and K_{wj} is the number of crossed walls of category j.

Table 2.4

Loss factors in indoor environment

	Material	Loss factor (dB)
	Brick	2.5 to 6.0
	Plasterboard	1.3 to 2.9
Wall type (L_{fi})	Concrete	10.8
	Light wall	2.31
	Thick wall	15.62
	Medium floor	23.62
Floor type (L_{jw})	Normal floor (offices)	14.6
	Reinforced floor	55.3

The loss factors for the inner, external, and ground walls have been obtained from field measurements. An approximate indoor coverage value can be based on the outside signal level, calculated by averaging the signal power around the building at about 2m above street level [23].

Example 9: Indoor pathloss attenuation

Consider a mobile receiver moving in a building and receiving signals from a transmitter placed outside the building (e.g., at the level of the first floor). In the first case, the pathloss can be calculated with the mobile also situated on the first floor and separated by an external wall (thick wall) and by three internal walls (light walls). In the second case, the mobile is located in the basement and is separated from the transmitter by an external wall.

In Figure 2.30, the parameter values are $f = 900$ MHz, $L_0 = 37$ dB, and $n = 2$.

2.4.2 Exact Methods and Ray Tracing/Launching

Empirical methods of calculating pathloss values do not consider the real environment (i.e., positions of obstacles or reflections). They consider parameters that characterize the environment in a statistical way (e.g., urban or suburban contexts). Alternatively, exact or approximate methods are based on accurate databases of the propagation environment. Optical geometry is one of the more practical exact methods that can be implemented using 3D *ray launching* or *ray tracing* algorithms. It is also possible to solve Maxwell's equations numerically with limit conditions representing the architecture and physical properties of the environment. This last method is very complicated and not commonly used.

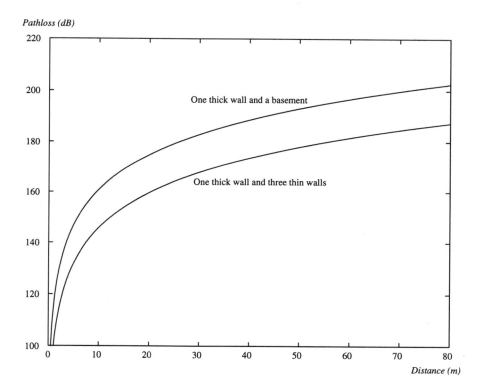

Pathloss (dB)

Figure 2.30 Indoor propagation pathlosses.

2.4.2.1 Exact Methods

The ideal method to simulate radio propagation consists of solving Maxwell's equations numerically [24]. Solving these differential equations in a given context requires the selection of a certain number of points, where the power of the received signal (having undergone a pathloss, a reflection, or a refraction, for instance) will be determined. A systematic method consists of tracing a grid on the area and calculating the solutions at different points. The size of each mesh is a fraction of a wavelength of the carrier frequency. An increase in frequency leads automatically to an exponential increase in the number of spots. Consequently, powerful computers with large memories are required for the calculations. This method is rarely used to predict propagation in a macrocellular context.

2.4.2.2 Ray Tracing and Ray Launching Methods

The *ray tracing method* is an entirely deterministic technique. It is based on a simple approach to the multipath propagation problem, derived from the geometry of optics [25]. The ray tracing method uses very precise geographical data that describes the local environment, particularly the characteristics of buildings. It fits well in urban environments where the number of reflectors and obstacles is significant and also make the empirical models less precise (e.g., Okumura-Hata). Ray tracing provides several advantages over traditional prediction methods as it can:

- Give accurate field strength prediction,
- Give angle of arrival information,
- Include 3D antenna patterns,
- Predict multipath time-delay information,
- Simulate wideband channel impulse responses.

Ray tracing consists of looking for valid ray paths between one transmitter and one fixed receiver. It is a point-to-point algorithm. For coverage calculations, the receiver has to be moved within the prediction area.

The *ray launching method* consists of shooting into the entire 4π stereoradians of space, straight rays that propagate from the transmitter in accordance with geometric optical theory to provide full coverage directly. A group of rays spaced at regular angles are traced from the site of the transmitting antenna. These rays are followed as they propagate through the environment medium. Paths linking the transmitter to a receiver are searched for systematically. This model implies the combination of the following phenomena: reflection, diffraction (whether horizontal or vertical), scattering, and going into and through vegetation [26]. These different phenomena are based on the geometric theory of reflection for the reflected rays and on Fresnel and Beckmann theories for the diffracted rays. Antenna patterns are considered by directly weighting the rays. In the simulator, the tracing of a ray is continued until one of the following conditions is met:

- Its power has fallen below a given threshold,
- It has left the simulation area,
- It has undergone a maximum number of reflections. A maximum number of reflections (between 5 and 10) per ray is taken into account because any increase over this number has little impact on the total power of the received signal. Also most paths take on a limited number of obstacles. Around 7, and at most 15,

diffraction pathlosses can be calculated and added to the free-space pathloss (one per obstacle and one between every obstacle).

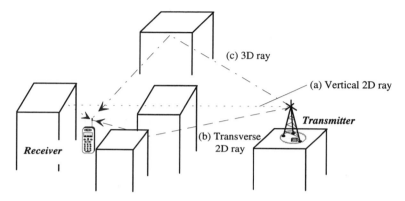

Figure 2.31 Ray tracing model.

If a ray meets the receiver, then a path is considered possible for a radio wave. The impulse response of the channel between the transmitter and the receiver is then determined by the superposition of the amplitudes, phases, and delays of the received rays. The reflection and transmission coefficients for the perpendicular and parallel polarized waves are determined compared with the angle of arrival of the initial ray and of the electric parameters of the wall (Figure 2.31).

Example 10: A ray tracing automatic algorithm
This example is of an algorithm used for most important transmitter/receiver propagation links. Three sub-models can be defined, with each being considered as dominant according to the particular situation. The three models used are:

- A propagation model for rays only in the vertical plane,
- A propagation model for rays in the horizontal plane, and
- A propagation model for rays in three dimensions via reflections from buildings and the ground.

The first model takes into account up to 16 rays (reflected rays, simple and double with combinations). This model is dominant in the case of macrocells. The diffracted rays, whether simple or double, are traced with the second model. This model is dominant when the transmit antennas are placed below roof level.

The problems linked to ray-tracing and ray launching techniques are the following:

- Strong dependence on the accuracy of the geographical database,
- A calculation time that is likely to become very significant for large areas.

Therefore, ray tracing and ray launching techniques are generally used only for indoor or microcellular environment propagation, not for macrocellular cases.

2.4.2.3 Knife-Edge Diffraction Attenuation

During its propagation, a radio wave can be blocked by obstacles such as mountains, buildings, or trees. In this case, the signal that reaches the receiving antenna may arrive as a diffracted ray.

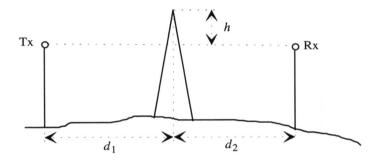

Figure 2.32 Knife-edge obstacle.

One of the principal disadvantages of statistical methods (Okumura-Hata or Cost 231) is that they do not take into account the real contours of the studied ground. For example, the propagation pathloss on a circle of radius R centered on an omnidirectional transmitting antenna and whose height is fixed will be the same in all points, whatever obstacles are met by the signal between a receiver and the transmitting antenna.

The effect of the obstacle represented in Figure 2.32 (see also Figure 2.33) depends on the parameters h, d_1, and d_2. The attenuation in dB due to an obstacle of this type depends on

$$V = h \sqrt{\frac{2(d_1 + d_2)}{\lambda \cdot d_1 \cdot d_2}}$$

The value of the parameter h can be negative if the edge is placed beneath the direct path between the transmitter and the receiver. The diffraction loss is then given by the relation:

$$L = 10\log_{10}\left|\frac{E}{E0}\right|^2 = 20\log_{10} F \qquad\qquad (2.38)$$

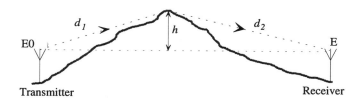

Figure 2.33 Knife-edge diffraction: first example.

In practice, the diffracted wave meets many edges. A few techniques have been defined to calculate the multidiffraction loss such as the methods of Deygout [27] or Giovanelli [28] and are presented in the following sections.

Multiple diffraction propagation prediction: The Deygout method. This method developed by Deygout introduces a correction factor that takes into account the different obstacles encountered by the radiowave between the transmitter and the receiver, and on which the radio signal is going to diffract.

The basic principle of this method is derived from Huyghens' concept, and the formulas are based on Fresnel integrals. A radio wave propagating by diffraction from point A toward point B, goes around a certain number of obstacles that can be considered as transverse and opaque screens. These factors are taken into account according to their importance with the help of Fresnel's calculation method.

Reflections that happen at the different obstacles are not considered and if a direct path exists between points A and B diffraction is not taken into account. Between A and B obstacles met by the radiowave are considered in an order that depends on their geographic situation. Note that the results are, however, independent of the direction of processing (i.e., A toward B or B toward A). Consider the case represented by Figure 2.34.

For every spot on the radio path, the ratio h/r is defined where:

h: height of the point in relation to the line AB (positive if above AB, otherwise negative),

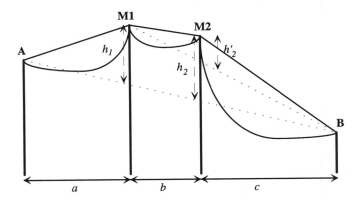

Figure 2.34 Multiple-edge diffraction (second example).

r: radius of the first Fresnel ellipsoid of path AB, measured at the vertical of the point.

The diffraction loss a_m resulting from the highest value of the ratio h/r (for $M1$ in the example) has the value of

$$a_m = \begin{cases} 6+12\left(\dfrac{h}{r}\right) & \text{for} \quad -0.5 < \dfrac{h}{r} < 0.5 \\ 8+8\left(\dfrac{h}{r}\right) & \text{for} \quad 0.5 < \dfrac{h}{r} < 1 \\ 16+20\log_{10}\left(\dfrac{h}{r}\right) & \text{for} \quad \dfrac{h}{r} > 1 \end{cases} \tag{2.39}$$

If $h/r < -0.5$, and taking $a_m = 0$, The h/r value is given by the following expression:

$$h\!\!\!/_r = \frac{h_1}{\sqrt{\dfrac{\lambda.a(b+c)}{a+b+c}}} \tag{2.40}$$

where λ is the wavelength.

The calculation is repeated on each side of the edge $M1$. For example, for $M2$, the loss depends on the ratio:

$$h\!\!\!/_r = \frac{h'_2}{\sqrt{\dfrac{\lambda.b.c}{b+c}}} \tag{2.41}$$

The diffraction loss can be added to the propagation loss computed by a formula such as that of Okumura-Hata [27].

2.5 CONCLUSIONS

The fundamental difference between mobile radio communications systems and wired systems lies in the transmission characteristics. The transmission problems encountered in mobile radio are the main limitation to the capacity of these systems.

Propagation phenomena are very difficult to define, particularly in urban and indoor environments. The complexity of the mobile radio channel in these conditions makes it impossible to derive models that can precisely estimate the level of radio signal strength at a given point. The models available to designers of mobile radio systems are either statistical and able only approximately to reflect reality, or they require substantial computational power to give more exact answers. It is only possible to apply the more exact methods if the propagation environment is precisely known (e.g., positions and types of structures). The use of such models is, however, fundamental in the system planning process and to that of cellular systems, in particular. This is why empirical models are most often used and calibrated by measurements taken in the field.

In Chapter 4, the main techniques used to resolve transmission problems in mobile radio are introduced. The means to increase the capacity of a system, particularly in cellular, are dealt with in Chapter 6.

APPENDIX 2A

NOISE THAT CAN AFFECT RADIO RECEPTION

2A.1 Internal Noise Sources

2A.1.1 Thermal Noise

This originates in the thermal agitation of electric charge carriers in a resistive material. These induces, even in the absence of an electric field, a random fluctuation of the instantaneous value of the observable voltage (*Johnson noise*). Thermal noise gives rise to a random voltage $n(t)$. The thermal noise power is proportional to the resistance temperature and to the frequency band under consideration (Nyquist theorem).

Note: QUALCOMM Company has tested receiver preamplifiers cooled at temperatures of $-193°C$ to reduce the thermal noise by a factor of 6 dB [29] in order to improve the MS-BS radio link quality.

2A.1.2 Shot Noise

Shot noise is due to the statistical fluctuations of electric charge carriers when they cross a potential barrier. It relates directly to the average current value and adds to it. It is white noise that follows a Poisson distribution process.

2A.1.3 Flicker Noise (1/f)

This kind of noise depends mainly on the technology of the equipment used. Its origin is not well known and is characterized by a spectral density that is a function of $1/f^{\alpha}$ ($\alpha > 0$).

2A.2 External Noise Sources

2A.2.1 Atmospheric Noise

Its main origin is electric storms. The level varies as a function of frequency, the geographic region, the season, meteorological conditions, or the time of day, for example. Atmospheric noise level decreases with the increasing latitude and has a significant contribution for frequencies below 20 MHz.

2A.2.2 Galactic noise

This includes all kinds of noise generated outside the Earth and its surrounding atmosphere. The main sources of galactic noise are the sun and cosmic radiation. It has significant influence between 15 MHz and 100 GHz.

2A.2.3 Man-made Noise

The most important noise for the radio communications systems is that generated by human activity. Two kinds of man-made noise are significant: automotive noise and industrial plant noise. In the mobile radio environment automotive noise is dominant; next are power-generating facilities; and the third is industrial equipment [8]. Other noise such as that from consumer products, lighting systems, medical equipment, electrical trains, and buses is lower and can generally be ignored. Man-made noise is much more important in urban areas than in suburban areas. The difference between the noise power in the two environments can be as much as 16 dB. For instance, in urban areas the noise generated by ignition or by electrical equipment is the most important and is a significant parameter in the degradation of mobile communications performance. Power line noise and industrial noise is present within the spectrum from the fundamental generation frequency of either 50 or 60 Hz, up into the UHF range.

APPENDIX 2B

DECIBELS

The *bel*, named after the inventor of the telephone, uses logarithms to base 10 to express the ratio of power transmitted to power received. The resulting gain or loss of a circuit is given by

$$B = \log_{10}\left(\frac{P_0}{P_1}\right)$$

where B is the power ratio in bels, P_0 the output or received power, and P_1 is the input or transmitted power.

The bel was used for many years to categorize the quality of transmission on a circuit because the human ear has a logarithmic response to signal volume change. Gradually, the bel was replaced by the use of the *decibel* (denoted by dB), which is a more precise measurement unit, as it represents one-tenth of a bel. The decibel is also a unit that describes a ratio using a logarithm to the base of 10. It is used in measurements and calculations that imply power ratios.

To define the dB, let P be the power level at a point of the system and P_0 be the reference power level with which P must be compared. The power ratio P/P_0 is defined in decibels by the following formula:

$$10.\log_{10}(P/P_0)$$

For example, a power ratio of 2 corresponds to 3 dB and a ratio of 10 corresponds to 10 dB. In the power domain, if the input power of a system is 1W and the output power is 2W, the system has $10.\log_{10}(2W/1W) \cong 3\ dB$ gain [12]. If for another system, the input power is 1,000W and the output power is 1W, the gain is $10.\log_{10}(1/1,000) = -30\ dB$ and is then a loss.

As a decibel is defined in terms of ratios or of relative units, it does not indicate the absolute level of the measured parameter (e.g., power). Derived decibel units can be used for that purpose. It is therefore possible to express P in dBW (decibels relative to 1 watt) using the formula $10.\log_{10}(P/1\ W)$ or in dBm using the formula $10.\log_{10}(P/1\ mW)$.

APPENDIX 2C

DETERMINATION OF THE PLANE EARTH PROPAGATION FORMULA

In the plane Earth propagation case, formula (2.6) becomes (Norton and Bullington) [13]:

$$P_r = \frac{P_t G_r G_t}{[4\pi(x/\lambda)]^2}[1+R.\exp(j\delta)+(1-R)A.\exp(j\delta)+...]^2 \qquad (2C.1)$$

where the first term in the summation represents the direct path, the second term represents the reflected path, the third term represents the surface wave, and the following terms represent the inductive field and secondary effects of the ground.

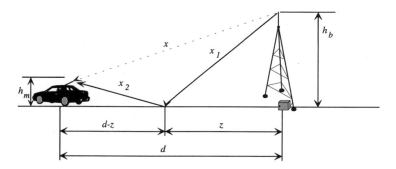

Figure 2C.1 Propagation paths over a plane Earth.

The reflection coefficient of the ground R depends on the angle of incidence θ, the wave polarization, and the ground characteristics. R is given by the following formula:

$$R = \frac{\sin(\theta) - z}{\sin(\theta) + z} \qquad (2C.2)$$

where $z = \sqrt{(\varepsilon_0 - \cos^2(\theta))} / \varepsilon_0$ for vertical polarization, $\varepsilon_0 = \varepsilon - jx60 \times \sigma \times \lambda$, ε is the dielectric constant of the ground relative to free space, σ the conductivity of the Earth in mhos/m, and λ the wavelength. The phase difference between the direct and reflected paths is given by δ. According to Figure 2C.1, the direct path distance value is given by the formula $x = \sqrt{d^2 + (h_b - h_m)^2}$.

In terms of phase units (i.e., 2π radians per distance λ) the following is obtained:

$$\phi(x) = \frac{2\pi d}{\lambda} \sqrt{1 + (\frac{h_b - h_m}{d})^2} \qquad (2C.3)$$

The reflected path includes two segments, x_1 and x_2, with $x_r = x_1 + x_2$; that is $x_r = \sqrt{h_b^2 + z^2} + \sqrt{h_m^2 + (d-z)^2}$. With $tan(\theta) = h_b/z = h_m/(d-z)$, it becomes $z = d.h_b/(h_b + h_m)$. Simplifying the previous formulas obtains the following:

$$\phi(x_r) = \frac{2\pi d}{\lambda} \sqrt{1 + \left(\frac{h_b + h_m}{d}\right)^2} \qquad (2C.4)$$

therefore $\delta = \phi(x_r) - \phi(x) = \frac{2\pi d}{\lambda} \sqrt{1 + \left(\frac{h_b + h_m}{d}\right)^2} - \frac{2\pi d}{\lambda} \sqrt{1 + \left(\frac{h_b - h_m}{d}\right)^2}$

If d is large relative to the antenna heights, then δ can be approximated to $\delta \approx \frac{4\pi h_b h_m}{\lambda d}$.

The effect of surface waves is significant only in the region located a few wavelengths above ground. Thus, it cannot be taken into account in mobile communications problems. With this hypothesis, and with $R = -1$, (2C.1) becomes (using the Euler relation, $exp(i\alpha) = cos(\alpha) + i \sin(\alpha)$ and series expansions of $sin(x) \approx x$ and $cos(x) \approx 1$) :

$$P_r \approx P_t G_t G_t \left(\frac{h_b h_m}{d^2}\right)^2 \qquad (2C.5)$$

REFERENCES

[1]   Collin, R. E., "Antennas and Radiowave Propagation," *McGraw-Hill International Editions*, Electrical Engineering Series, 1985.

[2]   Elbert, R. B., *Introduction to Satellite Communication,* Artech House 1987.

[3]   Deygout, J., *Données fondamentales de la propagation radioélectrique*, Eyrolles, 1994.

[4] Remy, J. G., Cueugniet, J., and Siben, C., *Systèmes de Radiocommunications avec les Mobiles*, Collection CNET-ENST, Eyrolles 1992.

[5] Proakis, J. G., *Digital Communications*, McGraw Hill, New York, 1989.

[6] Pichna, R., and Q. Wang, "Power Control," *The Mobile Communications Handbook*, Editor-in-Chief Jerry D. Gibson, CRC Press, 1996.

[7] Pahlavan, K., and A. H. Levesque, *Wireless Information Networks*, Wiley, 1995.

[8] Lee, W. C. Y., *Mobile Communications Design Fundamentals*, éditions Howard WAMW, Indiana, 1986.

[9] Gallager, R. G., "The Use of Information Theory in Wireless Networks," *Multiaccess, Mobility and Teletraffic for Personal Communications '96* (Paris), 20 May 1996.

[10] Hashemi, H., "The Indoor Radio Propagation Channel," *Proceedings of the IEEE*, Vol. 81, No. 7, pp. 943–967, July 1993.

[11] Frullone, M., G. Riva, P. Grazioso, and G. Falciasecca, "Advanced Planning Criteria for Cellular Systems," *IEEE Personal Communications*, pp. 10–15, Dec. 1996.

[12] Freeman, R. L., "Telecommunication Transmission Handbook," *Wiley Series in Telecommunications*, 1991.

[13] Hess G.C., *Land-Mobile Radio System Engineering*, Artech House, 1993.

[14] Rappaport, T. S., R. Muhamed, and V. Kapoor, "Propagation Models," *The Mobile Communications Handbook*, Editor-in-Chief Jerry D. Gibson, CRC Press, 1996.

[15] Gudmundson, M., "Analysis of Handover Algorithms," *Proceedings of the IEEE Vehicular Technology Conference'91*, Saint-Louis, MI, pp. 537–542, May 19-22 1991.

[16] Xia, H. H., H. L. Bertoni, L. R. Maciel, A. Lindsay-Steward, and R. Rowe, "Radio Propagation Characteristics for Line-of-Sight Microcellular and Personal Communications," *IEEE Transactions on Antennas and Propagation*, Vol. 41, No. 10, pp. 1439–1447, Oct. 1993.

[17] Okumura, Y. et al., "Field Strength and Its Variability in VHF and UHF Land-Mobile Radio Service," *Review of the Electrical Communication Laboratories*, Vol. 16, Nos. 9 and 10, pp. 825–873, 1968.

[18] Hata, M., "Empirical Formula for Propagation Loss in Land Mobile Radio Services," *IEEE Transactions on Vehicular Technology*, Vol VT-29, No. 3, pp. 317–325, 1980.

[19] GSM 03.30, Annex B, Version 4.3.0., ETR 103, Feb. 1995.

[20] Cost, "Urban Transmission Loss Models for Mobile Radio in the 900 and 1800 MHz Bands," COST 231, TD (91) 73, 1991.

[21] Walfish, J., and H. L. Bertoni, "A Theoretical Model of UHF Propagation in Urban Environments," *IEEE Transactions*, Al-38, pp. 1788–1796, 1988.

[22] Ikegami, F. et al., "Propagation Factors Controlling Mean Field Strength on Urban Streets," *IEEE Transactions*, AP-32, pp. 822–829, 1984.

[23] Cichon, D. J., T. C. Beeker, and M. Dottling, "Ray Optical Prediction of Outdoor and Indoor Coverage in Urban Macro- and Micro-Cells," *Proceedings of the IEEE Vehicular Technology Conference '96*, Atlanta, GA, pp. 41–45, Apr. 28-May 1, 1996.

[24] Lee, J. F., "Numerical Solutions of TM Scattering Using a Obliquely Cartesian Finite Difference Time Domain Algorithm," *IEE Proceedings H*, 140, No. 1, pp. 23–28, 1993.

[25]   Bergholm, P., M. Honkanen, and S.-G. Häggman, "Simulation of a Microcellular DS-CDMA Radio Network," *Proceedings of the International Conference on Universal and Personal Communications'95*, ICUPC'95, Tokyo, Japan, pp. 838–842, Nov. 6-10, 1995.

[26]   Bic, J.-C., D. Isner, M. Juy, and P. Metton, "Les spécificités de la propagation en zone urbaine," *REE*, N° 7, pp. 11–24, July 1996.

[27]   Deygout, J., "Evaluation de l'affaiblissement de propagation des ondes décimétriques en zone rurale," *L'Onde Electrique*, Vol. 72, No. 3, pp. 56–63, May-June 1992.

[28]   Giovanelli, C. L., "An Analysis of Simplified Solutions for Multiple Knife-Edge Diffraction," *IEEE Transaction on Antennas and Propagation*, Vol. AP-32, No. 3, pp. 297–301, Mar. 1984.

[29]   Wallace, S., D. G. M. Gruickshank, and J. Urwin, "An Independant Investigation Into the Commercial Opportunities Afforded by CDMA," *Proceedings of the International Symposium on Spread Spectrum*, Hanover, Germany, pp. 1053–1057, Sept. 1996.

SELECTED BIBLIOGRAPHY

[1]   "Coverage Prediction for Mobile Radio Systems Operating in the 800/900 MHz Frequency Range," *IEEE Transactions on Vehicular Technology*, Vol. 37, No. 1, Feb. 1988.

[2]   Babich, F., G. Lombardi, and E. Valentinuzzi, "Indoor Propagation Characteristics in DECT Band," *Proceedings of the IEEE Vehicular Technology Conference '96*, Atlanta, GA, pp. 574–578, Apr. 28-May 1.

[3]   Bultitude, R.J.C., "Measurements, Characterization and Modeling of Indoor 800/900 MHz Radio Channels for Digital Communications," *IEEE Communication Magazine*, Vol. 25, No. 6, June 1987.

[4]   Corazza, G. E., V. Degli-Esposti, M. Frullone, and G. Riva, "A Characterization of Indoor Space and Frequency Diversity by Ray-Tracing Modeling," *IEEE Journal on Selected Areas on Communications*, Vol. J-SAC 14, No. 3, pp. 411–418, Apr. 1996.

[5]   Guérin, S., "Indoor Wideband and Narrowband Propagation Measurements around 60.5 GHz in an Empty and Furnished Room," *Proceedings of the IEEE Vehicular Technology Conference '96*, Atlanta, GA, pp. 165–169, Apr. 28-May 1, 1996.

[6]   Har, D., H. L. Bertoni, and S. Kim, "Effect of Propagation Modeling on LOS Microcellular System Design," *International Journal of Wireless Information Networks*, Vol. 4, No. 2, pp. 113–123, 1997.

[7]   *IEEE Transactions on Antennas and Propagation*, Vols. AP-17, No. 3, May 1969 and AP-22, No. 1, Jan. 1974.

[8]   Kraus, J. D., *Antennas*, McGraw-Hill International Editions, 1989.

[9]   Rizk, K., J. F. Wagen, and F. Gardiol, "Ray Tracing Based Path-loss Prediction in Two Microcellular Environments," *Proceedings of the International Conference on Universal and Personal Communications'94*, ICUPC'94, The Hague, Netherlands, Sept. 1994.

[10] Valenzuela, R. A., "Antennas and Propagation for Wireless Communications," *Proceedings of the IEEE Vehicular Technology Conference '96*, Atlanta, GA, pp. 1–5, Apr. 28-May 1, 1996.

CHAPTER 3

ACCESS—RADIO CHANNEL DEFINITIONS AND RESOURCE ACCESS

In a mobile radio system, mobile terminal access to the resources (radio channels) must necessarily be shared. As the users are mobile and the radio resource is limited, it is clearly impossible (for capacity reasons) to allocate a permanent channel to each user on every radio site in the network (Figure 3.1). It is therefore necessary to define a method for dividing the available bandwidth into several channels, and define the protocols that the mobiles will use to gain access to these channels. For instance, within a cellular system (see Chapter 6), the allocation of a channel to a mobile goes through three steps. First, the radio spectrum has to be split into many channels according to the chosen multiple access method. The second stage—which can be modified by the operator—consists of allocating the channels to the radio sites (this step is addressed in detail in Chapter 7). The third stage—realized dynamically—consists of a mobile gaining access to a channel for the duration of a communication session.

In the first part of this chapter, the three main multiple access methods currently used in most mobile radio systems are examined. These are *frequency*, *time* and *code division multiple access* methods. In many mobile radio systems, a combination of these methods has been used.

In the second part of this chapter, the main random access protocols currently in use in mobile radio communications and similar systems are introduced. They allow different mobiles to make random access to the radio channels while minimizing contention.

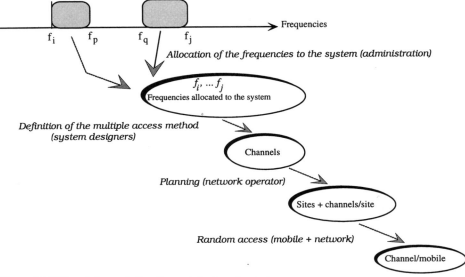

Figure 3.1 Different stages of allocation and sharing of resources.

3.1 MULTIPLE-ACCESS METHODS

3.1.1 Definitions—Narrowband and Wideband Systems

Definitions

Communication within a mobile radio system uses the limited frequency band allocated to that system. This frequency band must be used in an intelligent way to ensure that the maximum number of communication sessions can be served. It is split into channels that are allocated on demand to allow the exchange of information between a mobile terminal and the network or between mobiles. The definition of a communications channel is dependent on the multiple-access method.

FDMA-TDMA-CDMA. The three principal multiple access techniques are the following (Figure 3.2):

- Frequency-division multiple access (FDMA),
- Time-division multiple access (TDMA),
- Code-division multiple access (CDMA).

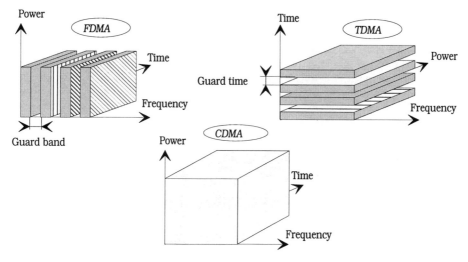

Figure 3.2 Schematic representation of the three multiple access techniques.

The efficiency of the different multiple access methods in terms of capacity has been the theme of much research. The efficiency is very dependent on the conditions within which the system is implemented and thus huge differences may appear between systems using different multiple access methods. The effect of multipath, interference, frequency reuse, or signal processing techniques, for instance, considerably influences the capacity of each system. Chapters 6 and 7 introduce some techniques that improve system capacity.

Nevertheless, keep in mind that in information theory the theoretical system limit depends on the occupied bandwidth, B, and on the signal to noise ratio, R. This is given by the *Shannon theorem*, which defines the theoretical system throughput by the following formula:

$$D = B.LOG_2(1+R) \tag{3.1}$$

Example 1: Calculation of the theoretical channel throughput

For a 200 kHz bandwidth channel and a signal to noise ratio (SNR) of 15 dB (i.e., a power ratio of 31.62), the maximum theoretical rate is:

$$D = 200.10^3.LOG_2(1+31.62) = 1,006 \ KBPS$$

Uplink and downlink channels. For communication from the mobile to the base station, the radio channels are known as the *reverse link* or *uplink* channels, and from the base station to the mobile they are called *forward link* or *downlink* channels.

TDD and FDD duplexing. In systems where communication can be bidirectional, *full-duplex* links by using *frequency-division duplex (FDD)* or *time-division duplex (TDD)* techniques can be introduced.

In the FDD technique (Figure 3.3), the two communicating points use different transmission frequencies and transmit simultaneously. The main drawback of the FDD technique is the loss of usable frequencies within the band due to limitations in RF filtering necessary to maintain separation between co-located transmitters and receivers. This mode is particularly interesting in macrocell systems (see Chapter 6).

Figure 3.3 FDD duplexing principle.

In the TDD technique, the two communication endpoints use the same carrier but transmit at different times. The TDD mode (Figure 3.4) is optimal for microcells because of the short guard times. It may also be more efficient than the FDD mode in cases where available spectrum is limited.

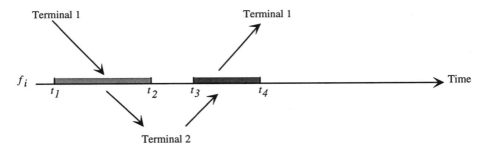

Figure 3.4 Downlink (BS toward MS) and uplink (MS toward BS) channels in TDD.

Combining the two duplexing modes (Figure 3.5) is often used in mobile radio systems, and in particular, in digital cellular systems such as GSM. This allows several duplex logical channels to be carried on a single pair of RF carriers.

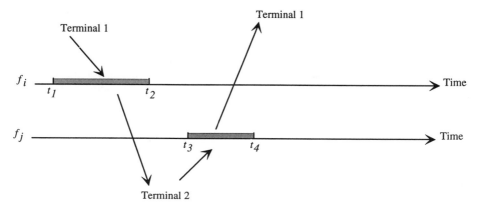

Figure 3.5 Combined FDD and TDD.

Traffic and signaling channels. Communications channels can be classified into two categories: *traffic* channels and *signaling* channels. Traffic channels (also called data channels) are used to carry user information such as voice and data. Signaling channels (also called control channels) carry system information (general information transmitted by the system to one or several mobiles, or commands exchanged between the network and mobiles). These two types of logical channels can be mapped on to the same physical channels, for example, in the same *slot*—the time unit during which a burst of data can be transmitted—of the same carrier and on different frames or on different slots within the frame or even on a different carrier.

Signaling can be transmitted *in-band*; that is, on the same logical channel as user information, and can lead to short communication interruptions, especially in analog systems. In addition, signaling can be transmitted *out of band*; that is, in a separate channel independent of user information, and thus the user's communication is not affected. Some special cases, such as handover, may contradict this rule. Generally, the portion of the spectrum occupied by signaling is low compared with that of user information (e.g., approximately 1/8 or less in GSM).

Narrowband or Wideband Systems

Mobile radio systems may be classified by the multiple access technique used and also distinguishes *narrowband* systems from *wideband* systems. In narrowband architectures, the whole frequency band is divided into several narrowband sub-channels, each of which can only be used by one or several users. In wideband architectures, the whole or a large portion of the spectrum is accessible to all the users simultaneously. Thus, FDMA intrinsically introduces a narrowband architecture (as the objective is to divide the available bandwidth in a maximum number of channels). On the other hand, CDMA induces a wideband architecture (as the whole allocated bandwidth is used). The TDMA technique can be implemented either as a narrowband system if the defined channels are narrowband, or as a wideband system. One disadvantage of narrowband systems is their sensitivity to the fading phenomenon. When a channel experiences fading, all information transmitted on this channel is affected. A definition of wideband and narrowband systems by function of the channel coherence bandwidth is given in Chapter 2.

3.1.2 Frequency-Division Multiple Access

The oldest multiple access method is FDMA (Figure 3.6). It is mainly used in analog systems and is combined with TDMA in most of the digital systems. In analog systems each channel, or carrier, is used to carry a unique call and in one direction at a time (uplink or downlink). According to system capacity and signaling requirements, one or several control channels are used.

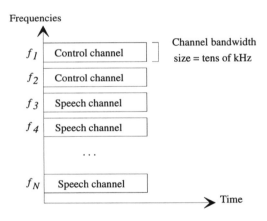

Figure 3.6 FDMA architecture.

The main characteristics of the FDMA method are as follows:

- *A unique circuit per carrier*: Each FDMA channel is defined to carry one single communication (a bidirectional full-duplex communication requires two carriers).
- *Continuous transmission*: When the communication channels have been allocated on the uplink and downlink, the two communicating entities can transmit in a continuous and simultaneous way.

- *Low bandwidth*: FDMA channels are relatively narrow, generally 30 kHz or less, because they are used for only one circuit per carrier; trends have been towards the design of narrower channels (10 to 15 kHz).
- *Symbol duration and binary throughput*: With a constant envelope digital modulation, binary data throughput is approximately 1bit/Hz or less. In a 25-kHz channel, with a 25 kbps transmission rate and one bit per symbol, the symbol duration is about 40 μs. The intersymbol interference is thus very low because multipath propagation delays are rarely higher than 5 μs (see Chapter 2).
- *Low mobile terminal complexity*: FDMA transmission mode does not require complex equalization or framing and synchronization related to the burst transmission, such as in TDMA systems, because the information is transmitted and received synchronously without interruption.
- *Low transmission overheads*: As the transmission is continuous, only a few bits of overhead are required. Control information, such as handover commands, is transmitted on the user information channel.
- *High fixed equipment costs*: One FDMA disadvantage comes from the fact that it requires the implementation of more transmission equipment at the base station than in the TDMA case. This is because there is only one channel per carrier.
- *Use of duplexers*: As the transmitter and receiver must operate simultaneously, the mobile must use a duplexer to avoid interference between them. This equipment adds additional costs.
- *Handover complexity*: Because of the need for transmission continuity, processing a handover from one channel to another is more difficult to manage than in a TDMA system where switching between two channels can be realized during the time interval between two consecutive transmission slots.

3.1.3 Time-Division Multiple Access

The TDMA technique may be regarded as the first alternative to the FDMA technique. It has been implemented in current digital systems in preference to the CDMA method. It allows transmission of larger information rates than in an FDMA system. The carrier (radio frequency) is divided into N timeslots (or simply, *slots*) and can thus be shared by N terminals. Each terminal uses a particular slot different from slots used by the other terminals. The number of slots per carrier is chosen according to various factors, such as the modulation technique or the available frequency band.

3.1.3.1 Basic Principles

The TDMA transmission is discontinuous. For example, the mobile transmits on *slot* 1, waits during *slot* 2, and receives on *slot* 3, waits again during *slot* 4, and transmits again on *slot* 1.

As a consequence in this type of transmission, the gross channel rate is not equal to the communication rate (i.e., the channel throughput seen by the user). The channel throughput must be faster by a factor of at least the number of slots in a frame.

Example 1: Transmission rate in a slot
Take the configuration represented in Figure 3.7, where each frame has 4 slots. A mobile transmits speech with a rate of $d = 16$ kbps (13 kbps with overhead) in slot number 2 of the frame. The transmission rate must thus be equal to $D = 4 \times 16 = 64$ kbps during this slot.

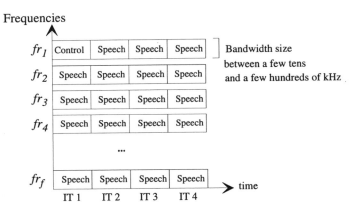

Figure 3.7 Example of TDMA architecture.

3.1.3.2 Characteristics

The different TDMA system characteristics are the following [1]:
- *Several circuits per carrier*: All TDMA systems must by definition multiplex at least two circuits or channels per frequency carrier. The GSM system, for instance, multiplexes eight circuits per carrier.
- *Burst transmission*: The transmissions are not continuous. At any moment, several mobiles (some with a high power level, others with a low power level) connected to the system, but in different cells are transmitting. The impact on the co-channel interference level varies significantly from one slot to another.
- *Narrowband or wideband*: The channel bandwidth required for TDMA systems is in the order of a few tens to a few hundreds of kHz. The bandwidth is determined by

modulation technique. For instance, GSM channel bandwidth is 200 kHz with a transmission rate of 271 kbps using GMSK modulation with a BT value of 0.3.

- *Binary rates and symbol duration*: The highest channel rates are between 300 and 400 kilosymbols per second (one bit per symbol). This makes intersymbol interference much more important than in an FDMA system. For example, for a 300 kilosymbol rate, the symbol duration will be 3.33 µs, which represents about the same duration as the observed delay spread in an urban environment. The intersymbol interference thus has an important impact on TDMA systems throughput.
- *Complexity of the mobile equipment*: The TDMA mobile terminal is more complex than an FDMA one, particularly in the digital processing part.
- *High overheads*: The burst transmission mode requires the receiver to synchronize at each burst. Furthermore, guard times are necessary to separate a burst from the preceding and following ones. The message overhead transmitted in TDMA systems can be as high as 20 to 30% of the total number of transmitted bits. To these overheads, a training sequence must also be added.
- *Cheaper base stations*: The main advantage of a TDMA system compared with a FDMA system is that the first one requires fewer radio channels than the latter, which implies a reduction in the amount of equipment at the base stations.
- *Handover complexity*: As the transmitter is inactive during certain slots, the mobile may process handovers (see Chapter 8) more efficiently than in an FDMA system. Furthermore, during the idle periods, it can scan neighboring base station channels during the handover preparation phase.

TDMA systems often use a complex hierarchical structure (e.g., GSM system). In Figure 3.8, a general case of the timing hierarchy that can be adopted in a TDMA system is shown.

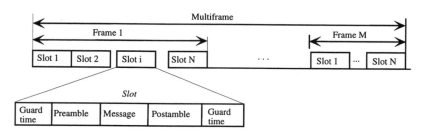

Figure 3.8 Typical TDMA timing hierarchy.

The different elements that can be identified in this structure are
- The *burst*, which is the information transmitted in one slot.
- The *slot*, which is the time unit in which a burst can be transmitted.

- The *frame*, which is a group of N consecutive slots that corresponds to the slots' periodicity associated with a communication.
- The *superframe*, which is a timing structure comprising a fixed number of consecutive frames. The superframe structure is used to organize system and network signaling.
- The *guard time*, which is the time interval between bursts used to avoid bursts overlapping.
- The *preamble*, which is the first part of a burst.
- The *message*, which is the part of the burst that includes user data, and whose size is larger than that of the other portions of the burst.
- The *postamble*, which is the last portion of the burst. It is used to initialize the following burst.

A burst may include a training sequence used by the equalization process (see Chapter 4).

Example 2: Channel organization in the GSM system (Figure 3.9).

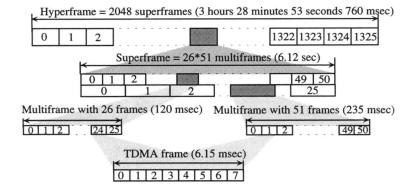

Figure 3.9 Frame organization in GSM.

GSM channel structure is organized in 5 levels:
- The slot level,
- The frame level—one frame has eight slots,
- The multiframe level; with two types of multiframes, a 26 multiframe (has 26 frames) and a 51-TDMA multiframe (has 51 frames),
- The superframe level; one multiframe contains 1,326 frames,
- The hyperframe; one hyperframe contains 2,048 superframes.

3.1.3.3 Timing Advance and Guard Time

The implementation of a TDMA structure requires the mobile transmissions to be synchronized to ensure that the transmitted messages do not overlap and do not go beyond the limit of their timeslots. Mobile synchronization can be realized globally with each mobile and the system locked to universal time. Nevertheless, mobiles can lose their synchronization (e.g., in GSM the accuracy is in the order of the tens of microseconds).

A mobile is synchronized after its first access to a different base station on the network. Transmission delay time adjustment is implemented by the base station to which the mobile is connected. In the case represented in Figure 3.10, two mobiles A and B are located respectively at distance d_A and d_B from the base station, with $d_A \ll d_B$. Without A and B transmission synchronization, their messages arrive at the base station with different delays. These delays Δ_A and Δ_B are such that

$$\Delta_A\left(=\frac{2d_A}{c}\right) \ll \Delta_B\left(=\frac{2d_B}{c}\right)$$

which may cause A and B messages to overlap, corrupting the transmissions.

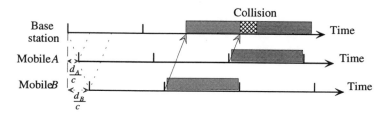

Figure 3.10 Transmission without timing advance.

Prior to full mobile synchronization the base station receives a message from mobile i, which allows it to determine Δ_i. The mobile is able to adjust the start of its transmission after the reception, from its base station, of the value of Δ_i. This parameter is called *timing advance*. By determining the value of this parameter for each transmitting mobile, the transmitted bursts will fit exactly in the timeslots when they arrive at the base station (Figure 3.11).

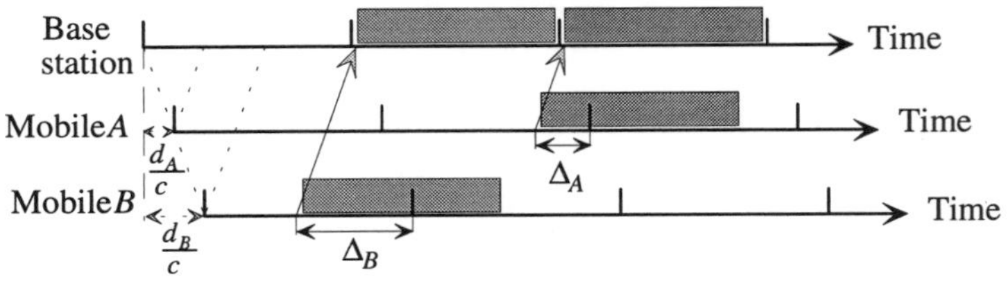

Figure 3.11 Transmissions with timing advance.

To take into account the small changes of Δ_is, which are due to variations in the mobiles and their altering of position, a *guard time* is introduced at the beginning and at the end of each timeslot. Nevertheless, the problem remains with the initial access of the mobile when it has not yet received its value of Δ_i. For this reason, the GSM system (Figure 3.12) defines an access burst the format of which is different from that of all other bursts. This burst is only a short burst (77 bits instead of 156 for the normal bursts). This ensures that even for a mobile located at the boundary of a cell whose radius is the maximum acceptable in GSM (and thus for which the value of Δ_i is also maximum), the transmitted burst will not overlap with the following timeslot. The maximum value of Δ_i is 233 μs, which allows a maximum cell radius of

$$233 \times 10^{-6} \times 3 \times 10^{5}/2 \cong 35 \text{ km}$$

3.1.4 Code-Division Multiple Access

CDMA architecture is based on the *spread-spectrum* technique. Use of this technique started with military systems because of the advantages it offers in tactical environments. This technique is used, for instance, in the GPS (see Chapter 8) and joint tactical information distribution system (JTIDS) [2]. It was in 1978 that spread-spectrum techniques were proposed for the first time for high-capacity cellular mobile communications systems.

In 1991, the Qualcomm Company proposed the use of CDMA techniques for cellular mobile communications. The EIA/TIA/IS-95 standard (see Chapter 12) is the result of this proposal and was published in July 1993. The first commercial CDMA network was opened in Hong Kong in September 1995.

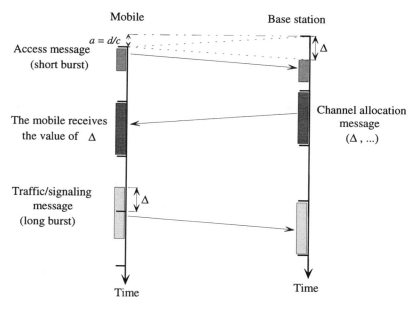

Figure 3.12 Mobile synchronization in GSM.

The definition of spread-spectrum techniques can be summed up by the following characteristics:

- It is a transmission mode where the transmitted data occupies a larger bandwidth than that required to transfer the communications data. The name "spread-spectrum" comes from the fact that the signal occupies more spectrum than in FDMA and TDMA systems.
- Spread-spectrum is realized before transmission by addition of a code that is independent of the data sequence. The same code is used at the receive end, which must operate synchronized with the transmitter, to despread the received signal in order to recover the initial data.

The CDMA technique is based on spread-spectrum modulation and has the peculiarity of being an access method where many terminals are authorized to transmit simultaneously and in the same frequency band.

Example 3: CDMA—an analogy

For a good understanding of the way the CDMA technique operates, an analogy can be made between a system using this access method and a room with many people talking to each other in pairs. In this room, each person A_i can communicate with his or her party B_i, because he or she can track the sound of B_i's voice and extract it from the overall interference created by other people's voices. If each pair speak a language (a code in the case of the CDMA technique) that is different from that used by the other couples, extracting the desired voice is made easier because the correlation between this voice (the signal) and the others (the interferers) is lower. Similarly, the codes used in the CDMA technique are defined so that they have a low correlation.

Two basic spread-spectrum techniques are defined, *fast frequency hopping CDMA* (*FFH-CDMA*) and *direct sequence CDMA* (*DS-CDMA*). Hybrid methods exist as well as variants of these two forms of CDMA that have properties similar to the first two. Nevertheless, DS-CDMA seems to offer some advantage in terms of spectral efficiency [3].

3.1.4.1 Principle

In the CDMA technique, each station is allocated a random sequence or code. This sequence must be different and orthogonal or quasi-orthogonal (i.e., decorrelated) from all other sequences. In FFH-CDMA, this code is used to generate a unique hopping sequence. In DS-CDMA, this code is used to make a quasi-random (i.e., similar to noise) high-rate transmitted signal combined with the information to spread the spectrum.

Example 4: Spreading gain calculation

The information rate D to transmit with an R baud modulation is $R = 1/T$ (T: symbol duration) $\approx D$. If a direct sequence spread-spectrum technique is used, the information rate D is transmitted at a rate $N.R$. (N large, from a few tens to a few thousands) and is therefore approximately N/T. The spreading gain (in dB) is thus $10\log_{10}(N)$.

3.1.4.2 Characteristics

The main characteristics of this type of system are the following:

- *Large number of channels per carrier*: CDMA systems use a single or very few carrier frequencies. Each carrier is theoretically able to serve tens of users simultaneously.
- *Bandwidth*: Present CDMA systems require significant bandwidth; that is a channel bandwidth between 1 to 10 MHz. Nevertheless, the transmissions have a limited spectral density (i.e., transmitted power per Hz), which as a consequence reduces the jamming effects on other communications.
- *Binary throughput and symbol duration*: Because of the high gross bit rate, the symbol duration is very short. With a 1 Mbps throughput, each symbol has a duration of about a microsecond (in the case of BPSK modulation). This property improves the timing resolution, which becomes proportional to $1/NT$ instead of $1/T$.

This is useful for mobile-base station timing measurements and in the recovery of multipath signals.

- *Mobile complexity*: The digital processing of received and transmitted information is more complex than in other kinds of systems due to the additional level of coding.

Antijamming and discretion properties. Spread-spectrum modulation techniques were quickly adopted for military applications because of their interesting properties (Figure 3.13).

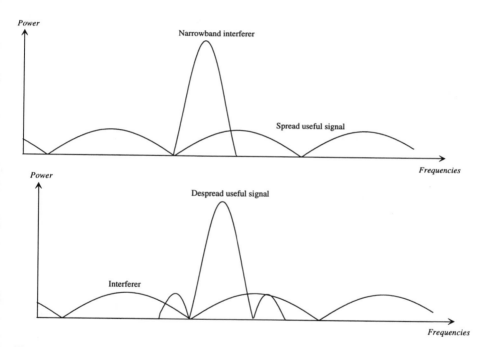

Figure 3.13 Discretion and protection against narrowband jammers.

The advantages they present for the military applications are the following:

- *Low probability of intercept*: the energy spreading within a wideband minimizes the probability of intercept by an eavesdropper,
- *Confidentiality*: receivers that do not have the appropriate despreader (which is difficult to determine) and cannot demodulate the signal; only authorized mobiles are able to demodulate the transmitted information,

- *ti-jam*: resistance to enemy efforts to jam the communications.

The despreading operation at the receiver side effectively spreads the spectrum of all discrete jammers (whether deliberate or not) allowing the communication to take place even with interference on the channel. This discretionary property is introduced because the signal occupies a bandwidth much larger than the minimal bandwidth required to transmit the signal. As a consequence, the transmitted signal also appears similar to noise. The transmitted signal is able to propagate through the channel without being detected by an eavesdropper.

Power control requirements. One of the main disadvantages of the CDMA method is the necessity of having a very efficient power control mechanism. Fast and accurate transmitter power control is required at the mobiles to maximize the number of users simultaneously communicating in the system.

The power control mechanism can be achieved either in an *open loop* or in a *closed loop*. The open-loop method is based on the hypothesis that power loss is similar on the uplink and the downlink. The mobile terminal keeps the sum of the received and transmitted powers at a constant level (typically −73 dBm). Transmitted power is adjusted inversely according to the level of received power. In the closed-loop power control method, the base station senses the received power of the signals transmitted by the mobile and sends commands to the mobile to make it adjust its power in fixed steps. In the IS-95 system, for example, adjustments are possible 800 times per second in steps of 1 dB.

Seamless handovers. The CDMA technique theoretically allows each base station to use the whole bandwidth allocated to the system. From the system point of view, this allows the realization of *seamless* or *soft handovers* (see Chapter 8) which minimizes the probability of link loss or quality of service degradation. This also allows the benefit of macrodiversity (see Chapter 4), where each mobile can simultaneously communicate with several base stations.

3.1.4.3 Direct Sequence Spread-Spectrum

The basic principle of DS-CDMA spread-spectrum transmitter/receiver chain is reproduced in Figure 3.14. At the transmitter level, the radio frequency carrier is modulated by a binary flow with rate R. The modulated data is then input in a spreader and modulated at a much higher chip rate W. In practice, the two modulations are combined.

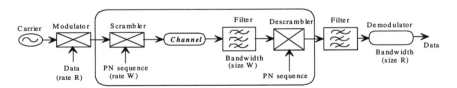

Figure 3.14 Transmitter and receiver for direct sequence spread-spectrum.

At the receiver, the signal is modulated again using the same pseudo noise spreading code (see definition in the following section) as in the spreader. The signal energy spread over W Hz on the channel is collected by the despreader on a band of only R-Hz wide. The spectral density of the additive white noise is not affected by this process, as can be seen in the following.

First of all, a PN sequence must be specified. It is defined as a coded sequence of I's and 0's possessing certain autocorrelation properties. Used sequences are generally periodic (the I's and 0's sequence is repeated with a known periodicity). Such sequences have long periods and require simple equipment. A maximal length sequence of this type is always periodic with a period $N = 2^m - 1$, where m is the length of the shift register used. PN sequences are chosen because of their inherent security features (ciphering) and for their autocorrelation properties.

The data sequence $b(t)$ is used to modulate a wideband PN sequence $c(t)$ by applying two sequences to a product modulator (or multiplier). For this purpose, the two sequences are represented by their polar form; that is by two levels equal in amplitude and opposed in polarity (e.g., -1 and $+1$). The PN sequence has the role of a spreading code. By multiplying the narrowband signal $b(t)$ by the spreading code $c(t)$, each information bit is "cut" in several timing slots called *chips*. For the baseband transmission, the product $m(t)$ represents the transmitted signal: $m(t) = c(t).b(t)$. The received signal $r(t)$ is the sum of $m(t)$ and additive interference denoted by $i(t)$: $r(t) = m(t)+i(t)$.

To recover the original sequence $b(t)$, the signal $r(t)$ is again presented at the input of a demodulator followed by a lowpass filter. The multiplier uses a PN sequence that is the exact copy of that used at the transmitter. The receiver must therefore be perfectly synchronized to the transmitter, which means that the PN sequence is exactly aligned with that on the transmitter. The resulting demodulated signal is thus

$$Z(T) = C(T).R(T) = C^2(T).B(T)+C(T).I(T) \qquad (3.2)$$

Therefore, the required signal is multiplied twice by $c(t)$, whereas the jamming signal is multiplied only once.

As $c(t)$ consists of a succession of $+1$ and -1, this succession is disrupted when squared: $c^2(t) = 1$ for all t where $z(t) = b(t)+c(t).i(t)$. The sequence $b(t)$ is then recovered at the output of the multiplier with a term representing the additive interference.

The signal $b(t)$ is narrowband, while the signal $c(t).i(t)$ is wideband (as $c(t)$). Thus, applying a baseband filter (lowpass) with restricted bandwidth so that the signal $b(t)$ can be recovered, the product $c(t).i(t)$ also becomes a narrowband signal which reduces its effective power.

To summarize, the general direct sequence spread-spectrum transmission/receiving scheme can be represented by Figure 3.15. In Figure 3.15, $\{w_1^i ...w_N^i\}$ represents the N chips spreading sequence. If $\{w_n^i\}$ is random enough then the throughput increases and the spectrum is spread with a factor $T_S/T_C = N$.

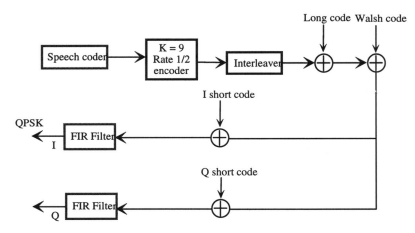

Figure 3.15 The IS-95 transmission chain.

Example 5: DS-CDMA transmission/reception operating principle

In Figures 3.16 and 3.17, the schematic operation of DS-CDMA transmission/reception is represented.

In Figure 3.16, only E_1 transmission takes place. The S_1 sequence that corresponds to the bit value of *1* is thus input (after supposed error-free and interference-free propagation) into the despreader of R_1 and R_2. As R_1 only knows the code S_1 and is synchronized to it, it can extract the signal and recognizes the bit *1* initially transmitted. At the R_2 side, the received signal is considered as noise and nothing is presented at the receiver input.

In Figure 3.17, transmitters E_1 and E_2 simultaneously transmit on the channel. Reception of each of them is affected by the interference created by the other transmitter. R represents the signal resulting from the transmission of the two signals coming from E_1 and E_2. If R is presented at the input of R_1 despreader and correlated to the sequence S_1, the resulting signal allows recovery of the bit *1* transmitted by E_1.

It is obvious that the higher the number of transmissions on the same carrier, the higher the error probability at the receiver. To obtain *1*, the output of the correlator is compared with a threshold and a decision on the received data can be made.

Conclusions. DS-CDMA systems are relatively complex and were only of interest until recently for specific applications (generally *satellites* or *military*). With the tremendous progress of technology, this method is presently used in an increasing number of cellular systems (IS-95 standard) in the world and in some wireless LANs (see Chapter 13).

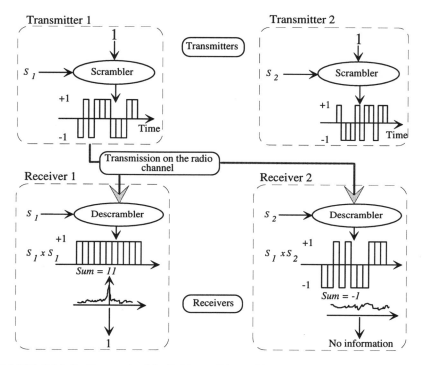

Figure 3.16 S_1 (S_2) is the binary code of the first (second) transmitter.

3.1.4.4 Fast-Frequency-Hopping Spread-Spectrum

The second main category of spread-spectrum techniques is *frequency hopping* (*FH*). This technique, as the preceding one, has for a long time been implemented in military systems as a solution to force hostile jammers to cover a larger spectrum. It consists of making the modulated carrier hop randomly from one frequency to another. This results in the spectrum of the transmitted signal spreading sequentially instead of instantaneously. Two frequency hopping modes are defined.

- *Slow Frequency Hopping* (*SFH*), where the symbol rate R_s of the signal is an integer multiple of the hop rate R_h. In this case, several symbols are transmitted at each frequency hop. This form of frequency hopping can also be used in TDMA systems (see Chapter 6).

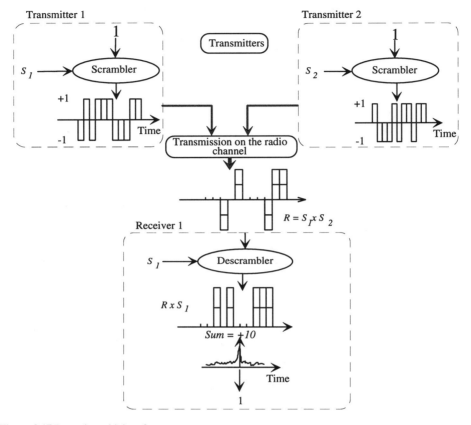

Figure 3.17 Reception with interference.

- *Fast Frequency Hopping* (*FFH*), where the hopping frequency is a multiple integer of the symbol rate. This means that the carrier frequency changes (hops) several times during the period of each burst transmission.

Considering the CDMA method, the focus here is on FFH.

The basic principles of the transmission chain in the FFH-CDMA technique is reproduced in Figure 3.18.

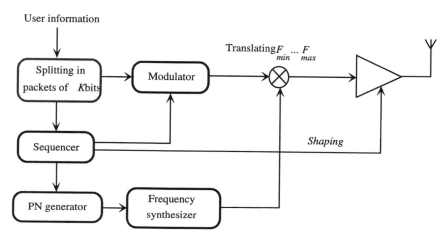

Figure 3.18 Basic principle of the transmission chain in FFH-SS.

The receiver takes the binary data at the input of the modulator. The resulting modulated wave and the output of the digital frequency synthesizer are then applied to a mixer that consists of a multiplier followed by a filter. This latter is designed to select the output of the multiplying process to transmit it. In particular, k bit segments of a PN sequence direct the frequency synthesizer which allows the carrier frequency to hop on 2^k different values.

Frequency hopping systems are well suited for working in tactical environments (where very powerful jammers only jam part of the bandwidth). The antijamming property is obtained after detection of the jammed bursts (channel/frequency) and by correction of the codes. It can therefore be qualified as a *passive* type. In FFH-CDMA (Figure 3.19), the transmission, and thus the number of transmitted bits, on the carrier lasts just a fraction of time on each frequency. This technique allows the information loss to be limited to when the channel experiences a selective fade or collision with a jammer.

At the receiving side, noncoherent detection is used. Nevertheless, the detection procedure is somewhat different from that used in slow frequency hopping. In particular, two procedures can be considered. First, for each symbol, separate decisions are made on the received chips using a simple rule based on majority voting to estimate the received symbol. Then, for each symbol, the likelihood functions are calculated as ones for the total received signal of the K chips and then the most dominant is selected.

Example 6: FFH on four levels

Assume an FFH system which uses a four-level M-ary frequency shift keying (*MFSK*) modulation. Figure 3.20 illustrates the transmission process on the different sub-carriers.

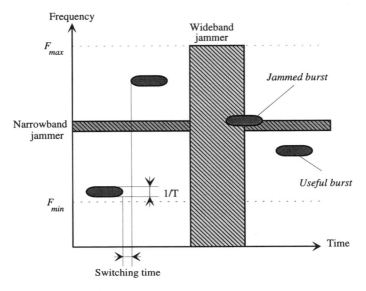

Figure 3.19 FFH-SS transmission example.

In this example, 3-bit chip sequences are used, which can take eight possible values (000, 001, ... 111). The FFH carrier of the hop is obtained by FFH modulation on one of the eight frequencies. The sub-frequency on this carrier is determined by MFSK modulation (four possible levels).

3.1.4.5 Mixed Systems

Mixed systems combine the two previous methods. The basic principle generally consists of defining frequency hopping using only one symbol per burst. The modulation becomes M-ary and not simply binary as in the direct-sequence spread-spectrum technique. It thus leads to an intrinsic improvement of the performance by gathering the bits into m bit packets. Therefore, the value of a packet (from 0 to 2^{m-1}) corresponds to a waveform among M ($M = 2^m$). A unique decision can be taken for the m bits. Furthermore, by the use of interleaving (see Chapter 4) to scatter error bursts and error detecting codes systematic error correction can be included.

Presently, mixed systems are the most efficient. Indeed, they are designed to offer resistance to all kinds of jamming and to the problems of interception/intrusion. Detecting the presence of a transmission and initial synchronization are, however, more complex to achieve than for the previous methods. This is because of the difficulty in detecting wideband signals with low SNRs. These systems are, for these reasons, more complex and

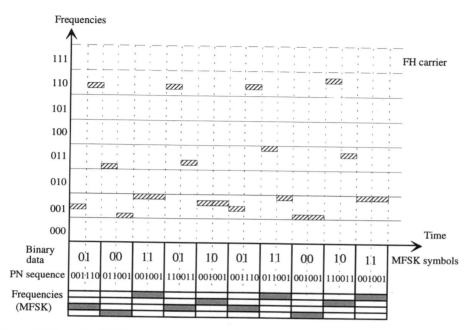

Figure 3.20 Example of FFH transmission with a four-level MFSK modulation.

more expensive, and are used for very specific applications such as JTIDS (AWACS military aircraft communications system), or MILSTAR (satellite wideband system).

3.1.4.6 Power Control

In general, the reverse link is not synchronized, which makes the arrival time of the codes different from one mobile to another. The base receiver for each mobile then receives interference from other mobile transmitters as the codes are not orthogonal. The number of served mobiles in the same cell is limited by this interference [4].

Consequently, power control is an essential function to implement on the reverse link to minimize interference. Without power control, the transmissions of the mobiles situated far from the base station are not decoded because of the interference generated by the mobiles close to the receiver. If all the signals arrive with the same power, the tolerance to the CDMA interference is proportional to a parameter that is called the system processing gain. Alternatively, the processing gain can be defined as the factor by which the receiver's input signal to interference power ratio is multiplied to yield the SIR at the output of the receiver's correlation detector [2]. The obvious way to increase CDMA system capacity is thus to reduce the interference level through the implementation of a power control mechanism.

3.1.4.7 Performance

The CDMA technique is a very hot topic for the mobile radio community (researchers, manufacturers, and network operators). The capacity it promises to offer is claimed to be greater than that of TDMA or FDMA techniques. Nevertheless, experiments have been carried out that show the technical feasibility of this kind of system, but so far without giving any indication about its behavior at high load (most critical conditions).

3.1.5 Conclusions

The multiple access method allows the system designers to define how the frequency band is split into several channels. The next step consists of defining the way mobiles make access to these channels. The random access protocol allows a group of mobiles to have access to the communication channels with minimum contention without the need for central control.

Combining the three access techniques (FDMA, TDMA, and CDMA) has been considered and will be used for third-generation systems. Combined FDMA/TDMA access methods show good spectral efficiency and the access methods combining FDMA and CDMA offer protection against co-channel interference (Figure 3.21).

Note: The *space division multiple access* (*SDMA*) method is a technique that has seen a growing interest since the beginning of the 1990s. It relies on the splitting of the space into several regions (by antenna processing techniques), which allows a spatial sharing of the spectrum. It is more a method based on the interference reduction that is mainly used in radar systems. This method is addressed in Chapter 4.

3.2 RANDOM ACCESS PROTOCOLS

In mobile radio systems, the number of mobiles located in the area covered by a base station are constantly changing. Contrary to the case of the fixed network where each terminal has its own communication channel, at least from the user terminal point to the first network equipment, the mobile terminal must have access to a channel shared by all the mobiles located in its area. The use of an access method is thus necessary.

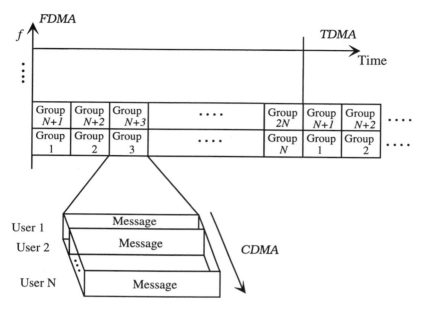

Figure 3.21 Combining the three multiple access methods.

Access protocols can be classified in four main categories:

- Random access protocols (e.g., ALOHA, S-ALOHA, CSMA),
- Centralized access control demand protocols,
- Distributed access control demand protocols (e.g., FIFO, token),
- Adaptive access protocols.

The constant changes in the mobile population identities within an area make it very difficult to use tokens or priorities (as in the case of distributed access control protocols such as are used on LANs). Also, base stations would ignore the number and the identities of the mobiles present within their cell coverage should they require access to the resource. Thus, a centralized access demand protocol is not usually adopted. In contrast, random access methods allow efficient and flexible access to the channel. One of the consequences of this kind of method is that the mobiles will contend (i.e., compete) to have access to the channel. This contention can give rise to collisions among the messages transmitted by different mobiles. These access methods are often called *contention schemes*.

The access problem in mobile radio systems is complicated because of the combining of two important phenomena:

- *Hidden nodes* exist because of the mask effect (Figure 3.22),
- *Capture effect* or *near/far effect* shown in Figure 3.23.

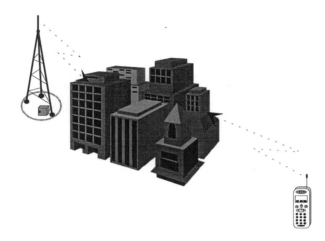

Figure 3.22 Mask effect (MS and BS are hidden stations, one for the other).

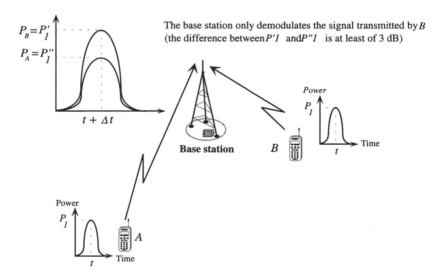

Figure 3.23 Capture effect.

3.2.1 Protocols Nonslotted and Without Carrier Sensing

The simplest random access algorithm is certainly the *ALOHA protocol* or *pure-ALOHA*. This protocol owes its name to the ALOHA system developed by N. Abramson [5] for radio communication at several sites on the Hawaiian islands.

The principle of this protocol is very simple. When a station has a message to transmit, it transmits it on the radio channel without taking note of the channel occupancy. Of course, as users send their packets arbitrarily collisions among packets will occur when the transmissions overlap.

Figure 3.24 represents the operating principle of this protocol. The message *1* is transmitted without problem but the end of message *2* is jammed by the transmission of message *3*. Both messages are thus lost. Message *3* transmitter generates a random waiting delay τ and retransmits its message after this period because there has been no reception of an acknowledgment. Message *2* will similarly attempt a retransmission but at a different random time.

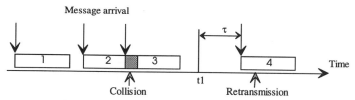

At time *t1* , no acknowledgment for message 3 is received:
determination of a random waiting period before new access attempt

Figure 3.24 ALOHA message-arriving process (collision of messages 2 and 3).

Example 8: Nonslotted ALOHA protocol throughput calculation

We assume the following hypotheses: *n* stations are waiting; packets length is equal to *1*; τ is the waiting time, which is a random variable exponentially distributed with the probability density $x.e^{-x.t_i}$ (x is an arbitrary parameter that can be seen as the throughput of the retransmission attempts of the station; t_i is the duration of the interval between the beginning of the ith and the $(i+1)^{th}$ transmission attempts); and λ is the arrival rate that follows a Poisson process.

The train of retransmission attempts follow a Poisson process with the law $G(n) = \lambda + nx$. The ith attempt will succeed if t_i and t_{i-1} are greater than *1* (packet length). The probability of success is: $Prob(Success) = e^{-2G(n)}$.

With $G(n)$ being the transmission attempt rate, then there is a throughput (defined as the number of successful transmissions per time unit), which is a function of *n*:

$$D(n) = G(n).e^{-2G(n)} \tag{3.3}$$

The maximum throughput of ALOHA is obtained for $D(n) = 1/2e$. The major drawback of the ALOHA protocol stands in its low performance at high load (instability and limited throughput) because of the uncoordinated mobile transmissions.

3.2.2 Carrier Sensing Protocols

Sensing the carrier before transmission is the basic principle of *carrier sense multiple access* (CSMA)-type protocols. The performance is for this reason higher than that of ALOHA. The collision problem can be reduced by previous sensing, but as a consequence there is a loss of capacity due to this sensing period before transmission.

3.2.2.1 1-Persistent CSMA

The simplest basic version of the CSMA technique is called *1-persistent CSMA*. Each station listens to the channel before transmitting, and the transmission of a message is authorized only if the channel is idle. If the channel is busy, the station keeps listening until the end of the current transmission in order to transmit as soon as the channel is detected idle. The name of *1-persistent CSMA* comes from the fact that the station transmits with a probability *1* when the channel becomes idle (Figure 3.25).

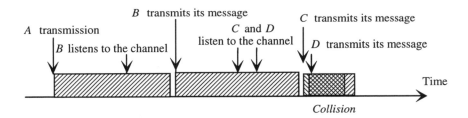

Figure 3.25 1-persistent CSMA.

In the case of a collision (by the lack of an acknowledgment received at the sender after a predetermined period), the originating station retransmits after a time interval (of fixed or random duration).

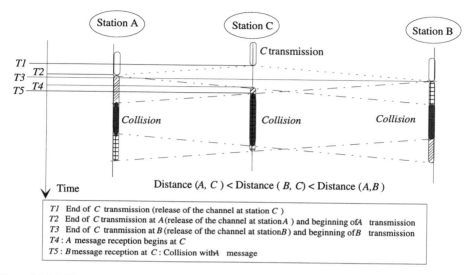

Figure 3.26 Collisions problem with the CSMA 1-persistent protocol.

The drawback of this method lies in the fact that the distance between the stations induces a propagation delay, called the *vulnerability* period, during which collisions may occur. Here, Figure 3.26 represents the state of the channel seen by different stations in the network. The channel does not become idle at the same time for all the stations, which may cause collisions in the case of two stations wishing to transmit as soon as each of them senses the channel is idle. To solve this problem, the nonpersistent version of the CSMA protocol has been proposed.

3.2.2.2 Nonpersistent CSMA

In the *nonpersistent CSMA* version, a station that detects the channel is busy delays its transmission attempt by a random time (Δ in the example of Figure 3.27).

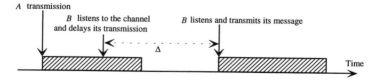

Figure 3.27 Nonpersistent CSMA.

The use of this variant eliminates a large number of collisions. Therefore, *nonresistant CSMA* gives rise to better performance than the *1-persistent CSMA*. Nevertheless, at low load, the random waiting time before new access attempt makes the transmission delays longer and introduces a degradation compared with that of the *1-persistent CSMA* version.

Example 8: Nonpersistent CSMA protocol throughput calculation

Let a = *propagation delay/packet transmission duration*. The throughput of nonpersistent CSMA is thus defined in the function of G (system load):

$$S = \frac{G \cdot e^{-a.G}}{G(1+2a) + e^{-a.G}} \tag{3.4}$$

3.2.2.3 p-Persistent CSMA

The *p-persistent* version of CSMA combines the advantages of the two previous versions by introducing the notion of timeslot (or slot). Note that the slots used here are not the same as those defined in the TDMA method.

The duration of a slot is equal to the maximum propagation duration (duration between the transmission and the reception of the two farthest stations the network). A station that has a message to transmit listens to the channel and if idle it transmits with the probability p . It delays its transmission in the next slot (i.e., introduces slippage) with the probability $q = 1 - p$ (Figure 3.28). If the channel is busy the station remains sensing the channel and repeats the previous procedure as soon as the channel becomes idle.

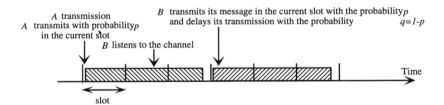

Figure 3.28 p-persistent CSMA

The performance of this protocol is highly dependent on the optimization of the parameter p .

3.2.2.4 CSMA With Collision Detection

The CSMA/CD (collision detection) reduces the jamming period by stopping the transmission in case of a collision. This protocol assumes that each transmitting station is also listening to the channel (*listen-while-talk*). A transmitting station that detects the collision of its message automatically stops transmitting and a *jamming packet* is transmitted to compel the other colliding stations to stop transmitting (Figure 3.29). The additional advantage of the simultaneous transmission/reception is that the messages do not require acknowledgment from the addressees.

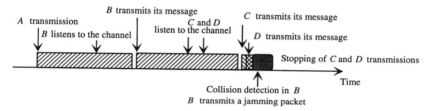

Figure 3.29 CSMA/CD.

This option is preferred as the unused channel is freed up sooner than if the colliding messages are allowed to continue and each sending mobile is waiting for an acknowledgment. The principle of collision detection can be implemented with all the variants of the CSMA protocol. Nevertheless, the maximum throughput of CSMA does not exceed 50%.

3.2.2.5 CSMA in the Mobile Radio Context: The BTMA Protocol

In the mobile radio context, the hidden stations problem means that the CSMA protocols cannot be used just as they are. In Figure 3.30, station A sees the channel idle and transmits a message although the channel is engaged because of the transmission of station B. The reception of the two stations at the third station C is jammed.

The use of a signaling channel called *busy-tone channel* (BTC) partially solves this problem. This protocol uses, in addition to the data transmission (data channel, DC), a BTC channel to transmit a signal (busy tone) during the transmission of a message on the DC. An ongoing data transmission (the DC is busy) is indicated to the other stations by transmitting a busy tone on the BTC. There are two possible cases, either it is the transmitting station that sends the busy tone during the transmission of its message, or more common it is the station which demodulates these signals that transmits the busy tone (in this latter case, the busy tone is used as an acknowledgment).

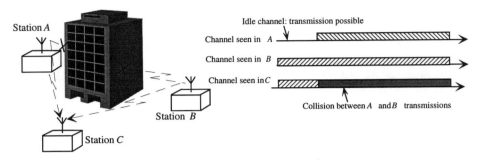

Figure 3.30 Hidden-station problem in the CSMA protocol.

The Busy Tone Multiple Access (BTMA) protocol is therefore similar to CSMA in the mobile radio context. Numerous studies [6, 7] have shown the superiority of the BTMA protocol compared with CSMA and ALOHA protocols in pure form. All variants of the CSMA method can be implemented in mobile radio systems using the BTMA principle (Figure 3.31). This protocol has been implemented in many packet radio networks.

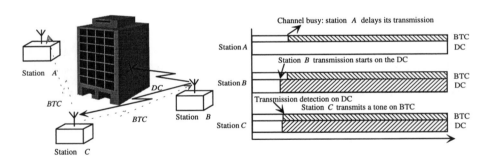

Figure 3.31 BTMA protocol principle of operation.

3.2.2.6 DSMA

The *digital/data sense multiple access* (DSMA) protocol is also a random access protocol based on the CSMA method and used very widely in data radio transmission systems such as CDPD and ARDIS [3] (see Chapter 13).

DSMA is used in a centralized context where the base station includes in its control channel message a flag indicating the state (busy or idle) of the uplink channel. The mobile stations listen to this flag before deciding to make access. If it indicates that the channel is idle, the transmission may be tried on the next slot. As soon as the transmission start is detected, the base station sets the flag to the busy state until the end of the transmission.

Example 9: DSMA/CD protocol operation in the cellular digital packet data (*CDPD*) system

The nonpersistent slotted DSMA/CD protocol used in the CDPD uses mini-slots, which consist of 10 Reed-Solomon symbols; that is, 60 bits or 3.125 ms [8]. Only the stations that detect the idle channel can transmit. When the channel is busy, access transmissions are delayed (Figure 3.32).

The downlink channel carries two indicators used to access to the uplink channel:

- *Channel state*: indicates the state (*I* for idle, *B* for busy) of the uplink channel. A mobile with a message to transmit delays its transmission until it detects a value of the flag equal to *I*.
- *Decoding state*: indicates if the message from the station which accessed successfully has been correctly decoded by the base station. In case of collisions, the transmitting station is informed by this indicator that data has been corrupted.

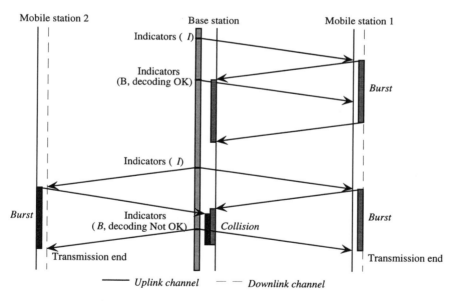

Figure 3.32 Transmission with the DSMA/CD protocol.

3.2.3 Non-Sensing Slotted Protocols

One problem occurring in the ALOHA protocol is that the transmission of n successive colliding packets generates the loss of n packets and the sterilization of the channel during this period.

The *S-ALOHA* (*Slotted-ALOHA*) protocol [9] consists of allowing the start of transmissions at a specified time. The channel is slotted in timeslots identical to those used in the TDMA method where the beginning of the timeslots are the same for all stations (and thus requiring the necessity to synchronize all stations). The messages are limited to have a maximum length of the duration of one slot.

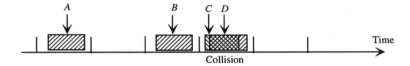

Figure 3.33 Slotted ALOHA.

Therefore, when a collision between two packets occurs (Figure 3.33) the channel is unused for a period of at most one slot (instead of two or more slots in the case of the pure-ALOHA protocol). The maximum throughput of this protocol is determined by using the same hypotheses as previously, and the following formula is obtained (at equilibrium $\lambda = G.e^{-G}$):

$$S=G.e^{-G} \tag{3.5}$$

Its maximal value is obtained for $G = 1$. The maximum throughput of the S-ALOHA protocol is therefore (See Figure 3.34):

$$S = 1/E \cong 0.368$$

Example 10: Throughput of the S-ALOHA protocol with capture effect

Split the users into two groups. A group (denoted by H) of transmitters close to the receiver and having a received signal level of P_H at this receiver, and a group (denoted by L) of transmitters located far from the receiver and at which the received power is at level P_L, with $P_L \ll P_H$ [10].

Let S_H (S_L) be the mean number of messages from the group H (L) received successfully in each slot, and G_H (G_L) be the mean number of messages transmitted per slot for the group H (L).

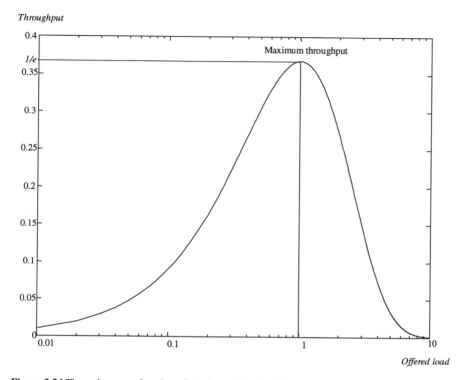

Figure 3.34 Throughput as a function of the channel load in ALOHA.

Assuming that the number of transmitters is very large and that the arrivals are Poissonian, the formula $S = G.e^{-G}$ can be applied and then

$$S_H = G_H. e^{-G_H} \tag{3.6}$$

The successful transmission of a message from group L requires that any other message from groups L and H is transmitted at the same time (i.e., with the probabilities e^{-G_L} and e^{-G_H}). The two transmissions are assumed to be independent, thus the successful transmission probability is equal to $e^{-(G_L+G_H)}$. G_L is equal to the number of new (i.e., transmitted for the first time) messages of group L plus the number of messages from group L that are retransmitted, that is:

$$G_L = S_L + G_L.PROB[RETRANSMISSION] = S_L + G_L.(1- e^{-(G_L+G_H)}) \tag{3.7}$$

Therefore

$$G_L = G_L \cdot e^{-(G_L + G_H)} \tag{3.8}$$

The total throughput S is the sum of the throughputs S_L and S_H and is therefore equal to (Figure 3.35 for the corresponding curves):

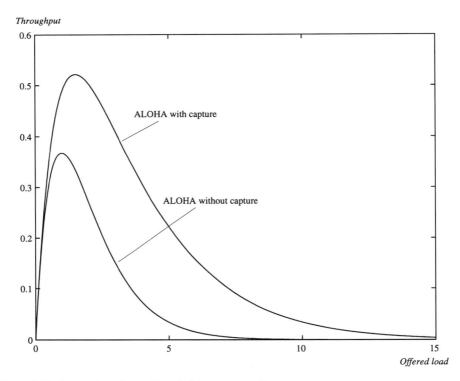

Figure 3.35 ALOHA throughput with and without capture effect.

$$S = G_H e^{-G_H} + G_L \cdot e^{-(G_L + G_H)} = e^{-G_H} \cdot \left(G_H + G_L \cdot e^{-G_L}\right) \tag{3.9}$$

Assuming that $G_H = G_L = G/2$, then

$$S_{Tot} = e^{-G/2} \cdot \left(\frac{G}{2} + \frac{G}{2} \cdot e^{-G/2}\right) = \frac{G}{2} \cdot e^{-G/2} \cdot \left(1 + e^{-G/2}\right) \tag{3.10}$$

3.2.4 Reservation-Based Framed Protocols

Reservation-based random access protocols can be classified into two categories: *Reservation-ALOHA (R-ALOHA)*-type protocols and *ALOHA-Reservation*-type protocols. Although there are no predefined reservation slots for the first type of protocols, the second type define some reserved slots per frame. These slots are only used for access requests. *R-ALOHA*-type protocols are simple and totally distributed. *ALOHA-Reservation* type protocols need control from the base stations to determine the number of information and access slots allocated in accordance with demand and the required quality of service.

The S-ALOHA protocol previously presented is efficient especially for short messages (i.e., messages that could be transmitted within a limited number of slots). On the other hand, the channel sensing protocols do not use a frame structure and have as a drawback the introduction of access delays. This particularly is the case when the transmitting station transmits a long message. The use of frames allows, therefore, better sharing of the channel among several users having messages of different priorities and lengths to transmit.

3.2.4.1 The R-ALOHA protocol

The *R-ALOHA* protocol is derived from the S-ALOHA protocol and uses a frame structure allowing N simultaneous users to transmit their messages on the same channel.

Several versions of the R-ALOHA protocol exist. Some versions make the distinction between the slots (used for the transmission of user data) from the *minislots* (used to transmit the access demands to the slots). The version presented here [11] is the simplest and uses an implicit reservation mechanism that does not require a central control process.

The terminals continuously evaluate the number of idle slots detected in the last frame, each frame being composed of N slots (in the example represented in Figure 3.36, N = 5). When a station has a message to transmit it processes the following algorithm. First, it randomly chooses a slot among the N' estimated as idle. Let j be the chosen slot. It then transmits its packet in slot j of the following frame. At the end of the transmission, the station senses the channel. If no other transmission is detected, slot j is now considered as reserved for that station. Otherwise it has to repeat the procedure.

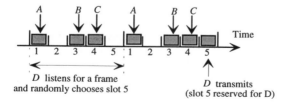

Figure 3.36 Reservation mechanism in the R-ALOHA protocol.

Two versions of this protocol are generally defined. Either the end of the transmission is indicated in the last packet or no indication for the end of the transmission is given. The latter implies the loss of a slot at the end of each transmission. The R-ALOHA protocol has been implemented within satellite systems and is also implemented for wireless LANs (see Chapter 13).

3.2.4.2 The PODA Protocol

The *priority oriented demand assignment* (*PODA*) protocol is used for satellite transmission systems. It is based on a frame structure with two parts, namely an *information field* and a *control field*. The information field consists of two sub-parts. The fixed-PODA(F-PODA) is used to transmit long messages in reserved channels and the contention-PODA (C-PODA) used to transmit short messages. The access mode to the C-PODA part uses the S-ALOHA protocol. The control field is used for contention in order to have access to the F-PODA part (Figure 3.37). The sizes of the different subparts are adjustable according to the traffic (e.g., in accordance with the traffic distribution between short and long messages).

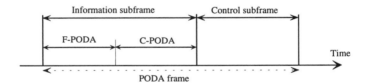

Figure 3.37 PODA frame.

3.2.4.3 The PRMA Protocol

The *packet reservation multiple access* (PRMA) protocol [12] can be considered a version of R-ALOHA within a centralized control system such as a cellular network. It uses the BTMA principle with indication of channel occupation made by the receiving station.

The state of the slots is indicated by the base station on the downlink channel: I for *idle* and R for *reserved* (Figure 3.38). A station that wants to have access to the uplink channel must first listen to the downlink channel, notice the idle slots (as in the R-ALOHA protocol case) and transmit on the chosen slot starting in the next frame. If the base station correctly receives this first packet, the slot (previously idle) becomes busy and acquires the state R. This informs the transmitting station that this slot is reserved for it and that its access attempt has been successful.

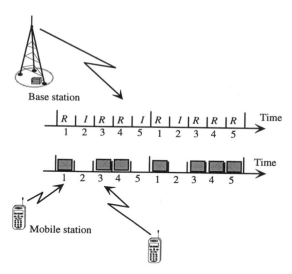

Figure 3.38 PRMA protocol behavior.

3.2.4.4 The RAMA Protocol

The *resource auction multiple access* (RAMA) protocol [13] is a protocol based on an access policy that goes through several stages (see also the access protocols used in the HIPERLAN and IEEE 802.11 systems, Chapter 13). It may be used for the allocation of channels in FDMA, TDMA, or CDMA systems.

Each terminal is identified uniquely by a number with several digits that is used during access. Each base station establishes a list of available or busy resources, and the idle resources are auctioned. After the auction process, each terminal is allocated a resource. The procedure is repeated until no demands or resources remain free.

During the access stage, the terminals transmit the digits of their number one after the other. Two different digits are orthogonal between each other; that is, if they are transmitted at the same time, the base station can demodulate both the signals. Note that a priority can be established among the services and according to the waiting period to access the system. This priority avoids too long access delays. We have detailed in Figure 3.39 the main stages of the access process in the RAMA protocol.

In Figure 3.39, T_s also denotes the transmission delay of one digit by a mobile station, and t_d is the waiting delay before a new transmission on the uplink channel (t_d is in practice taken equal to the maximum propagation duration in the cell plus the processing time at the base station.)

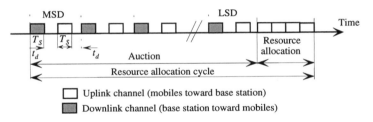

Figure 3.39 RAMA protocol behavior.

All the terminals transmit the first digit of their identifying number (*MSD* for—*most significant digit*). The terminals whose digit is the highest (e.g., 8 among 3, 6, and 8) are held by the base station for the next step during which they transmit the following digit of their identity number until the last digit (*LSD* for—*least significant digit*). This selection policy of the terminals by the highest digits could be replaced by a different policy if desired. In the case where several terminals having the same digit collide, the base station, even if it cannot demodulate the signal of each colliding station, will be able to detect a nonzero level of energy on the corresponding frequency and deduce the presence of stations.

Simulations [13] have shown that for a loss probability of 1%, the RAMA protocol allows a multiplexing gain of 2.63 with a fast voice-activity detector mechanism and of 2.28 in the case of a slow voice-activity detector mechanism.

3.3 CONCLUSIONS

In this chapter, the three main multiple access methods were presented that are used in current mobile radio systems and random access protocols, some of which are already implemented in real systems. All these methods defined for the radio resource access and sharing can be combined one with another. In general, ALOHA medium access control protocols are used to access FDMA, TDMA, or CDMA circuit-switched systems. When designing a packet data access system, other protocols must be used, because of the relatively low performance of pure-ALOHA at high load. In this case, CSMA or a variant such as DSMA may be implemented with an FDMA/TDMA method.

APPENDIX 3A
TDMA SYSTEM MEASURES OF EFFICIENCY

Three main types of efficiency measures used for a TDMA system can be defined. *Frame efficiency* (η_F): Percentage of used frame used for the messages:

$$\eta_F = 1 - \frac{(S + \sum_{i=1}^{n}(G_i + P_i + Q_i))T_S}{F} \tag{3A.1}$$

where:

F = frame duration (in μs),
S = number of symbols in the synchronization bursts,
G_i = ratio between the guard time and slot i symbol duration,
P_i = number of symbols in slot i preamble,
Q_i = number of symbols in slot i postamble,
T_S = symbol duration (in μs),
n = slot number.

Burst efficiency (η_B): Ratio between the number of useful transmitted information bits and the total number of transmitted bits:

$$\eta_B = \frac{rM}{P + M + Q} \tag{3A.2}$$

where

r = codec coding rate,
M = number of symbols in the message,
P = number of symbols in the preamble,
Q = number of symbols in the postamble.

System efficiency (η_S): Ratio between the useful capacity (user traffic) and the real capacity. For a Gaussian channel, we have

$$\eta_S = B = \log_2(1 + \frac{E_S T_S}{N_0 B}) \tag{3A.3}$$

where

B = channel bandwidth,
E_S = energy per symbol,
T_S = symbol duration,
N_0 = spectral density of a white Gaussian noise.

APPENDIX 3B

THROUGHPUTS OF SOME RANDOM ACCESS PROTOCOLS

The channel throughput of the different variants of ALOHA and CSMA can be simply expressed as follows [3]:

Table 3B.1

Throughput of some random access protocols

Protocol	Throughput
Pure-ALOHA	$S = G.e^{-2G}$
Slotted-ALOHA	$S = G.e^{-G}$
Nonslotted 1-persistent CSMA	$S = \dfrac{G[1 + G + a.G(1 + G + aG/2)].e^{-G(1+2a)}}{G(1+2a) - (1 - e^{-aG}) + (1 + a.G).e^{-G(1+a)}}$
Slotted 1-persistent CSMA	$S = \dfrac{G[1 + G - e^{-aG}].e^{-G(1+a)}}{(1+a) - (1 - e^{-aG}) + a.e^{-G(1+a)}}$
Nonpersistent nonslotted CSMA	$S = \dfrac{G e^{-a.G}}{G(1+2a) + e^{-a.G}}$
Nonpersistent slotted CSMA	$S = \dfrac{a.G.e^{-a.G}}{1 - e^{-aG} + a}$

where

a is the ratio of propagation delay/packet transmission delay,
G is the offered traffic (or load) in number of packets generated per packet duration,
S is the number of well-received packets.

REFERENCES

[1]   Yacoub, M. D., *Foundations of Mobile Radio Engineering*, CRC Press, 1993.

[2]   Scholtz, R. A., "The Origins of Spread-spectrum Communications," *IEEE Transactions on Communications*, Vol. COM-30, No. 5, May 1982.

[3]   Pahlavan, K., and A. H. Levesque, *Wireless Information Networks*, Wiley, 1995.

[4]   Thompson, J. S., P. M. Grant, and B. Mulgrew, "Smart Antenna Arrays for CDMA Systems," *IEEE Personal Communications*, pp. 16–25, Oct. 1996.

[5]   Abramson, N., "The ALOHA System: Another Alternative for Computer Communications," *Fall Joint Computer Conference AFIPS Conference Proceedings*, Vol. 37, 1970.

[6]   Cidon, I., and R. Rom, "Hidden Terminal Performance in Two Interfering Channels," *Proceedings of the IEEE Infocom'85*, Washington, DC, pp. 222–224, Mar. 1985.

[7]   Sidi, M., and A. Segal, "A Busy-Tone-Multiple-Access-Type Scheme for Packet Radio Networks," *The International Conference on Performance of Data Communications Systems and Their Applications*, Paris, France, pp. 1–10, Sept. 1981.

[8]   Sreetharan, M., and R. Kumar, *Cellular Digital Packet Data*, Artech House, 1996.

[9]   Namislo, C., "Analysis of Mobile Radio Slotted ALOHA Networks," *IEEE on Selected Areas in Communications*, Vol. SAC-2, No. 4, pp. 583–588, July 1984.

[10] Saadawi, T. N., M. H. Ammar, and A. El Hakeem, *Fundamentals of Telecommunications Networks*, éditions Wiley & Sons, 1995.

[11] Lam, S. S., "Packet Broadcast Networks - A Performance Analysis of the R-ALOHA Protocol," *IEEE Transactions on Computers*, Vol. C-29, No. 7, pp. 596–603, July 1980.

[12] Goodman, D. J., R. A. Valenzuela, K. T. Gaylard, and B. Ramamurthi, "Packet Reservation Multiple Access for Local Wireless Communications," *IEEE Transactions on Vehicular Technology*, Vol. 37, No. 8, pp. 885–890, Aug. 1989.

[13] Amitay, N., and S. Nanda, "Resource Auction Multiple Access (RAMA) for Statistical Multiplexing of Speech in Wireless PCS," *IEEE Transactions on Vehicular Technology*, Vol. 43, No. 3, pp. 584–596, Aug. 1994.

SELECTED BIBLIOGRAPHY

[1]   Barton, S. K., "Introduction to CDMA," *Annales des Télécommunications*, 48, No. 7-8, pp. 384–389, 1993.

[2]   Fukasawa, A., R. Fisher, T. Sato, Y. Takinawa, and T. Kato, "Wideband CDMA System for Personal Communications," *Proceedings of the IEEE International Conference on Universal Personal Communications'95*, Tokyo, Japan, pp. 833–837, Nov. 6-10, 1995.

[3]   Goodman, D. J., and A. A. M. Saleh, "The Near/Far Effect in Local ALOHA Radio Communications," *IEEE Transactions on Vehicular Technology*, Vol. VT-36, No. 1, pp. 19–27, Feb. 1987.

[4]   Goodman, D. J., and S. X. Wei, "Efficiency of Packet Reservation Multiple Access," *IEEE Transaction on Vehicular Technology*, Vol. 40, No. 1, pp. 170–176, Feb. 1991.

[5]   Goodman, D. J., and S. X. Wei, "Factors Affecting the Bandwidth Efficiency of Packet Reservation Multiple Access," *Proceedings of the IEEE Vehicular Technology '89*, San Francisco, CA, pp. 292–299, May 1989.

[6]   Guilhoussen, K., I. Jacobs, R. Padovani, A. Viterbi, L. Weaver, and C. Whealey, "On the Capacity of a Cellular CDMA System," *IEEE Transaction on Vehicular Technology*, Vol. 40, No. 2, pp. 303–312, May 1991.

[7]   Habbab, I. M. I., M. Kavehrad, and C.-E. W. Sundberg, "ALOHA With Capture Over Slow and Fast Fading Radio Channels With Coding and Diversity," *IEEE Journal on Selected Areas in Communications*, Vol. 7, No. 1, pp. 79–88, June 1989.

[8]   Ramamurthi, B., and A. A. M. Saleh, "Perfect Capture ALOHA for Local Radio Communications," *IEEE Journal on Selected Areas in Communications*, Vol. SAC-5, No. 5, pp. 806–813, June 1987.

[9]   Rom, R., and M. Sidi, *Multiple Access Protocols - Performance and Analysis*, Springer-Verlag, New York, 1990.

[10] Tobagi, F. A., and L. Kleinrock, "Packet Switching in Radio Channels: Part II - The Hidden Terminal Problem in Carrier Sense Multiple Access and the Busy-Tone Solution," *IEEE Transactions on Communications*, Vol. Com. 23, No. 12, pp. 1,417–1,433, Dec. 1975.

[11] Whipple, D. P., "The CDMA Standard," *Applied Microwave & Wireless*, pp. 24–39, winter 1994.

[12] Zander, J., "Radio Resource Management - an overview," *Proceedings of the IEEE Vehicular Technology '96*, Atlanta, GA, pp. 16–20, Apr. 28-May 1, 1996.

CHAPTER 4

PROTECTING AGAINST CHANNEL IMPERFECTIONS

Signals that propagate through the mobile radio channel are degraded in several ways, adversely affecting communication quality. The performance measure used to quantify signal degradation in digital systems is the bit error rate (BER). While fixed digital networks can provide BER values of less than 10^{-8}, and even 10^{-12}, mobile radio communications system BER is usually between 10^{-2} and 10^{-4}. In other words, transmission in the mobile radio environment has to face from ten thousand to more than one million times more errors than fixed link transmissions.

As one of the main objectives of a transmission system is to provide communications with acceptable quality, mobile radio systems must be designed so that they can operate even under very high BER conditions. In practice, when designing a mobile radio system, the BER considered and simulated, can be between 5 and 25% which reflects as closely as possible, the real conditions of a transmission system. In the very particular case of military mobile radio communications systems, signals with a BER of 50% may have to be decoded. Such systems are required to ensure communications with acceptable integrity, even in cases where half the transmitted bits are corrupted. It is necessary, therefore, to provide algorithms and mechanisms to detect and correct errors that may occur during transmission over the mobile radio interface. These means are arbitrarily classified into two categories:

- Mechanisms that are part of the transmission chain (mainly coding, equalization, and interleaving),
- Mechanisms and algorithms that are more at the physical level (e.g., diversity, fading margin, frequency hopping, power control, adaptive antennas).

This chapter is concerned with the description of some of the mechanisms used to improve the radio link quality. Some of the mechanisms, used to increase transmission system capacity, are presented in Section 6.3. Communication quality and system capacity are closely dependent on each other and the distinction between techniques is arbitrary. The reader will find other works adopting a different viewpoint.

4.1 MECHANISMS IMPLEMENTED IN THE TRANSMISSION SYSTEM

In this section, mechanisms aimed at minimizing error rates will be considered. Presentation of these mechanisms is oriented around their contribution as transmission problem countermeasures. For a more detailed discussion, the reader should consult the references listed at the end of this chapter.

4.1.1 Review of the Transmission System

The purpose of a telecommunications system is to transmit information between two or more remote points. Signal processing is performed before transmission to allow restoration of the transmitted information in the most reliable way possible while minimizing the use of the resource (e.g., radio channel). The main steps in the transformation of a signal in the transmitter and receiver are presented in Figure 4.1.

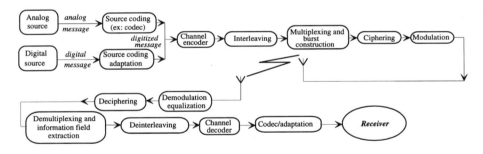

Figure 4.1 Block diagram of a typical communication chain.

Here the emphasis is on digital transmission. In this case, a converting device is used to change analog signals (e.g., speech, music or video) to a digital form and/or to adapt the input transmission rate to the channel rate when the source delivers digital signals. Next, *source coding* is applied to the output source signal. The main purpose of source coding is to compress the source data. A second coding process, called *channel coding*, is then applied to introduce redundancy, which is used for the detection and correction of transmission errors at the receiver.

Transmission errors that occur in the mobile radio channel generally happen in bursts. The transmitted data is interleaved to spread any errors that occur during transmission(in a fade). The information is then formatted into *packets*. In cases where ciphering is used, the information is treated (see Chapter 5) so that it can be detected and interpreted correctly only by authorized users. The signal, appropriately modulated (e.g., amplitude, frequency, or phase modulation), is then transmitted through the radio channel.

At the receiver, these operations are performed in reverse. In addition, an equalization mechanism is implemented to correct any distortion (notably of phase and amplitude) that may have altered the signal during transmission.

To detect and reconstitute the original transmitted message, the receiver includes circuits for amplification, frequency conversion and demodulation (in cases where carrier transmission is used), filtering, then sampling and decision-making. Frequency conversion and demodulation allows recovery of the baseband signal from the modulated carrier. To minimize noise, the baseband signal is filtered and sampled at regular time intervals. Finally, a decision circuit identifies the value of the transmitted bits from the received samples. The choice made by the decision circuit is binary (0 or 1).

Later in this section, the focus will be on coding mechanisms, equalization and interleaving. The first section reviews the main modulation types.

4.1.2 Modulation

The digital message is a sequence of binary elements that is associated with an electromagnetic signal in the transmitter. This function is called *modulation*. It can be defined as the process during which the amplitude, the frequency, or the phase of a carrier signal changes in accordance with the information to be transmitted. Modulation associates each n bit code word (called *n-tuplet*) in the message to a signal $S_i(t)$, $i = 1, ... M$ of duration $T = n.Tb$, chosen among $M = 2^n$ signals.

Note: Baseband transmission and carrier modulation transmission
Baseband transmission is the use of a frequency band situated between a frequency equal to or near zero and a frequency F_i. This is the case for cable transmission. *Transmission via carrier modulation* uses a frequency band centered on a frequency f_0. To baseband transmissions, this is the type used for transmission over radio cable and is known as *line coding*, while in the case of modulated carrier transmission it is *modulation*.

The three basic digital modulation methods are amplitude modulation [*amplitude-shift keying (ASK)*], frequency modulation [*frequency-shift keying (FSK)*], and phase modulation [*phase-shift keying (PSK)*]. They are represented schematically in Figure 4.2. In the case of binary modulation, the number of states is 2.

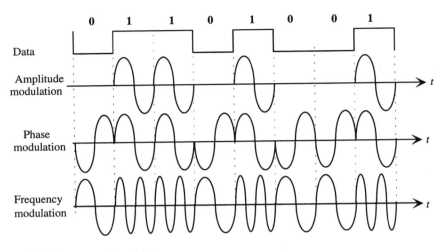

Figure 4.2 Different types of modulation.

Variations of signal amplitude due to Rayleigh fading make digital amplitude modulation practically unusable in mobile radio. Thus, the two modulation types which are generally used are based on phase modulation (e.g., PSK) or frequency modulation (e.g., FSK). In the following section, they are introduced together with the primary modulation methods.

4.1.2.1 On-Off Keying

On-off keying is the simplest modulation scheme. The carrier amplitude will be 1 or 0 depending on the value of the information to be transmitted (e.g., maximum amplitude to transmit 1 and zero to transmit 0, see Figure 4.3). This type of modulation is used in some radio communications systems such as wireless local networks (see Chapter 10). On-off keying modulation was also used in early systems for Morse code transmission.

4.1.2.2 Frequency-Shift Keying (FSK)

FSK is a very simple technique based on the use of a signal that can take two different tones. It can be implemented in a coherent or noncoherent form. In the case of coherent implementation, the impulse phase is known and the tones are orthogonal.

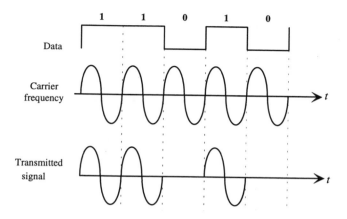

Figure 4.3 On-off keying.

When the tones are separated by a frequency value of 1/2T Hz (*T* is the duration in seconds of each bit), the modulation is called minimum (or sometimes median) shift keying (MSK). A particular case of this is Gaussian-filtered MSK (GMSK). This modulation is widely used in mobile radio communications systems and, notably, in the GSM system (see Section 4.1.2.6). It includes the addition of a premodulation filter with a Gaussian characteristic that reduces spectral spreading. Another variant, 4-ary FSK modulation, is employed in wireless LANs at 18 to 19 GHz and in some digital mobile radio systems operating in the low frequency bands. In the noncoherent case, the phase of the FSK signals is unknown and, therefore, it is necessary to separate the tones by a frequency value equal to an entire multiple of *1/T* Hz.

Figure 4.4 shows a representation of binary FSK in which the symbols *1* and *0* are represented by the transmission of sinusoidal waves which differ in frequency by a fixed value. In this case, a binary *1* is represented by the lower of the frequencies and a full cycle is required to transmit a single bit. Binary *0* is represented by two complete cycles of the higher tone. By this means, there are no phase discontinuities and the transmission is continuous phase. Disadvantages of FM systems is their relatively low bandwidth efficiency, receiver and transmitter complexity, and the necessity for the received signal power to be above a certain threshold for correct detection.

4.1.2.3 PSK Modulation

In PSK modulation, the phase of the signal is known and shifted by 180 deg (e.g., for *0*) or not (e.g., for 1), depending on the information to transmit (see Figure 4.5).

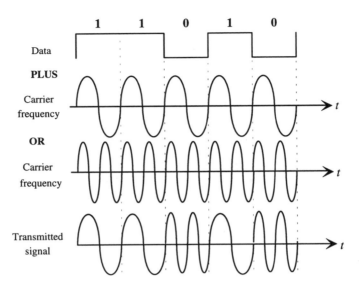

Figure 4.4 BFSK modulation.

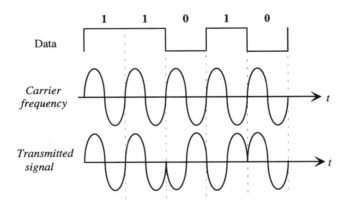

Figure 4.5 PSK modulation.

PSK modulation belongs to a family that requires the use of linear amplifier techniques, although in the case of binary (2 state) modulation it is usually ignored. Changes of phase bring with them changes in amplitude of the carrier, especially with higher order modulation schemes, and to avoid distortion of the signal and the generation of transmitter intermodulation products, linear amplifiers are used. Note that linear modulation offers good spectral efficiency but does require the additional cost and complexity of linear amplifiers. The main linear modulation methods are QPSK, D-QPSK, π/4 D-QPSK, PAM, and QAM.

4.1.2.4 Pulse-Amplitude Modulation

Pulse-amplitude modulation (PAM) consists of the variation of the transmitted signal amplitude inside a range of M discrete values (or levels). The symbol constellation (i.e., the representation on a circle $[0, 2\pi]$ of the transmitted symbols in amplitude and phase) contains four points. Hence, $M = 2^m$ allows the coding of m bits per symbol (in Figure 4.6, $m = 2$ and $M = 4$).

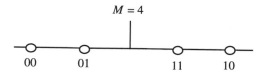

Figure 4.6 PAM (signal constellation for $M = 4$).

4.1.2.5 Quadrature Amplitude Modulation

Quadrature amplitude modulation (QAM) doubles the efficiency of PAM by also varying the amplitudes of the sine and cosine components of the carrier (Figure 4.7). The signal obtained consists of two sequences of PAM impulses in phase quadrature, hence its name QAM. If the number of constellation points is reduced to four and each symbol represents two binary values, then the modulation obtained is quadrature PSK. QPSK modulation is commonly used, in several variants, in mobile radio communications systems.

4.1.2.6 MSK and GMSK

Continuous phase modulation techniques overcome linearity constraints and allow reduced amplifier component cost. The modulation techniques, derived from this family of modulation, transmit at constant envelope with the drawback of a lower spectral efficiency.

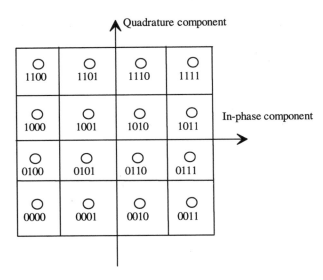

Figure 4.7 16-QAM (theoretical representation of symbols before combining with a code).

MSK is a form of FSK. The baseband signal is sinusoidally filtered to produce a smooth transition from one binary state to another with a continuity of the carrier phase at the binary transitions. The phase of the signal varies linearly with exactly ± 90deg with respect to the carrier during a period of 1-bit T_b.

As mentioned previously, GMSK is a special case of MSK modulation with a constant signal envelope. The Gaussian filter is used to limit adjacent channel interference. This technique gives fairly good spectral efficiency, constant signal amplitude and a robust modulation with a good C/I ratio. Other advantages are that the power consumption, weight, and cost of the transceiver can be minimized.

4.1.2.7 Conclusions

Modern techniques used in the mobile radio communications environment are based on linear modulation and are binary phase-shift keying (BPSK), quaternary PSK (QPSK), differential PSK (DPSK), and QAM.

At the end of the 1990s, orthogonal frequency division multiplexing (OFDM) modulation was considered for mobile radio communications. Classically, OFDM has been used for digital audio broadcasting (DAB). It significantly simplifies the equalization process and has the advantage of *soft* degradation in performance under poor signal conditions. In addition, because of the absence of equalization, OFDM presents lower complexity than other types of modulation.

The different modulation techniques used for mobile radio environment transmissions must use the spectral resources (transmit power and channel bandwidth) as efficiently as possible. To minimize transmit power, channel coding methods are generally used to the detriment of the signal bandwidth. In the case of systems having limited bandwidth, spectral efficient modulation techniques are preferred [1].

The choice between linear modulation and a constant envelope modulation is a function of the maximum spectral efficiency that has to be satisfied within the constraints of adjacent channel power and selectivity. More precisely, digital modulation techniques are chosen for mobile radio systems to satisfy the following properties [2]:

- Compactness of the power spectrum density: to minimize the effects of the interference on an adjacent channel, it is preferable that the level of power in the adjacent channels is between 60 and 80 dB below that of the transmit channel.
- Good performance in terms of BER: a low bit error probability must be obtained in the presence of cochannel interference, adjacent channel interference, thermal noise, and other forms of channel degradation such as fading or intersymbol interference.
- Envelope properties: mobile applications typically use nonlinear power amplifiers (class C amplifiers). If the modulation method contributes an amplitude component to the signal a nonlinear amplifier can affect the BER performance. To avoid the problems of intermodulation products caused by the nonlinear amplifier, the signal must have a relatively constant envelope.

4.1.3 Error Control: ARQ and FEC

Two stages of coding are performed before transmission: *source coding* and *channel coding*.

The purpose of source coding is to limit the size of the initial message by reducing the redundancy in the signal, and to transform it into a message in which successive symbols are independent and uniformly distributed.

Channel coding allows recovery of the signal at the receiver with a low error rate even in the presence of poor channel characteristics. It reintroduces redundancy to the message. This redundancy is, however, different from that contained in the source signal because the code is adapted to the channel characteristics (the redundancy that appears in the source message is totally independent of the channel). Channel coding is also known as *error correction/detection coding*.

Modulation and coding are logically independent processes but are firmly related to each other. An improvement in one or the other of the processes aims to improve spectral efficiency. A modulation scheme using multiple levels provides good spectral efficiency but requires a better SNR. For example, the SNR can be improved by using low-rate speech coding and efficient error correcting techniques before modulation.

Error control techniques used in mobile radio can be classified into three groups: error detection using block codes, the forward-error correction (FEC) codes using block and convolutional coding, and the automatic repeat request (ARQ) methods. The focus here is on channel coding designed to detect and correct transmission errors. Source coding is briefly discussed in the following section, taking as an example speech coders (*codecs*).

4.1.3.1 Source Coding: Speech Coding Example

An analog source is sampled, quantized, and converted to digital form before presentation to a source encoder, also called a *codec*. A digital data source can be processed directly by the source encoder. Speech coding, that is, using an analog source to be transmitted digitally, is the case considered here (see Figures 4.8 and 4.9).

Figure 4.8 Analog signal processing before the channel coding.

High efficiency coding algorithms play a very important role in high-density communication systems, where minimization of the data rate over the radio interface is crucial to save on bandwidth. Current coding algorithms allow toll quality voice at transmission rates from 4 to 16 Kbps against the 64 Kbps in use over land lines.

The objective of speech-coding is to transmit the signal with a certain level of quality while at the same time reducing the data rate. This reduction is obtained by taking advantage of signal redundancy and the limitations of the human ear.

Figure 4.9 Digital signal processing before channel coding.

After source coding, certain elements of the message which have little significance are removed. The message is then presented in its shortened form and formatted with a succession of mutually independent bits taking the values 0 and 1 with probabilities p_0 and p_1.

Remark: Evaluation of speech-coding algorithms

To evaluate the performance of a speech-coding algorithm, two criteria are used.

The first criterion estimates the speech quality by objective measurements (e.g., SNR) or subjective ones (mean opinion score). Here three levels of quality are commonly used: telephone, mobile radio communications, and military radio quality.

The second criterion used is an estimation of the algorithm complexity by calculating the number of operations per second necessary to code or decode the signal.

Speech encoders are generally divided into two groups: *waveform encoders* and *vocoders*. Hybrid encoders exist that combine the advantages of waveform encoders and vocoders.

Waveform encoders minimize the generated information rate by using the correlation properties between speech samples. The most advanced waveform encoders can achieve a quality that approaches PCM encoders at 64 Kbps, with rates less than 16 Kbps. Below this rate, the quality tends to decrease rapidly.

Among these, the best known is pulse code modulation (PCM). This type of encoder requires a high rate transmission and results in good quality. The PCM encoder used in fixed telephony is a typical example of such an encoder.

Vocoders are based on a simplified model of speech production consisting of an excitation sequence and a filter. They generate lower rates than waveform encoders and can offer better speech quality. This is the case for the *enhanced full rate* (EFR) encoder used in GSM, which has been shown to have a better subjective quality than PCM.

Hybrid encoders are very useful for mobile-radio telephony applications because of the low transmission rates they generate (see examples in Table 4.1). They remain efficient below 16 Kbps.

Example 1: Full-rate encoder regular-pulse excitation-long-term prediction (RPE-LTP) as used in GSM.

This encoder operates at 13 Kbps. The speech is sampled every 20 ms and encoded on 260 bits (giving the rate of 13 Kbps). In this type of speech coding, the most sensitive bits are protected by a cyclic redundancy check (CRC), a convolutional code of rate ½, and a length constraint of 5. The total coded binary rate per speech signal is 22.8 Kbps. To combat channel noise, 30% of the header (10.1 Kbps), which is reserved for guard time and synchronization, plus 0.95 Kbps for slow signaling are added before transmission. Thus, the total rate is 33.85 Kbps per channel and 270.8 Kbps for the eight channels. The spectral efficiency of the GSM system, using GMSK modulation and channel bandwidth of 200 kHz, is thus 1.35 bps/Hz.

As a function of their importance to the speech content, the bits transmitted by the speech encoder can be divided into several groups, as in the case of the GSM: *very important*, *important*, and *less important*, for example. These groups of bits are then coded differently by the channel encoder.

Table 4.1

Main speech encoders used in mobile systems

System	Speech coders	Rate after source coding (Kbps)	Rate after channel coding (Kbps)
IS-95	Vocoder (QCELP)	9.6; 4.8; 2.4; 1.2	28.8; 19.2
IS-54	Vocoder (VSELP)	8	13
GSM/DCS	Hybrid encoder (RPE-LTP)	13	22.8
PHS	Waveform encoder (ADPCM)	32	32
CT2/DECT	Waveform encoder (ADPCM)	32	32

4.1.3.2 Coding

Channel coding is specific to digital transmission and does not have an equivalent in analog transmission. It consists of the insertion of redundancy bits, generated according to a known procedure, into the bitstream-supplied from the source. Hence, channel coding results in an increase of the transmit bit rate. The channel decoder knows the coding procedure used at transmission and checks if the message is intact at the receiver. If this is not the case, it is able to detect the presence of transmission errors that, under some conditions, can be corrected.

The codes used for channel coding can be classified into two types: block coding, where the message is divided and transmitted in n bit blocks , and convolutional coding, where the messages are transmitted continuously.

Block coding starts by dividing the data into blocks (Figure 4.10). Each block is then processed, independently from other blocks, by a coding algorithm that adds redundancy and, hence produces a longer block. At the receiver, the decoder operates similarly but in reverse, and each block is processed individually. More precisely, block coding consists of the creation for each k bit of information, a block of n bits ($n > k$) called a *codeword* which includes ($n - k$) redundancy bits. These redundancy bits are called *parity check bits* or *parity bits*. The ($n - k$) parity bits of the codeword are calculated from the k information bits using a predetermined algorithm.

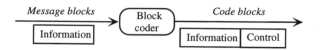

Figure 4.10 Principle of block coding.

Among the block coding techniques, cyclic codes have an advantage over other coding types because of their simplicity in the coding operation. They have a mathematical structure that allows very efficient decoding. A *cyclic code* is

- *Linear*: the sum of two codewords is also a codeword,
- *Cyclic*: a cyclic shift of a codeword gives another codeword.

Figure 4.11 illustrates the different steps in the construction of a block: the binary datastream is divided into k bit blocks, which are then coded using a block coding scheme to produce, at the encoder output, n bit codewords.

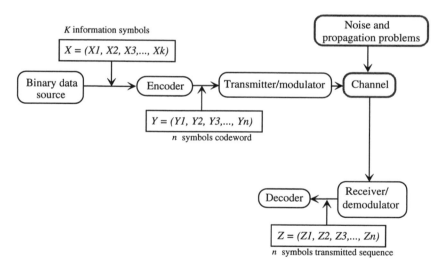

Figure 4.11 Block construction process.

To evaluate the correcting performance of a block code, the following definitions are necessary:

- The *Hamming weight* (*w*) of a codeword is the number of components different from *0*. For example, if $c = 10011$, $w(c) = 3$.
- The *Hamming distance* (d_H) between two codewords is the number of positions in which they differ. If $c_1 = 10011$ and $c_2 = 11000$, $d(c_1, c_2) = 3$. A codeword will be delivered without errors if the number of errors that it contains is strictly less than $(2.d_H - 1)/2$.
- The distance between any two codewords is the Hamming weight of the sum of the codewords.
- The *minimum distance* (d_{min}) of a linear block code is equal to the minimum weight of its codewords different from *0*.

A block code has the following properties:

- Detection of maximum *s* errors in a codeword $\quad d_{min} \geq s+1$,
- Correction of maximum *t* errors in a codeword $\quad d_{min} \geq 2.t+1$,
- Correction of *t* errors and detection of $s > t$ errors simultaneously in a codeword $d_{min} \geq s + t + 1$.

For a block code (n, k), the minimum distance is given by:

$$d_{min} \leq n - k + 1$$

The main cyclic codes are

- The *CRC codes* that are perfectly adapted to error detection.
- The *Golay codes*. The (23, 12) Golay code, for example, is a binary code that can correct any combination of up to three random errors in a 23-bit block.
- The *Bose-Chaudhuri-Hocquenqhem* (*BCH*) codes that constitute one of the most powerful block code classes.
- The *Reed-Solomon codes* (*RS codes*) that constitute an important sub-class of nonlinear BCH codes. The RS encoder differs from a binary encoder because it operates on symbols instead of on bits.

Convolutional codes are well suited to bitstream transmission. The message is operated on in a continuous and serial manner, generating *n* bits for *k* input bits of information (Figure 4.12). These *n* bits depend on the *k* input bits and also on the *m* previous blocks. Hence, convolutional codes have a memory of order *m*. The quantity (*m+1*) is called the *code constraint length*.

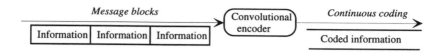

Figure 4.12 Illustration of convolutional coding.

Trellis codes are a category of convolutional code which combine a coding algorithm with modulation, giving the name *trellis-coded modulation (TCM)*. A trellis code is characterized by two important features. First, the number of points in the modulation constellation is greater than would be necessary for a format having the same binary rate. Second, these additional points allow the introduction of FEC redundancy coding without reducing the spectral efficiency.

Convolutional coding is used to introduce redundancy between successive points on the symbol constellation. This is because only certain patterns or sequences of points are allowed. They can be represented by a trellis structure (hence the name).

In Table 4.2 are summarized the main coding techniques implemented in mobile radio systems.

Table 4.2
Coding techniques used in mobile radio systems

Code Type	Characteristics
Hamming code	Length $=2.m$ 1, $m= 2, 3, 4, \ldots$
	$d_{min} = 3$
BCH codes	Length $=2.m$ 1, $m= 2, 3, 4, \ldots$
	d_{min} $2.t$ 1, t number of errors that can be corrected
	Parity bits number: $n-k$ $m.t$
Golay codes (23, 12)	$n = 23$, $k = 12$, $d_{min} = 7$, $t = 3$
Reed-Solomon codes (q-ary)	$q = p^m$, p prime and m integer, $N = q$ 1, $K = 1, 2, 3, \ldots N\text{-}1$;
	$d_{min}= N$ $K+1$
Walsh/Hadamard codes	$n=2^m$, $d_{min} = n/2$

Concatenated codes. To improve coding system performance, mobile radio systems use several types of code that can be concatenated.

Concatenating codes consist of associating two codes (sometimes more) such that the bits of the first encoder, called the *outer*, are applied to the input of a second encoder, called the *inner* (Figures 4.13 and 4.14).

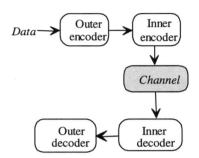

Figure 4.13 Concatenation of two codes.

To concatenate codes, the chosen codes must have complementary properties. For example, the inner code must have a good correction capability, while the outer code, which is in general more efficient than the inner code, corrects the residual errors. Actually, a widely used concatenation method uses a convolutional type inner code and a Reed-Solomon-type outer block code.

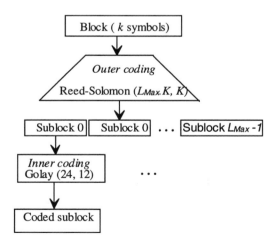

Figure 4.14 Example of code concatenation.

This type of coding is used in the GSM system. Convolutional block codes are implemented for error correction, a fire code for error correction, and a parity code for the detection of burst errors. In the channel coding process, the speech bits that are classified by the speech source coding scheme as *very important* and *important* are protected by a convolutional code.

4.1.3.3 Error Control

The control of errors (Figure 4.15) can be achieved by using ARQ (automatic repeat request) (Figure 4.16) and forward error correction (FEC), separately or in combination.

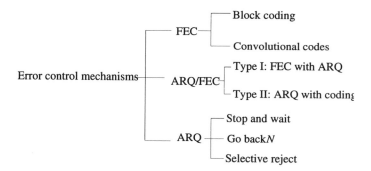

Figure 4.15 Error control techniques.

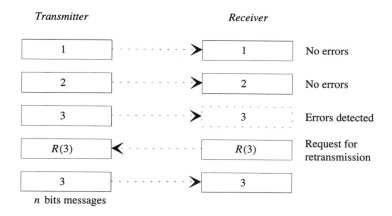

Figure 4.16 ARQ mechanism.

The main process of these techniques, as with channel coding, is the introduction of redundancy (Figure 4.17). Each method, however, is based on a particular principle.

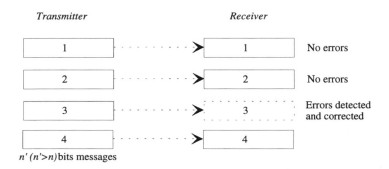

Figure 4.17 FEC mechanism

The ARQ technique uses redundancy for detecting errors. When errors are detected, a retransmission request is sent by the receiver. It is thus necessary to have a feedback communications link. The FEC technique uses redundancy directly to correct the errors at the receiver, without a need for retransmitting the message by the sender.

ARQ. ARQ combines block error detection coding with a mechanism of retransmission request. If a correcting code is added, the technique is called *hybrid* ARQ. ARQ methods work well when the error rate is low. The channel throughput can reach its maximum allowed by the transmission format. As the channel quality is adversely affected, because of propagation problems (e.g., fading, shadowing), the code blocks are frequently corrupted and the useful transmission bit rate decreases dramatically because of retransmission requests.

A distinction can be made between positive and negative acknowledgment systems. In a system with negative acknowledgment, the transmitter retransmits the data only in the case of an explicit request from the receiver. In a system with positive acknowledgment, the transmitter waits for correct reception confirmation from the receiver. In a system with positive acknowledgment, the throughput can be low, in particular if the round trip delay between the transmitter and the receiver is significant (e.g., a satellite link). The advantage of positive acknowledgment is an increase in the level of transmission security. In practice, the majority of ARQ systems use negative acknowledgment with positive acknowledgment as an option. The three basic ARQ techniques are: stop-and-wait (*SW*), go-back-N (*GBN*), and selective repeat (*SR*).

Stop-and-Wait (SW). The simplest strategy is called stop-and-wait: the transmitter stops transmitting after every data block and waits for either an acknowledgment message or nonacknowledgment message or the expiration of a timeout. If a positive acknowledgment is received (ACK), the transmitter sends the next data packet. If a negative acknowledgment (NACK) is received, the last packet sent is retransmitted. The performance of this mechanism is highly dependent on the round trip delay Td; that is, the time spent between the transmission of a packet and the reception of the corresponding ACK or NACK.

Stop-and-Wait is the most efficient of all the ARQ protocols if the round trip delay is very small. The main problem encountered is that its implementation requires a significant buffer size (theoretically infinite) to deliver packets correctly as these packets may be transmitted or received nonsequentially. This mechanism is not suited to speech services, which have real-time constraints because of the delays introduced by the propagation time in retransmission.

Go-Back-N. In the GBN mechanism, the channel is slotted (with a slot duration equal to τ) and the source continuously sends packets, each packet having a duration of m slots. By receiving an acknowledgment, one slot after the end of the transmission, the transmitter knows the result of the transmission. If a NACK is received, the transmitter terminates the transmission of the current packet, then retransmits both the erroneous packet and the packet which had just started before reception of the NACK. At the receiver, the packet received after an errored one is discarded, regardless of error rate.

In this mechanism, the acknowledgment is normally of the negative type. When a received block contains errors, the receiver requests the retransmission of the N last blocks (the most common values for N appear between 4 and 6). The advantage of this system is that the blocks need not be numbered individually and the algorithms are hence simpler than the SR algorithm (see below). The GBN drawback is that the delay must be lower than the transmission duration of N blocks. Another problem is that a retransmission requires the sending of more data than is necessary.

Selective Repeat. In the GBN method, some error-free received messages are retransmitted, thus increasing the quantity of information transmitted on the communication channel. In the SR method, only those messages that contain errors at the receiver are retransmitted. In this case, it has to be accepted that messages will be received nonsequentially. In most applications, this method shows better performance than the two previous techniques. Its application is however, more complex, and it requires buffer management and message resequencing procedures. Its use, still limited, should develop with the increasing performance of electronic sub-systems.

In the SR mechanism, the ARQ is negative. In many noisy environments, these protocols have the best performance with regard to throughput. The blocks, however, have to be numbered. ARQ mechanisms are used successfully in data transmission systems to reduce the bit error and the frame error rates on different channels. Because of the inherent delay in these techniques and often the low transmission rate constraint of the feedback channel used for packet acknowledgment, the ARQ mechanism cannot be used for speech communications.

Forward Error Correction. Shannon showed, in 1948, that error-free transmission will be possible if the information rate is less than the so-called channel capacity. He also showed that a communication system can be made reliable if a fixed percentage of the transmitted signal is redundant. Nevertheless, the way this can be achieved was not indicated in Shannon's work. Later, through intensive research, a number of techniques allowing the correction of errors without a need for retransmission were developed. These techniques are known as FEC and are used in systems where a feedback channel for retransmission requests isn't available, where the retransmission delay is prohibitive, where the error rate generates a large number of retransmissions, or when a retransmission system would be very complex to implement.

In digital mobile communications systems, the FEC technique is one of the most powerful in overcoming the power limitation problem. In fact, in the mobile radio environment the use of FEC allows a reduction in the value of the ratio $(E_b/N_0)_{min}$ when traded against a wider bandwidth. It is assumed that, by the use of error correction coding, the information is correctly received at the first attempt.

The implicit FEC principle consists of adding redundancy to the information. At the receiver end, the degraded signal is processed by a decoder that uses this redundancy to reconstitute the original message. The FEC technique is implemented by using convolutional codes or block codes.

ARQ/FEC. ARQ eliminates blocks containing errors even though these may contain some useful information (i.e., error-free information). In the mobile radio context, the high number of errors often requires the elimination of a large number of data blocks. The message throughput may become zero if all the message blocks contain some errors. This is often the case in mobile radio communications, which is why these systems often use mechanisms that combine ARQ and FEC techniques.

Among the best known ARQ/FEC combinations is ARQ type I. In this method, error correction is performed with FEC. When errors cannot be corrected, a retransmission request is triggered by the ARQ mechanism. In type II, ARQ/FEC data blocks are first transmitted without FEC. When a block is received with errors, the retransmitted message is protected by FEC.

4.1.3.4 Selection of a Coding Technique

The most important criterion for choosing an error control technique is the necessary level of protection for the application. For example, speech applications are more tolerant to the incidence of errors than data transmission. In fact, in data transfer applications (e.g., files or messages), the presence of errors in message headers makes them totally useless.

The way errors occur on the channel are analyzed so that an adequate error control method can be provided. In outdoor environments and for fast moving mobiles (e.g., in vehicles), the rapid signal fluctuations cause short error bursts (see Chapter 2). For slow-moving mobile terminals, the signal fluctuates slowly and the duration of fade intervals is longer than in the previous case. FEC can give a satisfactory protection level in the first case, leaving ARQ to protect against the longer error bursts. In indoor environments, it was shown [3] that FEC was not appropriate, because of long error bursts, and that an ARQ technique had to be used. Another important aspect associated with FEC is that the delay introduced for the coding and interleaving operations increases with the block size. Fast digital signal processors (DSP) therefore become necessary, which can consume significant power. On the other hand, ARQ has the drawback of significant transit delays and the time for the receiver to process and acknowledge a block.

4.1.4 Equalization

In mobile radio, transmitted signals undergo irregular variations and distortion during propagation. One of the chief causes is multipath, which produces intersymbol interference. The problems of nonideal frequency response characteristics of the channel can be corrected by using a filter or equalizer in the demodulator. It is therefore necessary to design techniques that equalize the received signal to eliminate or reduce the distortion. Three classical structures are defined for equalizers: the transversal equalizer or nonrecursive equalizer (adaptive transversal equalizer), the recursive equalizer with feedback (decision feedback equalizer), and the maximum likelihood equalizer. A receiver optimally adapted to a particular transmission channel must have an idea of the behavior of that channel.

In linear equalization (Figure 4.18), synchronous sampling consists of sampling of the signal at the period Δ to give samples y_k that are then processed by the equalizing filter to be delivered to the output as signal c_k. This filter must perform the inverse operation of the transmission channel. When the additive noise is nil ($v_k = 0$), the signal at the output of the equalizing filter is equal to the transmitted signal ($c_k = a_k$). In practice, the additive noise is nonzero and complex equalization is required. Therefore, a filter that minimizes the difference between c_k and a_k must be designed. The decision element consists of a simple uniform quantizer.

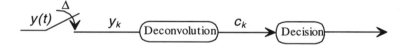

Figure 4.18 Main operations of a linear equalizer.

4.1.4.1 Linear Equalizers

The channel equalizer type, which is the most frequently used for intersymbol interference reduction, is a linear transversal filter with adjustable coefficients θ_i (Figure 4.19). Initially, the characteristics of the frequency response are unknown and are time varying, so the channel characteristics have to be measured, and the parameters of the equalizer adjusted. Once the parameters of the equalizer have been adjusted, they remain unchanged during transmission. This type of equalizer is called a *preset equalizer. Adaptive equalizers* periodically update their parameters according to the received data, and are therefore able to follow slow variations of the channel (see also Section 4.1.4.3).

Figure 4.19 Structure of a transversal equalizer.

A transversal equalizer is a transversal filter containing adjustable coefficients and followed by a decision-circuit with thresholds. It can be made recursive, but this option may not yield significant improvements and, in addition, can present stability problems. The majority of linear equalizers are nonrecursive.

4.1.4.2 Decision Feedback Equalizers

A decision feedback equalizer consists of a direct transversal filter and a recursive transversal filter, the inputs of which are the previously estimated symbols (Figure 4.20).

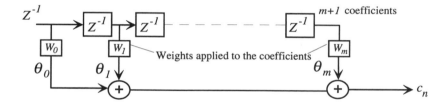

Figure 4.20 Structure of a decision feedback equalizer.

Unlike a linear equalizer, a decision feedback equalizer can compensate deep fades without increasing the noise. However, interference canceling by the recursive part, is based on the hypothesis that the decisions taken are correct. If this is not the case, the decision error propagates and introduces more errors. This error propagation affects the performance, in particular, under high error rate conditions. Nevertheless, in spite of these drawbacks, the decision feedback equalizer performance is notably higher than those of linear equalizers.

4.1.4.3 Adaptive Algorithms

To adapt an equalizer to the fluctuating characteristics of the transmission channel, adaptive algorithms can be implemented. Among the most commonly used algorithms, there are the "zero forcing" algorithms, the stochastic gradient algorithm, and the least mean square algorithm.

In an adaptive equalizer, the transversal filter coefficients, of a finite length, are periodically updated so as to minimize some specific criterion. This updating is generally performed after each decision; that is, at a frequency equal to the speed of modulation. Optimization of the equalizer coefficients according to the minimum error probability criterion would be extremely complex. This is because the dependency between the minimum error probability and the equalizer coefficients is itself complex. This is why, in practical applications, minimization of the least mean square error is preferred.

The basic problem in equalization is the algorithm for initial convergence. In a certain number of applications, this problem is solved by including a training sequence in each message.

Three working modes for an adaptive mode equalizer are defined:

- *Channel training*, it is implemented by the transmission of *training sequences*, known or unknown (the case of blind algorithms), by the receiver. The equalizer filter can converge from any initial condition to an optimal filter.
- *Channel variation tracking*, when the channel is variable.

- *Blind equalization*, when a training sequence is not possible (the case of transmitters sending messages to several receivers that may come on line at different times); blind equalization is accomplished without training sequences.

4.1.4.4 Viterbi Equalizer

The Viterbi equalizer uses a mathematical model of the transmission channel and calculates the most probable transmitted data. In order to do this, a training sequence is included in the data burst. By comparing this known sequence (*S*) in a correlator at the receiver with the received sequence (*S'*), the equalizer calculates a channel model that can be applied to the remainder of the message.

A binary sequence is entered in the channel model and the output is compared with the received sequence. By comparing the two bursts, the Viterbi equalizer selects the most probable transmitted sequence, which is then again introduced into the channel model (Figure 4.21). The process is repeated until a sufficiently close binary pattern is found.

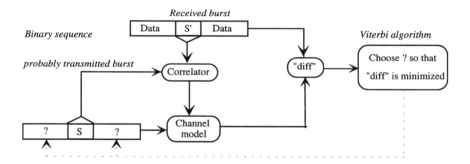

Figure 4.21 Viterbi equalizer.

4.1.5 Interleaving

In the mobile radio environment, errors can occur in bursts because of deep fades. These fades can affect a group of consecutive bits in the transmitted message. Unfortunately, channel coding is efficient only for the detection and correction of random errors or burst errors that are very short. Interleaving is the method of distributing the errors in a message so as to isolate or randomize them. This operation makes the errors simpler to detect and possibly to correct. Several interleaving levels can be realized. The lowest is at the bit level.

At a higher level, frame interleaving avoids the loss of important parts of a message (a message consisting of several frames). Interleaving changes the order of the transmitted bit sequence so that an error block produced in the transmission channel will be transformed into separated, independent random errors.

The simplest interleaving method at the bit level is realized using a matrix of L rows and n columns (Figure 4.22). Bits issued from the encoder are introduced into the matrix row by row and then delivered to the output by reading column by column (see, for example, the case of the Ermes standard, Chapter 11). For block codes, each row of the interleaving matrix contains, in general, a codeword ($N = n$). The parameter n is called the *interleaving factor* or *degree*. The deinterleaver, placed before the decoder, performs the inverse operation of the interleaver. When a packet of errors, of length $B \leq n$, is presented at the deinterleaver input, each group of n bits contains at most one error. The main drawback of the interleaving process is the delay introduced before transmission that can be, in some cases, inconvenient.

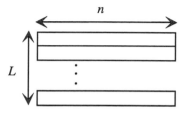

Figure 4.22 Example of interleaving at the bit level.

At the receiver, the matrix $L.n$ is regenerated after receipt of all the columns. Each row is then reconstituted using a decoder. L is called *interleaving depth*. An error of length less than or equal to L is dispersed over the codewords, contributing a unique error in each codeword.

4.2 DIVERSITY TECHNIQUES

The term diversity covers multitude of techniques widely used to combat fading in mobile radio systems. One of the principles, on which numerous diversity techniques are based, consists of sampling the received signal in several possible ways. The signals retrieved are either combined or classified so as to select the best one.

Without diversity, because of multipath effects (see Chapter 2), the transmitter must transmit with a significantly higher power level (fade margin) to protect the link during the short intervals where the signal fades deeply. In radio communications, the power available on the uplink is severely limited by the capacity of the terminal battery. For this, diversity techniques have a crucial role in reducing the transmission power [4]. Two diversity classes can be identified: The first one consists of techniques known as *explicit*, which uses transmission redundancy to exploit different channels. The use of a transmission with two signal polarizations results in this type of diversity. It is obvious that this class of technique leads to an increase in the transmitter power or a lower spectral efficiency. The second class uses the so-called *implicit* technique. The signal is transmitted from one place and at one time and the decorrelation effects in the propagation medium, such as multipath, are exploited to receive the signals through several channels.

The basis upon which diversity techniques operate is that the instantaneous SNR should be independent if measured at different places, frequency, or time. In practice, the spacing between measurement points of these parameters must be sufficiently large. A diversity of order L corresponds to the reception of L, independent copies of the transmitted signal.

Two types of diversity can be defined: microdiversity uses only one receiver and macrodiversity may use several transmitters and receivers located at different sites. In general, microdiversity techniques are used to combat Rayleigh fading while macrodiversity techniques are used to overcome masking effects.

4.2.1 Microdiversity Techniques

Microdiversity consists of the recovery of a transmitted signal at the same site (base station or mobile station). When the receiver collects several copies of the same information, the probability that all the signals have faded simultaneously is considerably reduced. Indeed, assuming that the same symbol is received on different receiver branches, each branch is subject to a particular random fluctuation. This reduces the probability that the received signal fades simultaneously on all other branches, which consequently reduces the bit error rate. Denoting the probability that a signal is the subject of a fade having a depth greater than a critical threshold as p, the probability that L copies of the fade are also below the threshold is then equal to p^L. A diversity of order L corresponds to reception of a symbol on L independent Rayleigh channels.

Microdiversity is widely implemented at cellular base stations. It can be implemented as follows:

- In *space*, by using multiple antennas ,
- In *frequency*, by receiving copies of the signal on different carriers,
- In *time*, by receiving copies of the transmitted signal at different arrival times.

4.2.1.1 Time Diversity

Time diversity is obtained by transmitting the same message at L different instants, which are separated by a time period greater than or equal to a duration $(\Delta t)_c$ called *channel coherence time*. This duration is equal to the inverse of the Doppler frequency (see Chapter

2). Time diversity is used for time selective channels. Its main advantage is use of a single receiver, making it appropriate for mobile stations (Figure 4.23), as well as for base stations. Its major drawback is that it leads to a reduction of channel throughput because the same block is transmitted several times, regardless of transmission channel performance.

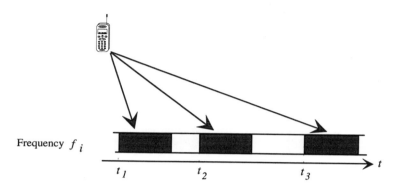

Figure 4.23 Time diversity.

In general, this type of diversity is used dynamically in data transmission systems through the ARQ or the FEC methods presented previously.

4.2.1.2 Space Diversity

Along with polarization diversity, space diversity is the most commonly used form. If two receiving antennas are used, spaced apart at a distance of between 0.5λ and 0.8λ, the signals received on each antenna can be considered to be uncorrelated. For systems using frequencies above 500 MHz, the separation distance can be less than 30 cm. At the base station end, a distance of 5 to 6m between antennas obtains gains of the order of 3 dB. The use of several antennas increases the received power without increasing the transmitted power. The inconvenience of this technique is the requirement to implement several receivers with similar characteristics and to process vector-type signals at the receiver.

In space microdiversity, several antennas are linked to a single multi-input receiver. This is different from macrodiversity schemes, where the different receiving antennas are linked to receivers at different sites and therefore have different network access points. An inconvenience of space diversity is that it requires the construction of robust towers (for fixing the different antennas) [5].

Finally, this technique is often used in urban environment cellular systems for the uplink (mobile to base station) because it reduces the effects of fading by selection of the best received signal or combination of signals. It contributes to link budget balancing (see

Section 6.2.4) by improving base station reception with a gain that increases (by 3 to 5 dB) with the density of the environment [6].

4.2.1.3 Polarization Diversity

A signal transmitted with orthogonal polarization has uncorrelated fading statistics (the correlation of the short-term envelope between the horizontal and the vertical antennas is very near to zero). Polarization diversity exploits the difference in the average power level that can reach values of between 6 and 10 dB.

The drawback of space diversity can be avoided by the use of polarization diversity. The antennas are physically situated in the same place and require less space than space diversity antennas. In some cases, three antennas with double polarization can have the same performance as those obtained with nine antennas in space diversity [6]. It is also often used for base station antennas in urban and suburban environments.

4.2.1.4 Frequency Diversity

The frequency diversity method (Figure 4.24) consists of transmitting the same signal on L carriers separated by a frequency bandwidth at least equal to the coherence bandwidth of the channel $(\Delta f)_c$. At 800 to 900 MHz, frequencies are separated by between 1 to 2 MHz. This type of diversity is used for frequency-selective channels.

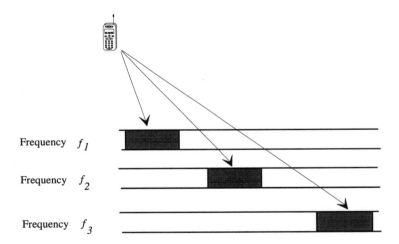

Figure 4.24 Frequency diversity.

Statistically, the same information signal will not fade at the same instant on different frequencies. Spread-spectrum techniques may be considered as frequency diversity techniques.

4.2.1.5 Signal-Combining Techniques

The received diversity signals must be processed to produce an output signal with an acceptable quality. The commonly used methods are:

Selection Combining. The selection combining method consists of the periodic selection of a single antenna from among those in use, according to one or several criteria. It consists of selection of the channel having the highest power or the one for that has the best C/I ratio. The performance of this technique is satisfactory for systems using two antennas. It is widely used in mobile radio communications systems.

Equal Gain Combining. This method is used in systems where it is difficult to obtain precise channel estimates (fast-frequency-hopping systems are typical) as in the case of a fast-varying channel, for example. All the input signals are amplified equally then phase matched and summed. The coherent summation of the wanted signal and the noncoherent summation of noise obtains a better SNR at the combiner output.

Maximum Ratio Combining. In this technique, the outputs of each antenna are weighted by an estimate of the C/I ratio received at each antenna. As with equal gain combining, the signals are phase matched and data decisions are made from the weighted summation of the demodulator outputs. This weighting method obtains an SNR higher than the one produced by an equal gain combiner, making it the most efficient of these three techniques.

4.2.1.6 Coding and Frequency Hopping Diversity

This technique combines the coding of signals with a frequency-hopping transmission mode (see Chapter 6), which provides transmission diversity. Diversity may be obtained by transmitting each symbol independently from the others. This is the case for GSM where interleaving on eight bursts is used by transmitting each burst on its appropriate frequency. This technique obtains excellent performance.

4.2.2 Macrodiversity Techniques

Macroscopic diversity or macrodiversity (sometimes called base station diversity) is a means to improve coverage and link quality, combat shadow fading and realize seamless handover (see Chapter 8). Macrodiversity techniques are just beginning to be implemented in public mobile radio communications networks, although they have been used for many years in public safety systems where diversity receive sites have aided the coverage for UHF hand-held portable equipment. For instance, CDMA-based systems like IS-95 (Chapter 12) and cordless systems like DECT (Chapter 10) use macrodiversity during the handover.

Macrodiversity allows a mobile to be connected simultaneously to several base stations. These techniques can be used in mobile radio networks with several network access points (e.g., cellular systems and cordless systems with several fixed sites). By allowing a mobile to receive and send signals to and from several stations at the same time, shadow effects, pathloss propagation, and fading characteristics experienced on the different links are independent of each other. Macrodiversity methods should allow reduction of the dynamic power control margin and also a reduction of the interference level in neighboring base stations. Due to the different propagation conditions along the separate paths between the mobile station and the base stations, the signals received at the different base stations on the uplink are not identical. A combination of the multiple received information results in better quality information in the majority of cases, which then results in a net gain. On the downlink, two options are possible: either the information is sent to the mobile by one base station at a time, after selection of the base station site that has the best propagation conditions with that mobile, or all base stations send the information simultaneously. The first scheme is called *selection macrodiversity* and the second is called *simulcast*. Statistically, three separate base stations can provide better coverage for a given region relative to the case when base stations are all located at the cell center.

4.3 ADAPTIVE ANTENNAS

Among the different antenna types discussed in Chapter 2, array antennas attract much attention because of their interesting applications in mobile radio. The use of intelligent antennas enables mobiles to be tracked on the move and can limit the interference generated by signals transmitted to a given mobile. They also help to reduce the level of interfering signals transmitted from other mobiles, as the base station antennas exhibit directivity only towards the mobile currently in communication. This concept is based on the array antenna structure. In the first part of this section, array antenna principles and later smart antenna strategies are introduced.

4.3.1 Array Antennas

An array antenna is composed of a set of identical transmitting elements that are arranged, in a regular manner, next to each other. Each element receives a signal, the amplitude and phase of which depends on its position. The resultant beamwidth of an array antenna has particular characteristics that are difficult to obtain with other antenna types. The signals received by each element are coherently combined by using an appropriate signal-processing algorithm that increases the power of the received signal. In fact, array antennas can provide a higher spectral efficiency and a better range than other antenna systems [7]. This range improvement is of high importance in the 2 GHz frequency band.

Array antennas have been used by radar systems and military communications; they allow detection and canceling of enemy jamming signals. In civil cellular systems, an array antenna receiver is essentially used to obtain acceptable bit error rates and SNRs for each user [8]. An antenna spot beam is realized, for instance, by using directional elements. These elements provide an antenna gain oriented in a selected direction, with the suppression of signals in others. For M elements, illumination of an aperture having $360/M$ deg can be formed. If used in a base station site, $360/M$ spot beams can be made to implement microsectorization.

4.3.2 Smart Antennas

Adaptive or "smart" antenna systems [9] constitute a multisubject technology that has increased rapidly during the 1980s and 1990s. They benefit from the increasing interactions between the radio and the signal-processing fields. The value of these systems is their ability to react automatically, in real time, to an environment the interference of which is unknown *a priori*. They reduce the side lobes situated in the direction of the interference, while keeping the main lobe in the useful signal direction. As the mobile station moves, the base station tracks the location and changes the beam direction accordingly. Usually, these systems are based on array antennas and a real-time adaptive receiver/processor that attributes weights to the array antenna elements to optimize the output signal according to predefined control algorithms. Thus, a smart antenna can be defined as an array that is able to modify its radiation pattern, its frequency response, and other parameters by using an internal feedback decision loop during antenna operation [10].

Adaptive antennas have been successfully used in several applications such as sonar and satellite communications. After some experimental research using adaptive antennas in mobile communications [11], products have recently become available from some manufacturers.

In traditional antenna systems, each transmitter sends its signal in an omni-directional or broad sectored beam (with an aperture width situated between 2π and $2\pi/3$) mainly because the receiver position is unknown [12]. This kind of transmission pollutes the electromagnetic environment by increasing global interference due to the radiation of the majority of power in useless directions. In contrast, smart antenna systems determine the mobile location and focus and transmit the energy only in the desired directions. In the ideal case, a smart antenna system enables the operation of a radio service with an efficiency that is similar to that obtained on cable transmission systems. The energy propagates directly to the receiver. This assumes, however, that the transmitter and the receiver are in line of sight, which is rarely the case in mobile radio.

Smart antenna systems introduce the *intelligent cell* concept using multiple antennas and advanced signal-processing techniques. Smart antenna systems can be divided into two categories [13]: *switched beam* and *adaptive array antennas*. Only the adaptive system however, obtains an optimum gain by identification and minimization of signal interference. This strategy is used in the adaptive system and the additional gains provided give improved performance and flexibility with respect to the more passive switched beam.

When smart antennas are used on the up and down links, global interference decreases. Channel capacity can hence be improved by using smaller frequency reuse patterns, or even by reusing frequencies inside the same cell.

4.3.2.1 Switched-Beam Systems

The switched-beam concept is an extension of cellular sectorization in which a cell site contains, for example, three macrosectors with 120 deg coverage (Figure 4.25). The switched-beam approach divides the macrosectors into several microsectors, *N,* for example. The system consists of *N* antenna beams from the *N* antennas covering the *N* zones. A beam selector is used to choose the right frequency in the right beam in order to reach the mobile concerned. Each microsector contains predefined fixed beams having a maximum sensitivity located in the beam center and a lower sensitivity elsewhere. In this type of system, fixed radiation patterns determined during installation are selected for operation. When a mobile enters a particular macrosector, the system selects the microsector that contains the most powerful signal. During the call, the system monitors the signal strength and will switch to other fixed microsectors when required.

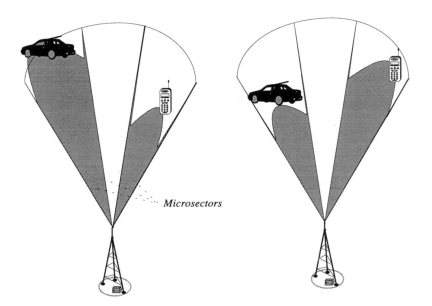

Figure 4.25 Sector coverage function of the mobile's position.

Compared with traditional sectorized systems, switched-beam systems increase the range of a base station from 20% to 200%, depending upon implementation. The additional coverage can save the operator on infrastructure cost.

In [14], two fixed-beam systems (the first with 12 beams of 30 deg aperture width and the second with 24 beams of 15 deg aperture width) are compared with a sectorized antenna system using dual diversity. Tests with the 24-beam antenna show gains of about 5 dB.

Switched-beam system limitations are mainly due to a degradation in communication quality caused if the switching of a mobile from one microsector to another is too slow. The other important drawback of these systems is caused by the system finding it difficult to distinguish a useful signal from the interferer. If the jamming signal is approximately in the center of the desired radius, the level of the interferer may increase more significantly than that of the useful signal. In this case, communication quality is adversely affected.

4.3.2.2 Adaptive Antenna Systems

Adaptive antenna systems, also called adaptive interference cancellation systems, adapt continuously to the radio environment and follow its variations. These systems are based on the use of sophisticated signal-processing algorithms that can distinguish useful signals from both the multipath and interfering signals and then calculate the corresponding directions of arrival. The precise tracking capacity and the interference rejection capability of the system allows several users to share the same channel within the same cell.

It is easy to eliminate the jamming or the jammed terminals outside the main beam by switching the beams of a multibeam antenna (Figure 4.26). Theoretically, an antenna containing M components can suppress $(M - 1)$ jamming signals by applying adequate weighting to the elements [15]. In practice, however, this suppression capacity decreases with the presence of multipath components. This technique can be improved by combining it with either CDMA, an adaptive equalizer, or polarization diversity.

Field experiments [16] have shown that adaptive antenna techniques allow gain improvements of 6 dB compared with those of sector antennas.

4.3.3 Space-Division Multiple Access Technique (SDMA)

The SDMA technique originates directly from the intelligent antenna concept. It can be used with all the conventional access methods (FDMA, TDMA, and CDMA). The necessary modifications are limited to base stations and do not involve the mobiles. This allows the introduction of SDMA in existing systems. Simulation results promise capacity gains, in the case of CDMA with trisector cells, of the order of 2 to 4 with $8 \leq N \leq 32$. Clearly, the greatest gains will be obtained with the highest complexity and cost of the array antenna [17].

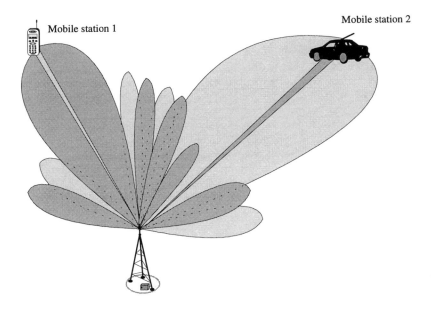

Figure 4.26 Adaptive antenna system: main beam directed to the useful signal.

The use of intelligent antennas allows the implementation of the SDMA technique. This technique has been used for several years in satellite systems, for example [8]. It may be considered as a special filtering operation obtained by using an adaptive antenna at the base station. An identical structure is adopted on the transmit side. This system is composed of an array antenna and a digital signal processor (DSP), the role of which is to process, in real time, the received signals and those to be transmitted by the antenna array. The N complex signals, obtained at the N antennas of the array, are input to the DSP, multiplying the signal of each antenna with an appropriate value W_i^* and performing, finally, the summation of all the terms. The output signal has the following form [17]:

$$y(t) = \sum_{i=1}^{N} w_i^* x_i(t)$$

The judicious choice of the weights of the vector $W = [W1, W2, ..., WN]$ defines, according to the wanted characteristics, the profile of the antenna radiation pattern (Figure 4.27). In particular, the vector W is determined according to an adaptive strategy. Its weights are regularly updated on the basis of the observation of the received data. To ensure good system operation, it is necessary that the adaption rate can compensate for the environment variations caused by motion of the sources and by the presence of multipaths.

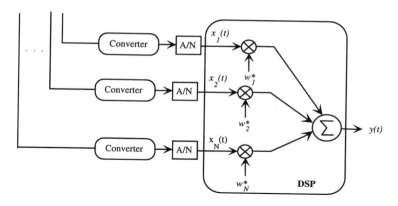

Figure 4.27 Structure of adaptive array antenna for reception.

Multiple antennas can be controlled in an adaptive manner by the base station to follow the mobiles [15]. In a cellular system, the most significant components of the received signal arrive with a low angular spread. Consequently, one possibility would be to precalculate a set of radiation patterns and to record the antenna weights, by switching between the diagrams during mobile movements. This process can be seen as interference canceling, in which the receiver adapts the weight of the elements to cancel the jamming signals totally. With M elements, $(M - 1)$ jammers can be eliminated (Figure 4.28). Thus, M users can share the same cell. In practice however, the improvement of performance is less than could be expected because of the many multipath components coming from the useful signal and from the interferers. To suppress the jamming signals, all the major multipath components must be controlled. These can arrive from a large set of angles, particularly in indoor environments.

In mobile radio, antenna patterns should be adjusted so that the beam pointing toward a mobile can follow its movement. This technique is known as *adaptive SDMA*. The approach has the following main advantages:

- Reuse of carriers and hence of the band, dynamically, in an adaptive manner inside the coverage area (more users can share the same frequency band).
- The beams are directed toward the user, multipath propagation is considerably reduced as well as the interference between channels.
- Energy consumption is reduced because it is beamed towards the user with reduced electromagnetic pollution.
- The confidentiality of communication is increased by the directional beams. The interception of a communication is not possible unless in the beam direction.

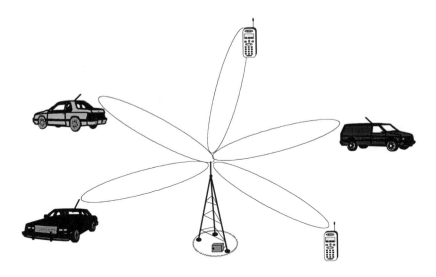

Figure 4.28 Principle of the SDMA technique.

4.3.4 Advantages and Drawbacks of Array Antennas

The following are the main advantages of systems using array antennas [8] that drive the use of the SDMA technique:

- An increase of the number of users for a given BER,
- A reduction of multipath fading,
- An increase in the BER for a fixed number of users in a cell,
- A reduction of the required C/I ratio for each antenna and hence a reduction of the necessary transmission power for the uplink,
- An increase of cell size because of the diversity gain,
- Use of efficient power control (in particular, in CDMA systems).

The main drawbacks of these systems are the following:

- Software and hardware complexity,
- The M receivers must be accurately synchronized in time to obtain acceptable performance,
- Processing complexity of the processing algorithms may be high,

- The size of the array is limited by the space available at the base station side. In general, the spacing between the antenna elements varies from one half to ten wavelengths,
- The antenna array can be affected by channel modelling errors, calibration of errors and noise correlation between antennas.

4.4 CONCLUSIONS

Signal-processing techniques and mechanisms used in a transmission system are essential in maintaining acceptable communications quality. The increase of system capacity and an improvement in communications quality are two closely linked concepts. Thus, an improvement of radio-link quality may be obtained by system-capacity enhancement techniques (e.g., frequency hopping, power control), or by traffic increase of a cellular network (e.g., cell splitting, sectoring, down-tilting), as presented in Chapters 6 and 7.

REFERENCES

[1]   Mundra, P. S., T. L. Singal, and R. Kapur, "The Choice of a Digital Modulation Scheme," *Proceedings of the IEEE Vehicular Technology Conference '93*, Secaucus (NJ), pp. 1–4, May 1993.

[2]   Stüber, G. L., "Modulation Methods," *The Mobile Communications Handbook*, Editor-in-Chief Jerry D. Gibson, CRC Press, 1996.

[3]   Wong, P., and D. Britland, *Mobile Data Communications Systems*, Artech House, 1995.

[4]   Paulraj, A., "Diversity Techniques," *The Mobile Communications Handbook*, Editor-in-Chief Jerry D. Gibson, CRC Press, 1996.

[5]   Dennis, T., "Airborne," *Mobile Asia-Pacific*, pp. 29-30, Aug./Sept. 1997.

[6]   Kamanou, P., "Outil de planification. Radio et méthodes associées," *Séminaire régional sur les systèmes cellulaires radiotéléphoniques mobiles*, ITU-BDT, Tunis, 9-13 Juin 1997.

[7]   Forssen, U., J. Karlsson, F. Kronesdet, M. Almgren, and S. Anderson, "Antenna Arrays for TDMA Personal Communications Systems," *Proceedings of the International Conference on Universal and Personal Communications'95*, ICUPC'95, Tokyo, Japan, 1995, pp. 382-386, Nov. 6-10.

[8]   Thompson, J. S., P. M. Grant, and B. Mulgrew, "Smart Antenna Arrays for CDMA Systems," *IEEE Personal Communications*, pp. 16-25, Oct. 1996.

[9]   Fernandes, J., O. Sousa, and J. Neves, "Impact of the Antenna Setup and Arrays on Mobile Radio Systems," *Proceedings of the International Conference on Universal and Personal Communications'95*, ICUPC'95, Tokyo, Japan, pp. 387-391, Nov. 6-10, 1995.

[10] Tsoulos, G. V., M. A. Beach, and S. C. Swales, "Application of Adaptive Antenna Technology to Third-Generation Mixed-Cell Radio Architectures," *Proceedings of the IEEE Vehicular Technology Conference '94*, Stockholm, Sweden, 1994, June 8-10.

[11]   Andersson, H., M. Landing, A. Rydberg, and T. Öberg, "An Adaptative Antenna for the NMT 900 Mobile Telephony System," *Proceedings of the IEEE Vehicular Technology Conference '94*, Stockholm, Sweden, 1994, June 8-10.

[12]   Nowicki, D., and J. Rouleliotis, "Smart Antenna Strategies," *Mobile Communications International*, pp. 53-56, Apr. 1995.

[13]   Lee, W.C.Y., "An Optimum Solution of the Switching Beam Antenna System," *Proceedings of the IEEE Vehicular Technology Conference 1997*, Phoenix, AZ, pp. 170-172, May 4-7.

[14]   Li, Y., M. J. Feuerstein, and D. O. Reulink, "Performance Evaluation of a Cellular Base Station Multibeam Antenna," *IEEE Transactions on Vehicular Technology*, Vol. 46, No. 1, pp. 1-9, Feb. 1997.

[15]   Winters, J. H., "Signal Acquisition and Tracking With Adaptive Arrays in the Digital Mobile Radio System IS-54 with Flat Fading," *IEEE Transactions on Vehicular Technology*, Vol. 43, No. 4, pp. 377-384, Nov. 1993.

[16]   Anderson, S., et al., "Ericsson/Mannesmann GSM Field-Trials With Adaptive Antennas," *Proceedings of the IEEE Vehicular Technology Conference '97*, Phoenix, AZ, pp.1,587-1,591, May 4-7, 1997.

[17]   Buracchini, E., F. Muratore Palestini, and M. Sinibaldi, "Performance Analysis of a Mobile System Based on Combined SDMA/CDMA Access Technique," *Proceedings of the International Symposium on Spread Spectrum*, Hanover, Germany, pp. 370-374, Sept. 1996.

[18]   Elbert, R. B., *Introduction to Satellite Communication*, Artech House, 1987.

SELECTED BIBLIOGRAPHY

[1]   Berlekamp, E. R., P. E. Peile, and S. P. Pope, "The Application of Error Control to Communications," *IEEE Communication Magazine*, Vol. 25, No. 4, pp. 44-57, Apr. 1987.

[2]   Bhargawa, V. K., and I. J. Fair, "Forward Error Correction Coding," *Mobile Communications Handbook*, Editor-in-Chief Jerry D. Gibson, CRC Press, 1996.

[3]   Bornkamp, B., A. Kegel, and R. Prasad, "Macro and Micro Diversity in Land-Mobile Cellular Radio Telephony Networks With Discontinuous Voice Transmission," *Proceedings of the International Conference on Universal and Personal Communications'95*, ICUPC'95, Tokyo, Japan, pp. 590-594, Nov. 6-10, 1995.

[4]   Comroe, R. A., and D. J. Costello, Jr., "ARQ Schemes for Data Transmission in Mobile Radio Systems," *IEEE Transactions on Vehicular Technology*, Vol. VT-33, pp. 88-97, Aug. 1984.

[5]   Corden, I. R., and M. Barret, "Adaptive Antennas for Second and Third Generation Mobile Systems," *RACE Mobile Telecommunications Workshop*, Amsterdam, Netherlands, pp. 728-732, May 17-19, 1994.

[6]   Falconer, D., "Signal Processing," *Proceedings of the IEEE Vehicular Technology Conference '96*, Atlanta, GA, pp. 11-15, Apr. 28-May 1, 1996.

[7]   Hanzo, L., R. Lucas, and J. Woodard, "Automatic Repeat Request Assisted Cordless Telephony," *Proceedings of the International Conference on Universal and Personal Communications'95,* ICUPC'95, Tokyo, Japan, pp. 934-938, Nov. 6-10, 1995.

[8]   Haykin, S., *Digital Communications,* Wiley, 1988.

[9]   Tangemann, M., C. Hoek, and R. Rheinschmitt, "Introducing Adaptive Array Antenna Concepts in Mobile Communication Systems," *RACE Mobile Telecommunications Workshop,* Amsterdam, Netherlands, pp. 714-727, May 17-19, 1994.

[10] Vaughan, R., "Polarization Diversity in Mobile Communications," *IEEE Transactions on Vehicular Technology,* pp. 177-186, Aug. 1990.

[11] Yamada, T., "Block Codes," in *Essentials of Error-Control Coding Techniques,* Academic Press, 1990.

CHAPTER 5

SECURITY

5.1 DEFINITIONS AND GENERAL PROBLEMS

Security in radio systems is a concept covering a wide range of quite different aspects and also requiring that protection mechanisms be involved at all levels. Security covers:

- *Integrity,* which ensures that a system can be accessed only by authorized users or entities,
- *Availability,* which ensures that the system responds efficiently,
- *System control,* where network managers must be able to control the access of the users to certain data,
- *Maintenance operations,* including, for example, the physical connections, communication equipment, and logical processes defined by the software and procedures.

More specifically, to ensure that the security of a radio system prevents intruders from reading or modifying the data being transmitted or stored as well as from gaining access to the resources or services of the system.

Network security can be analyzed and deployed under three major headings:

- *Organizational:* security rules and management priorities,

- *Technical:* the technical means and mechanisms to be implemented to ensure the network logical security,
- *Physical:* the physical security of the equipment, such as buildings and the communication methodology.

This chapter is mainly focused on the technical aspects of security issues, in which there are four main function groups. These are:

1. *Authentication:* helps to make sure that the person or body who attempts to be connected is genuine,
2. *Confidentiality:* ensures that the information relating to one person or entity will not be listened to or heard by a third party,
3. *Data integrity:* ensures that the exchanged information will not be modified during transmission,
4. *Nonrepudiation:* demonstrates that a given person or entity did transmit or receive the message.

This last function relates, more particularly, to financial or legal transaction services. In mobile radio systems, the main concerns about security are focused on authentication and the necessity to protect user data (see Section 5.2.2).

Other aspects, such as protection against intrusion (e.g., against messages that users would rather not receive) or secret transmission exchange (aiming to prevent the use of traffic analysis from learning that a message is being transmitted), are factors that should be taken into account when deciding a set of system security measures to be implemented.

In this chapter, protection measures are described for fixed and mobile radio systems. More specifically, concepts and methods are introduced that relate to confidentiality and authentication. These are the two major risks to which mobile systems are most vulnerable. The latter part of this chapter contains various examples showing the implementation of these techniques in cordless systems (refer to Chapter 10) and in cellular systems (refer to Chapter 12).

5.1.1 Complexity of the Problem

Security problems are getting more acute because:

- System coverage (the GSM system, for example, covers many countries worldwide),
- The number of accesses to the systems increase everyday and by a steadily growing number of individuals and entities,
- The complexity of the systems is increasing because of the sophistication of the services that are offered by the network operators and service providers. This is further exacerbated by the increase of intersystem interconnections.

With the increase of system coverage, the decentralization and distribution of decisions has been necessary both for system designers and operators. In the case of centralized systems (e.g., first-generation cellular systems), secure access control was relatively easy. On the other hand, for systems that are designed to be decentralized and where the data and equipment are distributed geographically, access to a system is multiplied both by system complexity and by location. (e.g., consider the case of international roaming, see Chapter 8) [1].

Moreover, systems become more and more complex because of the integration of network equipment acquired from different manufacturers, different working conditions, applications that are more tightly integrated, and the interconnection of heterogeneous networks from different operators. The use of intelligent networks and the handling of user data through various environments opens up possible new weak links that may make it easier to attack resources and sensitive information.

In such conditions, and in order for the integrity of the telecommunications system to be assured, it is mandatory to define a set of protection functions at all levels. It is obvious that in the case of mobile radio, the physical layer (radio interface) is the most vulnerable link in the system. As a result, a substantial part of this chapter is dedicated to the security mechanisms related, directly or indirectly, to this characteristic. At the data link layer level, ciphering can be used. At the network layer level, some software barriers—commonly known as *firewalls*—can be installed to prevent the inputting or outputting of messages [2].

5.1.2 Intrinsic and Extrinsic Protection

A communications system is protected by a set of techniques that may be subdivided into two categories. The techniques implemented by the designers and operators of the system belong to the *intrinsic* protection of the system. Subscription procedures (e.g., those defined and undertaken by the bodies providing the subscriptions) belong to the *extrinsic* protection of the system. These include subscription registrations and the checking of subscriber solvency and so forth.

Most problems relating to fraud cases are due mainly to loopholes found in the extrinsic protection system. For example, the GSM system is provided with strong intrinsic protection, as is shown later, but it is likely to suffer from a much higher number of fraud cases than an analog system such as NMT (provided with a notably weaker intrinsic protection system). This is due mainly to the international roaming that is offered by GSM as well as its more attractive services. In this chapter, we are interested in the different kinds of intrinsic protection.

5.1.3 Attacks and Origin of Security Problems

For mobile networks the use of a radio interface, if decoupled from the fixed physical links of the telecommunications network, leads to security and confidentiality related problems.

These are more complex than those met by wireline systems. The radio interface accessibility is much higher than that of the fixed network and augments the risks of passive (eavesdropping) as well as active attacks (identity violation). Broadly, security breaches of a mobile radio system can be categorized into two types: economic fraud and spying/sabotage.

5.1.3.1 Economic Fraud

Some intruders use the services and resources of the system for their own benefit or to assist others for financial gain. They benefit from network services without paying their bills (calls and subscription fees) to the service provider or to the network operator. These practices consist mainly by *rechipping, cloning, fictitious subscriptions* (or forged ones), *terminal theft*, and *fake devices* (which claim to be a different entity). Economic fraud can occur at two levels and can be divided into technical fraud and commercial fraud. The former has an impact at the system level, while the latter is independent of the system.

5.1.3.1.1 Technical Fraud

The main technical causes of fraud are *rechipping* and *cloning*. *Rechipping* consists of reprogramming the electronic chip of the terminal that contains the serial number (ESN, *electronic serial number*) and which identifies the terminal. The fraud operates by copying the serial number of an existing terminal of a subscriber whose account will be invoiced for the calls made through the rechipped terminal. This allows several terminals to be operated on a single network subscription. No totally reliable method to help fight against this kind of fraud exists. There are some methods that offer software traps that accept the new ESN but then stop the unit from operating after a limited number of calls. There is also another method that uses counters (see Section 5.3.3.5).

Cloning relies on the same principle as rechipping. It consists of devising a new terminal by using the identity (genuine) of another terminal that is already in use (Mobile Identity Number [MIN], and associated ESN). This process allows the user of the new terminal to make calls without being charged while the bills are issued in the name of the real subscriber. Cloning accounts for one of the major sources of fraud affecting the analog cellular networks (up to 90% of the losses are caused by these types of fraud [3] for some American operators). Due to the authentication algorithms in use (see Section 5.4.2.1.1), GSM SIM cards (see Chapter 12) seem to have not been cloned, at least up to 1999. Indeed, any attempt to replicate must be achieved by reverse engineering of the card itself, which requires bypassing the built-in high-level protection designed specifically to stop this [4].

Masquerades (e.g., fake base stations pretending to be base stations belonging to the real network) can be used to collect the terminal identities of units roaming nearby. The cloning of terminals, by use of these collected identities, allows intruders to act as if they were genuine subscribers to the real network and gain access to it.

5.1.3.1.2 Commercial Fraud

Commercial fraud occurs at two levels: first at the subscription level and second by use of stolen terminals.

For *fraudulent subscriptions* there are two common scenarios. Either a person subscribes under a false name and uses a false address (this practice is much more commonplace with cellular systems than fixed networks where the subscriber is connected at home), or the operator loses track of the subscriber who continues to use his or her subscription rights.

Terminal theft is generally accompanied by the theft of calls if the terminal is not protected by a password or has been stolen after the password has been entered by the user. Experience has shown that fraud always has a distinct appetite for following the cheapest and least path of resistance. It is thus more likely that external fraud will be of the subscription type, rather than sophisticated technical attacks that are likely to require special expertise and are expensive to implement.

5.1.3.2 Spying or Sabotage

Spying or sabotage attacks are used either to acquire confidential or private information, or to modify parts of it. Within these categories, two kinds of attacks can be identified: *passive* and *active*.

In the case of a *passive attack*, the information exchanged between the subscribers is *intercepted* by eavesdroppers (Figure 5.1). This kind of attack does not alter either the state of the system or messages passing through it.

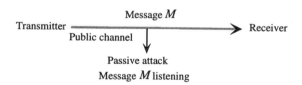

Figure 5.1 Passive attack.

In the case of an *active attack*, the intent is to alter or destroy information held on the network management database to the benefit (e.g., modification of his own profile, such as authorization for unlimited access to sensitive data) or detriment of a subscriber (e.g., to ban a subscriber from having access to crucial data). This attack could be the result of identity usurpation (use of a mobile with a false identity). An active attack may also be made by jamming the transmission channel to prevent a communication from being processed, or by *spoofing,* that is, modifying the content of the exchanged messages. The latter can be achieved by recording messages, modifying them, and sending them at a later time.

5.1.4 Security Services and Mechanisms as Defined in the OSI Model

In the Open System Interconnection (OSI) model, five main types of services are provided: authentication, access control, data confidentiality, data integrity, nonrepudiation. The corresponding security mechanisms are: encipherment, digital signature, access control, data integrity, authentication exchange, traffic padding, routing control, and notarization.

5.2 CONFIDENTIALITY PROBLEMS

The confidentiality of communications is defined by ISO as "the information property not to be used or deciphered by individuals, entities, or nonauthorized processes." This service could be considered both as value-added by the end user (e.g., such as the ciphering of voice communications on the radio links within GSM) and as a protection for the operators (notably for data management and signaling) [5].

5.2.1 Confidentiality Levels

Confidentiality of radio network communications can be assured in accordance with the following four-level classification [6]:

- *Level 0*: No confidentiality is activated. True identity of users is assumed without authentication. This is the case for analog radio where no ciphering is implemented and where eavesdroppers can tap into messages with suitable equipment (e.g., a scanner). The main advantage of this kind of protection is its simplicity and thus its cost, which is the lowest.
- *Level 1*: Confidentiality is similar to that of fixed networks. Ciphering is provided on the radio interface and authentication mechanisms are added to the identification process. This level of protection is one that is provided by the second-generation mobile radio systems. The rest of this chapter mainly focuses on this kind of protection.
- *Level 2*: Confidentiality is provided for professional transactions, such as communications related to stock market transactions, solicitor-client discussions, and contract negotiations. This level of protection requires end-to-end ciphering, not just air interface ciphering. Authentication may also be applied end to end between the communicating entities and not just between the system and each mobile entity,

as in level 1. This level of protection requires additional and specific mechanisms to be implemented at the communicating entities level (the higher layers in the OSI model), as well as those supplied by the communication system itself (e.g., use of passwords for connection to a particular service).

- *Level 3*: Confidentiality meets the constraints of a military or strategic user. In this case, the mechanisms used are much more specific than in the previous case with additional reinforced ciphering methods (user key encryption) and authentication processes.

5.2.2 Data to Be Protected

The act of communicating involves two types of data that should be protected:

- The *user data* exchanged with the network and that ensures transport to its final destination.
- The *signaling data* processed by the system for call management (e.g., subscription rights, subscribers' location).

Confidentiality should be ensured for various information exchanges or operations, the main ones including [7]

- The *users' calls* or entities to be connected.
- The *call establishment related information*: dialed number, calling card number, or kind of service requested, for example.
- *Subscriber location*: the confidentiality of the subscriber's location requires protection of the data exchanged between the terminal and the network, as well as the home database (see Chapter 8) within which the subscriber's location is stored.
- The *identity* of the user.
- The *profile of the calls*: the information that analyzes the user's operations (e.g., calling number, terminal frequency use, caller identity).
- *Financial transactions*: for example, credit card number.
- *Information flow or traffic analysis*: a passive attack would consist of collecting information that allows intruders to determine tactics and attack points. For instance, the intruder may, by analyzing traffic patterns (protocols, routing, time, size, and type of transmitted information), confirm known information and by inference determine confidential information.

5.3 PROTECTION METHODS

The main protection called upon to provide confidential access to the system and protection to exchanged information are methods (level 2 and higher protection levels) used by the communications system for *authentication* and *ciphering*. Cryptography is the main mechanism that facilitates the provision of these two services.

The message to be ciphered (called *cleartext* or *plaintext*) undergoes a certain transformation through a cipher algorithm. The outcome of the ciphering process (called *ciphertext*) is then transmitted (Figure 5.2).

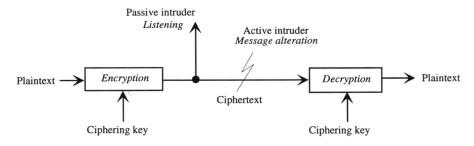

Figure 5.2 Basic ciphering principle.

Definitions: The ciphertext is the result of the transformation of the plain text through an encryption function characterized by a *key* that normally remains unaltered during the communication, hence its name *session key*. The key is a short string of characters that can be changed as often as required.

An eavesdropper is able to tap into or record the whole ciphered text. Yet, unlike the intended receiver, who is equipped with the deciphering key, the eavesdropper will find it very difficult to recover the plaintext.

The real secret of ciphering resides in the cryptographic key size, which is an element of paramount importance to determine the strength of the system. For example, a two-digit key allows 100 possible combinations with an exhaustive search (the simplest way but also the longest one). A three-digit key would offer 1,000. Thus, the larger the size of the cryptographic key, the more important the work factor becomes. This factor helps to quantify the amount of effort (in terms of typical computer operations) to be provided in order to break the ciphering method (called the *cryptanalyst method*).

Definition: The most basic attack used against a block cipher is to test each of the possible candidates for the key k until a match is found. If the key is b bits in length, then there are clearly 2^b possible keys. When designing a cipher algorithm, as a basic hypothesis, the fundamental rule should be to consider that the eavesdropper's cryptanalyst also knows the general encryption rules. The role of the keys is therefore a crucial one. These should be designed in such a way that exhaustive search methods should lead to astronomic computing times. It therefore follows that the main objective of a ciphering method is to deliver a cipher text that looks as random as possible in order to minimize the use of any shortcuts.

5.3.1 Communications Confidentiality: Ciphering or Encryption

Confidentiality allows a given subscriber to communicate without being subject to indiscriminate eavesdropping. Monitoring may be performed for two major reasons, either to understand the information content of the messages or to track down the movement of the users. The radio channel, being a broadcast medium, spreads in all directions and is, therefore, likely to be intercepted by whomever is provided with an appropriate device (a scanner most of the time). The public mobile radio networks, therefore, should systematically incorporate this type of protection to guarantee the confidentiality of messages for their customers.

Ciphering is the best technique available to avoid the interception of messages by eavesdroppers. It may also be used on a much wider scale for providing other security services such as integrity, authentication, or nonrepudiation of the received information.

Basic ciphering principles can be described as follows. Consider user A who would like to send a message M to user B. The ciphering procedure will modify the plaintext message M into a ciphered message $C=C_K(M)$ using a ciphering key K. To decipher C, user B must have the appropriate deciphering key K'. This key will help him or her to decipher C after processing by a reverse transformation $D_{K'}(C) = M$ (Figure 5.3).

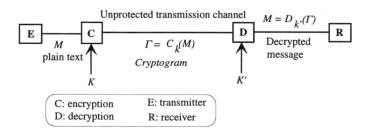

Figure 5.3 Ciphering system general principle.

In the case of an analog cipher, the transformation undergone by the message is carried out only at the signal level. This kind of cipher is used whenever the bandwidth is limited. For instance, in the case of speech, the signal is cut into segments of some tens of milliseconds, and then spectrally transformed in a pseudorandom way so that the ciphering can be obtained. This cipher technique is similar to the cipher by substitution method (see next section).

In the case of digital ciphering, the message is totally digitized and the ciphering is directly applied to the binary data. In this case, the obtained security level is higher than that of the analog techniques. The cipher techniques that are of interest in this chapter are the digital methods.

Digital ciphering systems can be subdivided into two large family groups:

- Systems characterized by a shared or symmetric key, and where ciphering and deciphering keys are the same ($K = K'$). In this case, the key exchange takes place through a confidential channel and the system is called a *secret-key cryptosystem*. These are implemented in digital cellular systems such as GSM and IS-54.
- Systems characterized by two keys, or asymmetric systems, and where ciphering and deciphering mechanisms are different (the knowledge of one key does not lead to the knowledge of the other); in other words, $K' \neq K$. The key K' cannot be simply determined from K (e.g., in the RSA algorithm, the determination of K' requires the factorization of a large integer number N) and the ciphering key can be issued from a register. These systems are called public-key cryptosystems. These cryptosystems are being considered for UPT and FPLMTS [8].

The symmetrical systems rely on three main techniques:

- Substitution,
- Transposition,
- Combinations of substitution or transposition.

5.3.1.1 Basic Ciphering Methods

Three basic and classic mechanisms are shown here:

Substitution cipher. This method, the simplest one, consists of replacing each letter or group of letters by another letter or group of letters. Nevertheless, the use of statistical characteristics of the communications language such as letter occurrence, syllables, and phraseology can help to make the breaking of this ciphering method easy.

Example 1: Example of substitution cipher using random letter allocation

   Initial letters
   Ciphered letters

   Message in clear: "this is an example."
   Ciphered message: "udej ej wg sowbnrs."

However, the simplest substitution method is the Caesar cipher in which the plaintext letter is replaced by the ciphertext letter several places down the alphabet. As there are only 26 Caesar alphabets, it is trivial to solve a Caesar cipher by exhaustive search.

Transposition cipher. In the substitution cipher, the order of symbols (e.g., letters) of the initial texts is maintained. The transposition cipher does not alter the letters but classifies them in an order that is different from the initial text. For this, it requires the introduction of a key (of size n) and requires the transcription of the text in the form of a table containing n columns. The cryptographic key is used to number the columns.

Stream/block cipher. Symmetric cryptosystems can be divided into *block ciphers* and *stream ciphers*. The difference between these two modes of ciphering can be summarized by the following definition from Rueppel [9]: Block ciphers operate with a fixed transformation on large blocks of plaintext data. Datastream ciphers operate with a time-varying transformation on individual plaintext digits.

Stream ciphers have the advantage of being very fast to operate. They are generally much faster than block ciphers. Each plaintext bit can be enciphered and serially output immediately. One of the properties of stream ciphers is that a transmission error in one bit of the ciphertext will only affect one bit of the resulting plaintext. There is no error propagation.

A block cipher used to encrypt the same plaintext block with the same cipher key at different times will give the same ciphertext block. A block cipher that operates on plaintext blocks of size n is called an *n-bit block cipher*. However, a single error in a block may cause the whole block to be corrupted if suitable error correction is not included.

5.3.1.2 Public-Key Algorithms

The problem related to cipher-key distribution has always been the weak link in most cryptographic systems. Whatever the cryptographic system, if the cipher key is stolen, the system loses all its security.

In order to solve this problem, Diffie and Hellman, in 1976, proposed a totally new encryption system [10]. In this system, the enciphering and deciphering keys are different. Furthermore, the deciphering key cannot be deduced from the enciphering one.

In the Diffie and Hellman proposal, the ciphering algorithm E and the deciphering algorithm D meet the following three characteristics:

- $D(E(P)) = P$,
- It is very hard to deduce D from E,
- E cannot be calculated even if the initial text (plaintext) is known.

User A initially chooses two algorithms, E and D, meeting these three characteristics. Both the ciphering algorithm E and the cryptographic key E_a are made public, hence the name *public-key cryptography*. Key E_a may be stored in a public file. User A is also required to issue the deciphering algorithm D. The only secret that A keeps is the deciphering key D_a.

When user A wishes to communicate with B, the public-key cipher method does not require that A and B have communicated with each other previously or even that they know each other. In other words, it is not necessary for A and B to have exchanged secret data (i.e., cipher or decipher keys) to be able to communicate their messages. The ciphering keys of A (E_A) and B (E_B) can be found in a public file. User A calculates $E_B(P)$, where P is the message to be transmitted knowing that the transmission is to B. User B will decipher it by applying his secret cryptographic key D_B. In other words, he or she calculates

$D_B(E_B(P)) = P$. No one, apart from B, will be able to decipher the message $E_B(P)$ because it is very difficult to deduce D_B from E_B.

In general, a public-key cryptography system supposes that each user U has two keys: a public key K, available to everyone and used for enciphering messages to be sent to U, and a secret cryptographic key K', used by U to decipher the messages that are received (Figure 5.4).

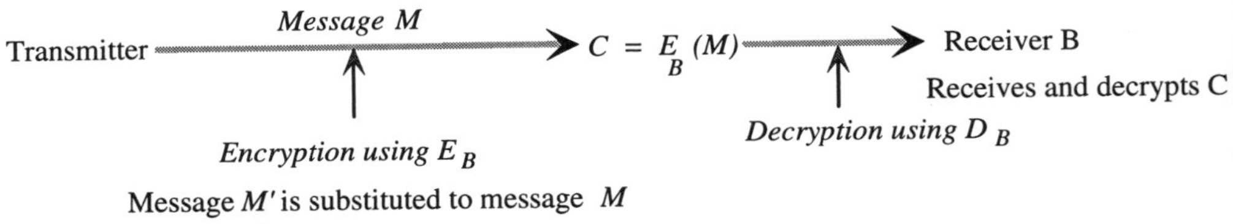

Transmitter ══════ *Message M* ══════▶ $C = E_B(M)$ ══════▶ Receiver B

Receives and decrypts C

Encryption using E_B *Decryption using D_B*

Message M' is substituted to message M

Figure 5.4 Public-key systems.

From a mathematical point of view, keys K and K' are interrelated by *trapp* functions. The major characteristics of these functions are their strength of calculation and a property making it impossible to reverse the process. In other words, it is impossible to calculate the secret cryptographic key without access to an extra piece of information (called the *trapp*).

Example 2: Signature with a public-key system
A signature is used to prove the identity of the transmitter and the authenticity of the message. The principle of the signature mechanism when using a public system cipher key is represented by Figure 5.5.

M is the message in clear. S is the message enciphered by the key D_A. User B receives a message couplet of (M,S) and using E_A (publicly available key) checks that $M = E_A(S)$. This is used to prove the identity of A and the integrity of M.

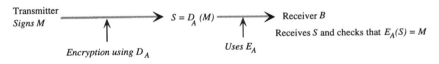

Transmitter
Signs M ══════▶ $S = D_A(M)$ ══════▶ Receiver B

Receives S and checks that $E_A(S) = M$

Encryption using D_A *Uses E_A*

Figure 5.5 Signature with a public-key cryptosystem.

The mechanism is based on the following characteristics:

- Only A knows D_A and is, therefore, the only one able to compute the signature S,

- No one can alter the message and pretend that it had been sent by A because no one is capable of calculating $S' = D_A(M')$.

Note that the two cipher and signature operations could be performed simultaneously. In this case, User B receives the pair (M, S) after having computed $S = D_B(C)$ and $M = E_A(S)$ (Figure 5.6).

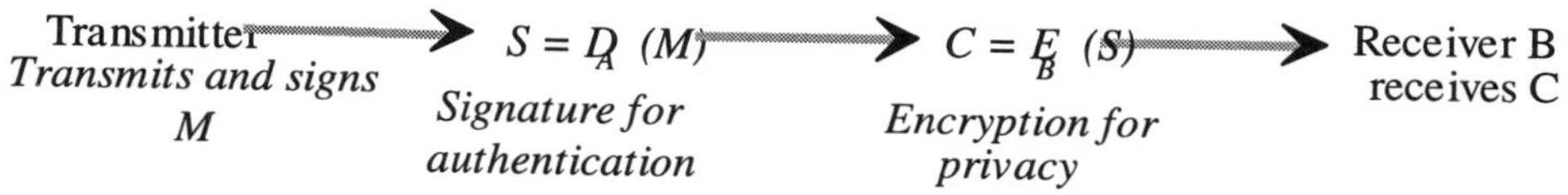

Transmitter ⟶ $S = D_A (M)$ ⟶ $C = E_B (S)$ ⟶ Receiver B
Transmits and signs M / Signature for authentication / Encryption for privacy / receives C

Figure 5.6 Signature and ciphering with a public-key system.

Existing mobile-radio communications systems such as GSM, DECT, or those operating in accordance with the IS-54 standard specification use the secret-key algorithm method only. The public-key cipher methods initially envisaged by the GSM standardization body were rejected because the protocols that were available at that time were not fast enough to provide satisfactory processing times. A public-key ciphering system could, however, be used to distribute keys used in secret-key systems. The complexity of the latter is notably lower than those of the public-key systems, and in turn, lends itself more easily to on-line ciphering.

5.3.1.3 Secret-Key Algorithms

Secret-key ciphering systems use the same key for encryption and decryption. They require the confidential exchange of keys, which is the most difficult task in these systems.

In the GSM system, for instance, each user's SIM card U has a secret key K_u and an algorithm for the cipher-key calculation that is also known by the network. When a ciphered communication is established between the user and the network, the network generates a random number—generally denoted by $RAND$—which is transmitted to the user mobile terminal. The terminal, with the aid of $RAND$, its secret-key K_u, and the cipher-key algorithm determines the ciphering key K_c. The network performs the same operation. Ciphered exchanges will then use the key K_c and the cipher algorithm (Figure 5.7).

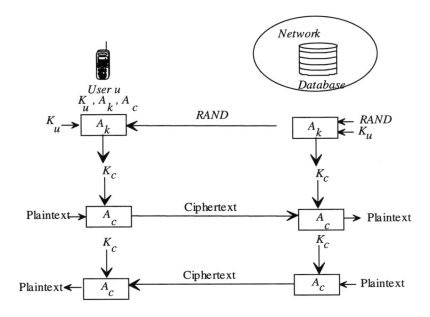

Figure 5.7 The basic principle of a secret-key cryptosystem.

In contrast to public-key cipher algorithms, the secret-key algorithms only require relatively low-complexity calculations. The major drawback of these systems is the need to possess a secret-key database. Thus, the weakness of these systems is that an internal intruder to the system might be able to obtain the ciphering keys of the subscribers to allow interception of all messages. Another drawback of these systems is in the initial supply of these secret keys that may be intercepted by a potential eavesdropper.

5.3.1.4 Comparison Between Secret Key and Public-Key Cryptosystems

The security of each algorithm depends on the size of the key. Use of a ciphering algorithm is conditioned by the nature of the information to cipher:

- If the number of messages and their size are large (e.g., file transfer, electronic mail), a secret-key protocol is more efficient than a public-key algorithm. For similar levels of confidentiality RSA is almost 1,000 times slower than DES and requires key sizes about 10 times larger.
- For key exchange and distribution (e.g., certificates, authentication, ciphering), the use of public-key systems is more efficient.

In most ciphering applications, a combination of these two algorithms is used. Session keys are generated randomly and are used in secret-key algorithms to protect transmitted messages. Public-key cryptography is used to protect and distribute these session keys.

5.3.1.5 Key Management

The ciphering method has an important impact on the architecture of the security regime. Systems using secret-key or public-key methods will be organized differently. The key management issue is very dependent on the ciphering method employed. We introduce some aspects of this problem in the following sections.

Key management is crucial in the sense that any ciphering system, however powerful it is, loses all its value as soon as the cryptographic key is known by a third party. The two important key-management methods are:

- Either *manual* (the keys are distributed by key fill units or similar), with the major risk of having a slow management system when the number of users is significant,
- Or *automatic* (the keys are distributed by the transmission network per se).

When the distribution method is automatic, the distribution system can be based either on symmetric or asymmetric algorithms. In Appendix 5A, details of some methods of cryptographic key distribution are outlined.

5.3.1.6 The RSA Algorithm

The Rivest, Shamir, and Adelman (RSA) algorithm is a public-key algorithm that was defined in 1977 [11]. Its origin relies on the factorization problem related to large integer numbers.

Presently, algorithms exist that are able to cope with factorization of integers of the length of about 100 digits. Even so, the computing power of a supercomputer is required to resolve the answer and this after several hundred hours. That said, these algorithms are unable to cope with processing numbers larger than 150 digits.

The objective of the RSA algorithm is to benefit from these technical constraints. Every user has two primary numbers p and q (e.g., each of 80 digits), the product of which is $n = p.q$.

The messages to be transmitted have to reach the set $0, ..., n-1$. Let a random number i between 0 and $n-1$, primary with $Pr(n) = (q-1)(p-1)$.

It is possible to find out a number j between 0 and $n-1$ so that the product $i.j = 1 \; mod \; Pr(n)$.

Numbers n and i are made public and numbers p, q, and j are kept secret.

A message M (between 0 and $n-1$) to be ciphered is raised to the power i. The rest of the division by n is noted C, $C = M$ (modulo n).

The ciphering of C is made by raising it to the power j (modulo n), C_j (modulo n) = M. The j key, therefore, remains secret because one has to know $Pr(n)$ to calculate it and this is virtually impossible to achieve.

The security of the method stems from the difficulty to factorize high numbers. If the cryptanalyst is able to factorize n (known), he or she can figure out p and q, and subsequently, $Pr(n)$. Knowing $Pr(n)$ and i, j can be calculated by using the Euclidean algorithm. Luckily, mathematicians who have tried to factorize these high numbers for the last 300 years have demonstrated that this problem is very difficult to resolve.

5.3.1.7 The DES Algorithm

Published in 1977 by the National Bureau of Standards (NBS), the Data Encryption Standard (DES) has been used as a national algorithm in the United States for the safeguard of federal agency data. It is a secret-key algorithm.

DES is based on the "Lucifer" cipher algorithm developed by IBM in the 1970s [10]. Since 1981, under the name of the Data Encryption Algorithm (DEA), the same algorithm has been adopted by ANSI. DES ciphers 64-bit blocks. The decipher algorithm can be easily deduced from the encipher one. They are both controlled by the same 64-bit cryptographic key (56 bits + 8 parity bits). As the cipher algorithm is publicly known, DES security is contingent upon keeping the key secret. An exhaustive search for the key would require 2^{56}, that is 10^{17} computations, an operation that was very difficult a few years ago with the computing power then available. To reinforce the safety of the algorithm, the use of a 112-bit cryptographic key has been suggested.

The DES ciphering method includes the following stages of calculation:

* Initial transposition of a 64-bit block to be ciphered,

* 16 calculation iterations,

* Final transfer, which reinverts the initial transfer.

More precisely, the plaintext is ciphered in 64-bit blocks to give 64-bit blocks of ciphertext as well. The algorithm is defined by a 56-bit cryptographic key in 19 stages (Figure 5.8). The first operation is a transposition that is independent of the key, and the last one is exactly the reverse of the former operation. The penultimate operation exchanges the 32 more significant bits with the 32 least significant bits. The remaining 16 operations are functionally identical but are determined by the key parameters. The algorithm had been defined to allow deciphering with the same key as used to encipher. The deciphering process stages are then followed in the reverse order of ciphering.

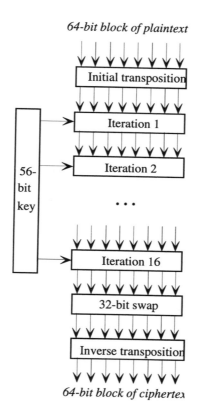

Figure 5.8 DES ciphering principle.

Today DES can be deciphered using powerful computers. Yet it is still being used on a wide scale for applications such as in cash point machines. For example, it has been demonstrated that to break DES would require about 3.5 hours work using a supercomputer. The most powerful attack requires some 2^{47} operations. This outcome shows that for high security applications a 56-bit key is now not enough, whereas it could perhaps be too powerful for low-level applications (e.g., speech communications in public mobile networks).

5.3.1.8 Additional Mechanisms

A technique that is commonly combined with the ciphering consists of introducing redundancy in the encrypted messages. In other words, information is added that is not necessary for the understanding of the message. Adding redundancy to information helps avoid third parties substituting encrypted messages for other messages with insignificant content to disturb the communication.

Another principle consists of using measures to avoid third parties, who record the messages and transmit them at a later stage, from introducing errors into the communication link. The method used consists of including a timeout after each message to ensure that messages which are overdue and could have been used for interception/ retransmission purposes by third parties are cancelled.

5.3.2 Location Confidentiality by Use of Implicit Addresses

Ensuring location confidentiality has many objectives:

- Ensure the confidentiality of the subscribers' operations,
- Meet the security demands of particular subscribers (e.g., VIP [7]),
- Allow a minimum confidentiality of messages by not revealing the identity of the mobile subscriber in case ciphering is not implemented.

Location confidentiality can be introduced at the level of the subscribers' addresses by use of *implicit addresses* or *aliases*, which are different from those used to identify each subscriber. An implicit address consists in a stream of pseudorandom bits that is incomprehensible except to the called person and the network. The implicit address should be totally different from the explicit address, which would help to indicate the subscriber's location.

The implicit addressing mechanism is a principle used in the GSM system. During the initial access, the mobile terminal identifies itself by using its fixed identity called the International Mobile Subscriber Identity (IMSI). Then the network allocates to the terminal, in ciphered mode if implemented, a temporary identity called Temporary Mobile Subscriber Identity (TMSI) to be used afterwards both by the mobile and the network for any interaction (e.g., messages transmission, location updating, paging of the subscriber). This identity is changed by the network either whenever the location area of the mobile changes (refer to Chapter 8) or periodically (see also Section 5.4.2.1).

5.3.3 Access Security: Integrity and Authentication

Integrity is the mechanism that allows a message transmitted through a nonsecure channel to resist active attacks aimed at modifying the messages or to substitute messages to original ones (Figure 5.9).

On the other hand network access security problems are addressed by implementing an *authentication* process. This is defined as being the mechanism that allows one party to prove his or her identity to the second. Authentication can take place:

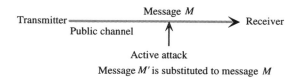

Figure 5.9 Data integrity problems.

- Either in one direction only, generally the case where the network requests the terminal to prove its identity, or
- In both directions, with mutual authentication of both the terminal and the network.

The network can request authentication at any time during communication (generally at the beginning of the communication and at each service request).

Authentication of users is the first form of protection against external attack. It relies on the use of an authentication protocol that might be simple or complex. The simplest protocols use a password chosen by the subscriber or given by the operator. The most secure protocols require exchanges of secret or random information that could be transmitted encrypted with the user's secret key.

The following criteria should be considered when designing an authentication protocol [12]:

- The authentication protocol should generate a minimum computing overhead. That is, the processing time must be as low as possible so that the call setup delay is not increased significantly.
- The authentication protocol should allow for domain separation. That is, secret or sensitive information used in the authentication process (user's secret key or password) should not be exchanged between the home domain and visited domain.
- The authentication protocol should be transparent to the user. That is, it must not depend, from the user interface point of view, on the network where it is processed (home or visited network).

In the main, an authentication procedure uses a secret key K, a function A (the access to which is controlled by a standardization body), and a random number $RAND$ (one per authentication). The larger the size of K, the safer the authentication but it requires heavier calculation in return (Figure 5.10). The choice of the function A is made such that the computation of the result $RES = A(K, RAND)$ is fast and that the determination of K from A, $RAND$, and RES is very difficult.

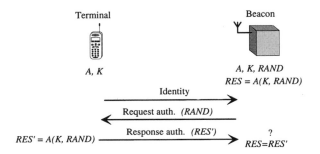

Figure 5.10 Terminal authentication by the network.

The calculation of *RES* is carried out in a different way in the network than in the mobile. From the network point of view, the calculation can take place at a central equipment level (i.e., *centralized* procedure at an authentication center) or at the level of the last network equipment, usually the base station, connected to the mobile (i.e., *distributed* procedure). The advantage of the first solution is its safety (only one site has to be protected). The drawback is the increased signaling load between the mobile and the central switch, something that generates extra costs and longer access delays. A distributed procedure helps obtain very short authentication delays because the exchanges take place at a local level but with the risk of having less security. This is because the secret data is transmitted through the network and stored at various points capable of implementing the authentication mechanism.

5.3.3.1 Authentication Based on a Shared Secret Key

Suppose that users A and B have already been provided with a common secret cryptographic key K_{AB}. The protocol is based on the principle found in several other authentication protocols. User A, for example, sends a random number to B who will transform it according to a certain algorithm and then send it back to A. These protocols are called *challenge-response* protocols. Suppose we have R_I's, the challenge is where i identifies the user who transmits the challenge. K_i is the key, where i indicates the owner of the key and K_s represents the session key.

First, A sends his identity in a nonciphered mode to B, who is unable to determine the genuine origin of the message (A or an intruder). B sends a challenge in the form of a random number R_B, which he sends to user A in clear. In his turn, A enciphers the message with the key K_{AB} and then sends the ciphertext $K_{AB}(R_B)$ back to B. Upon receiving the message, B is certain that it comes from A as the latter is the only one, along with B, to know K_{AB}. Moreover, as R_B had been chosen at random, it is highly unlikely that an intruder could have recorded the same random number R_B and its answer during a previous session.

User B is now certain of his communication with A, but it is not obvious for user A who, in his turn, chooses a random number R_A and sends it in clear to B. Upon receiving the answer $K_{AB}(R_A)$ transmitted by B, A is now sure that he is communicating with B. Finally, to set up a session key, A could choose one key, K_S, and send it to B using the cipher key K_{AB}.

The messages can be combined to help reduce the number of exchanged messages, as is reproduced in Figure 5.11.

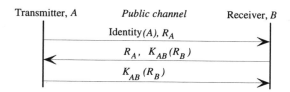

Figure 5.11 Mutual authentication with messages combination.

In this case A starts the exchange by initiating a challenge-response protocol. Equally, while responding to the challenge of A, B initiates its own protocol.

5.3.3.2 Authentication Using a Key-Distribution Center

To communicate with n persons it is necessary to have N cryptographic keys in order to ensure full confidentiality for all users. This would require the corresponding physical space (e.g., on memory cards) in each mobile. A different approach from that described above consists in introducing a *key-distribution center (KDC)*. With this method, each user is provided with a unique key shared with the KDC. The management of the authentication and of the session key is performed by the KDC. The simplest KDC authentication protocol is called the *wide-mouth frog* (Figure 5.12).

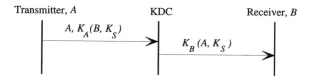

Figure 5.12 Wide-mouth frog protocol principle.

The underlying idea to this protocol is quite simple. User A picks a random session key K_S and indicates to the KDC that he would like to communicate with B by using K_S. This message is ciphered with the secret key K_A that A shares only with the KDC. The center deciphers the message by extracting the identity of B along with the session key. Then it can build up a message containing the identity of A and the session key to be sent to B. This ciphering is performed with K_B, the secret key that B shares with the KDC. When the message is deciphered B knows that A would like to communicate and is in possession of the correct session key.

5.3.3.3 Authentication Based on a Public-Key System

Mutual authentication can be achieved by means of a public-key cipher system. The setting up of the communication is reproduced by Figure 5.13.

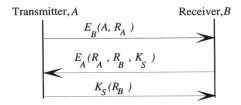

Figure 5.13 Mutual authentication using a public-key cryptography method.

User A starts by ciphering his identity and a random number $RAND$, by using B's public key, E_B. When B receives this message it is not known if it really comes from A. To make sure, B sends back a message containing $RAND$ and another random number $RAND'$, together with a session key K_S all enciphered using key E_A. Upon receiving this message, A will decipher it by using the key E_A. User A, therefore, finds the $RAND$ number in the message, which proves that only B, could have received and detected $RAND$. A responds by coding $RAND'$ with K_S. By means of this procedure, equipment can be authenticated without having to interrogate a database in real time. This characteristic helps reduce the system complexity.

5.3.3.4 Mutual Authentication Application

The authentication of the network by the terminals could be used to challenge decoy equipment that is trying to tap the identity of the terminals. From the terminal point of view, the principle is one of checking if the detected terminal is really part of its network. Mutual authentication is carried out as shown (Figure 5.14).

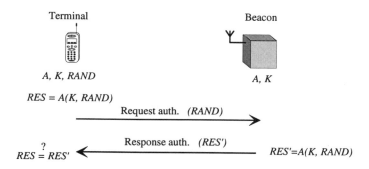

Terminal

Beacon

A, K, RAND

A, K

RES = A(K, RAND)

Request auth. *(RAND)*

Response auth. *(RES')*

$RES \stackrel{?}{=} RES'$

RES'=A(K, RAND)

Figure 5.14 Network authentication against fake base stations.

5.3.3.5 Authentication to Detect Perfect Clones

To protect the network against attacks from perfect clone terminals—in other words, terminals that are the identical copies of real terminals, both in terms of hardware and software—the authentication mechanism is not performed in a fixed way as described above. This is because it cannot help to distinguish clone terminals from the original terminal. The mechanism used becomes dynamic; a system that helps to add extra control. In this case, one may resort to a counter, the value of which is changed at each connection to the network. This method is used, for example, with PACS [13] (refer to Chapter 10).

In Figure 5.15, the network and the authentic terminal are both provided with a counter Z, whose value should always remain in synchronism at the network and the terminal. During the authentication phase the authentication center requests both parties to provide it with the value of this counter (*Z-Terminal*). If the value of *Z-Terminal* and that of *Z-Network* corresponds the authentication is successful. If not, this indicates the presence of a clone terminal and all accesses of the authentic terminal and its clone are banned.

Generally, during a call, counter *Z-Terminal* is incremented upon a command from the network. The network also increments counter *Z-Network*. If multiple handsets share the same identity, then the network will increment its counter *Z-Network* to exceed that of the legitimate user.

5.3.4 Personalization and Telepersonalization

The acquisition of the authentication and subscription data by the terminal (and in particular the key) can be performed in three different ways:

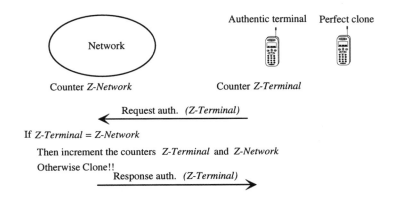

Counter *Z-Network*   Counter *Z-Terminal*

Request auth. *(Z-Terminal)*

If *Z-Terminal* = *Z-Network*

Then increment the counters *Z-Terminal* and *Z-Network*

Otherwise Clone!!

Response auth. *(Z-Terminal)*

Figure 5.15 Authentication using a counter.

- Personalization of the security identity module (the SIM card in GSM). The subscription and security data are registered on the SIM card (*subscriber identity module*, see Chapter 12). The former are loaded by the service provider and the latter are included either by the operator or the supplier of the SIM card.
- Manual personalization. Manual entry of subscription and authentication data. For example, the operator or service provider could send the secret data by mail and the subscriber manually inputs into the terminal through the keyboard.
- Telepersonalization. The secret data is sent at the first access of the terminal to the network over the communications path.

There are several kinds of telepersonalization:
- Hardware implemented key as illustrated by Figure 5.16,
- Predetermined sequence input by the user,
- Current key (transparent telepersonalization). The lifetime of this key is limited by the operator.

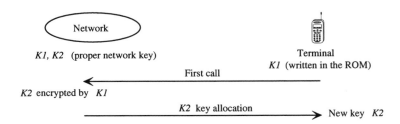

K1, K2 (proper network key)   Terminal
                                           K1 (written in the ROM)

First call

K2 encrypted by *K1*

K2 key allocation   New key *K2*

Figure 5.16 Telepersonalization.

5.3.5 Some Methods in the Struggle Against Fraud

With the development of the GSM networks and the international roaming service offered to the subscribers, fraud has acquired international proportions. This has compelled operators in different countries to cooperate, notably by exchanging information about different frauds and mobiles that have been reported stolen.

5.3.5.1 Data Exchanges

One of the best methods to help in the fight against fraud is the frequency at which the data history of the subscribers' calls is reported to the billing center of the subscriber's home network where it may be assessed. Hence, the latest generation of the "hot billing" systems helps to provide data in real time. The difference in gaining access to this information directly and in real time rather than receiving information a few days old helps considerably in reducing the losses from fraud. This is particularly significant with respect to fraud associated with international roaming on the GSM networks. To coordinate the identification of fraudulent equipment, a Central Equipment Identity Register (CEIR) has been set up in Dublin by the GSM MoU. This latter contains a database relating to virtually all mobile equipment being used together with details of manufacturer, model, and operator, for example. This initiative was to help implement a global database with which the members of GSM MoU may exchange a blacklist of terminals daily.

Implementing the principle of *electronic data exchange* among all the GSM networks for the recording of the calls of the roaming mobiles has proved to be a very efficient method to help in the fight against roaming fraud. This strategy allows the operators to detect the fraudulent use of SIM cards in a matter of a few days. Indeed, the acceleration of data transfer among GSM operators has already helped reduce the rate of roaming fraud by at least 50%. Operators belonging to the MoU are able to supply other operators within 72 hours, and sometimes in less than 24 hours, with details relating to subscribers who have used another network.

5.3.5.2 Security Modules

Use of modules such as the SIM card in GSM or the DECT authentication module (DAM) has the following advantages in terms of security:
- Allows production of these modules to be centralized and have the production sites (a few in number) more secure than the mobile station manufacturing facilities. (If the authentication module is built into the terminal, security measures have to be provided in all mobile station production sites.)
- Security modules are more tamper resistant. They can be physically protected against unauthorized attempts to read or modify their contents (e.g., ROM and EEPROM cannot be accessed directly, SIM files are protected by access conditions, K_i can neither be read nor modified).

5.3.5.3 Fraud Detection Systems

Finally, neural network computers are helping to detect potential frauds [3]. The use of profiles for each subscriber that identifies his or her normal behavior can help detect anything unusual by triggering an alarm. Equally, the subscribers can be gathered in groups of stereotypes, with any unusual activity triggering this alarm system if it does not fit within the set of these stereotypes. Based on these principles, fraud-detection systems (FDS) use two main criteria [4]:

- *A rules-based criterion:* the system generates a usage profile for each customer modeling their calling patterns and any unusual detected behavior of that user will trigger an alarm,
- *An event-based criterion:* a set of exceptional events is defined and any activity of the user falling within one of these events raises an alert.

Some conditions that may be included in the fraud-detection system could be as follows:

- *Pattern alarm:* a filter for single-call data records (e.g., a call to Australia longer than 25 min from a Paris cell),
- *Threshold alarm:* a filter for multiple occurrences of call data records during a short-term period (e.g., 15 calls to Lebanon from the Lyon area during the last 12 hours),
- *Accumulation alarm:* a filter for accumulated calling behavior during a short-term period (e.g., more than 150 calls, 65 international, total costs greater than US$ 5,000 within last 4 days),
- *Profile alarm:* a filter for accumulated behavior during long-term (7 to 30 days) period (e.g., more than US$8,000 turnover for international calls during last 30 days).

In many cases, related counters are driven by the billing system and produce a fraud or credit alert when a customer's behavior triggers one of these conditions. These limits are often arbitrarily set by the operator. The normal result of such monitoring can be the generation of many false alarms and resultant time-wasting. This is the reason why it should be useful to include self-learning components, generate a weighted database, and obtain information from fraudulent subscriber behavior to derive fraud probabilities.

One of the most difficult tasks to address is the design of triggering conditions. Indeed, cellular operators have observed that their worst subscription fraudsters and their best customers may have very similar usage profiles. Profile events include credit limit, suspect number, hourly and daily high-usage, long calls, international usage, off-peak, and suspect country.

As mentioned before, some FDSs are based on neural network techniques. Using these intelligent methods against fraud is motivated by the multiplicity and heterogeneity of the fraud scenarios. Neural networks are flexible enough to cope with this fraud diversity and can adapt the defenses to face new fraud scenarios. The FDS is thus able to detect abnormal usage patterns, which can appear in the billing records of the users and are often the result of a fraud. The neural networks engine acts by first classifying each user within a class of users according to his or her behavior and then defining a profile or training pattern to each of them.

5.3.5.4 Other Methods and Countermeasures

Another effective measure that can be used against subscription fraud (the main and to date the only crime affecting GSM [4]) is to introduce a random variation (i.e., a *moving target*) in the request for identity proof required from the customers when subscribing to the service.

Use of electronic money is also a useful mechanism to fight against fraud. It allows the subscribers to make a call only if their SIM cards contain prepaid units. This system has been used in Italy for a few years and more recently in France and the United Kingdom.

Phone specific fingerprinting allows real-time detection at the cell site. Newer analog telephones are issued with 16-digit individual security PINs. This number is transmitted automatically with the telephone number to authenticate the telephone, and if the PIN does not match the telephone number, the call is cut off [14]. Along the same lines, to fight cloning, the FCC has adopted a rule requiring that all cellular telephones produced after January 1, 1995, must be designed so that the factory-set ESN cannot be reprogrammed.

Finally, it should be mentioned that speech recognition algorithms can now provide new means of identifying the subscriber and can be used to augment security. However, these techniques will not be in use commercially for a number of years.

5.3.6 Conclusions

The security level provided by a ciphering service relies on two parts, the ciphering algorithm and the cipher/decipher keys. Even with the most powerful of procedures the security stream is equal to its weakest link. In practice, most of the frauds are not perpetrated through sophisticated cryptoanalysis or intelligent attacks, but through implementation errors or management failures.

One of the first tasks engineers in charge of design and implementation of security mechanisms have to consider is the related costs. For instance, information flow protection against traffic analysis (see Section 5.2.2) is only implemented in military systems. The security process consists of preventing any third party from detecting the presence or absence of a data exchange. This is an important constraint in the case of military operations. Some padding techniques, which consist of input even when there is no useful data to be processed, ensure that no curious eavesdropper is capable of making the difference between useful data ("real" message) and a padding message (absence of message). Padding consists of sending meaningless messages (a generator of messages that observes the frequency occurrence of letters of the alphabet and spurious data is necessary). This requires additional sophisticated mechanisms to be implemented.

In the case of the mobile radio systems, achieving a compromise between the level of confidentiality and the processing complexity is necessary. Mobile terminals should be operational over long periods of time with low capacity batteries. This requires the implementation of ciphering mechanisms with minimal energy consumption. These mechanisms may be implemented either at the hardware level (in special circuits), or at the software level (e.g., by using a Digital Signal Processor [DSP]). Studies show that special hardware can lead to better performances. The FEAL-32 [15] used to cipher 32 Kbps speech communications implemented on the hardware would entail a power consumption of about 1 mW, whereas in the case of a software implementation, energy dissipation will increase to about 70 mW.

5.4 EXAMPLES OF SECURITY FEATURE IMPLEMENTATION

In the following, we present how security features are implemented in cordless systems (which are described in detail in Chapter 10) and cellular systems (which are addressed in Chapter 12).

5.4.1 Authentication in the CT2 and DECT Systems

5.4.1.1 CT2 System

With the CT2 system, authentication operates by use of a key K and an authentication function A, stored at the terminal and at the network. At each authentication request the network transmits a random number $RAND$ to the terminal and calculates C' using A, $RAND$, and K (Figure 5.17). Upon reception of the authentication request message, the terminal computes C by means of A, K, and $RAND$. The outcome of this operation is then transmitted to the network, which checks whether C' is identical to C. The result C is also used as a ciphering key.

5.4.1.2 DECT System

With the DECT system the authentication of the mobile (called the *portable part*) by the network can be either *active* in the case when the mobile is located in its home network, or *passive* if the mobile is located in a visited network.

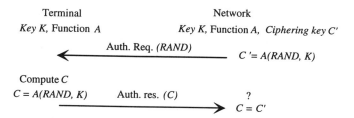

Figure 5.17 Authentication mechanism in the CT2.

For these two mechanisms (Figure 5.18), DECT uses two authentication algorithms, *A11* and *A12* and an authentication key *K* (128 bits). During an *active* authentication, the network transmits two random numbers R_S and *RAND-F* to the mobile. The calculation of the result, transmitted back to the network, is carried out in two phases. The mobile starts by calculating K_S (128 bits) using *A11*, R_S (64 bits), and *K*. Next it calculates *RES1* (32 bits) using *A12*, *RAND-F*, and K_S. The outcome *RES1* is sent to the network, which checks that it matches the result calculated by itself.

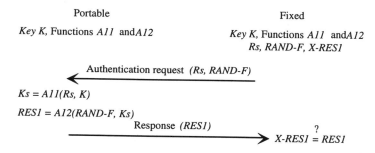

Figure 5.18 Active authentication in the DECT system.

In the case when the mobile is not in its home network, the authentication procedure is different (Figure 5.19). Only algorithm *A12* is common to both the home and visited networks. When challenging the mobile, the visited network calculates the first part of the authentication algorithm using *A11* and sends the result K_S with R_S and *RAND-F* to the mobile. The network also calculates *XRES* using *A12*, K_S, and the random number *RAND-F*. The mobile performs the second half of the active authentication procedure and transmits back the result *RES* using the network supplied value of K_S. This is compared with *XRES* by the visited network. This procedure helps decentralize the authentication procedure without disclosing secret information.

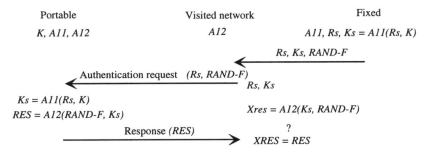

Figure 5.19 Passive authentication in the DECT system.

The initial distribution and installation of the keys in DECT can be achieved by different ways:

- *Over-the-air distribution and automatic installation.* The keys are distributed to the user equipment over the air and automatically installed into the equipment at the subscription registration.
- *Remote distribution and manual installation.* The keys are distributed remotely (by letter or by telephone call) to the user, who can enter the keys manually into the PP.
- *Local distribution.* The keys are locally generated and distributed to the Portable Part (PP) and Fixed Part (FP) (e.g., with a residential cordless telephone the user can choose the keys and enter them in the PP and the FP).
- *Installation at manufacture.*
- *Use of a* DECT authentication module (*DAM*). The keys are entered by the system operator along with other user data.

With cordless systems, network authentication can be initiated by a portable to detect fake FPs. In the case of CT2 the network authentication algorithm is the same as the one used for mobile authentication with a key that differs by 1 bit. In the DECT system, the algorithms used are different but the keys used are the same for both the authentication of the terminal and the network.

5.4.2 Authentication in Cellular Systems

Fraud vulnerability in cellular systems is highly dependent on the technology used. Analog networks are quite vulnerable to cloning. AMPS and TACS analog system operators, because of the increasing number of fraud cases, have designed and implemented new authentication techniques. For instance, some operators have introduced mechanisms that change the equipment serial number (ESN) of the mobile automatically after each call.

5.4.2.1 GSM System

With the digital systems, the security algorithms being used are quite numerous. For example, GSM uses the following procedures:

- *Authentication* of the SIM card. The authentication procedure is similar to that used in CT2.
- *Identification* (IMSI, TMSI). The IMSI is a permanent identity that is exchanged as rarely as possible on the radio air interface to avoid its interception by some eavesdroppers. The TMSI is used as a temporary identity, varying in time to provide confidentiality about the subscriber's location.
- *Ciphering.* A new ciphering key is calculated with each identification.

The security information in use is the *A3* algorithm for authentication, the *A8* algorithm for the ciphering key K_c calculation, the *A5* algorithm for ciphering, and key K_i for the computation of the authentication result (S') and of the cryptographic key K_c. Table 5.1 shows the sizes of the different security parameters used in GSM.

Table 5.1

Size of the different security parameters in the GSM

Parameter	Size
K_c	64 bits
COUNT	22 bits
BLOCK1	114 bits
BLOCK2	114 bits
K_i	128 bits
R or *RAND*	128 bits
S or *SRES*	32 bits

Authentication in GSM. The subscriber home database (called the *HLR*; see Chapters 8 and 12) is able to compute a series of triplets in form of (K_c, R, S), where R is a random number and S is the result of the function *A3* applied to K_i and R. These triplets are transferred to the visitor database (called *VLR*) in the visited network to allow it to undertake the authentication and ciphering operations at its Authentication Center (AuC). A ciphering key sequence number (*CKSN*) is used to avoid the possible lack of homogeneity between the key K_c located at the network level and the key K_c located at the mobile level. Each mobile has a sequence number stored with K_c in the SIM and in the subscriber profile at the MSC/VLR level. If the sequences are different, the MSC/VLR triggers an authentication before setting the ciphering mode. If CKSN has a value of zero, this means that no K_c has been allocated.

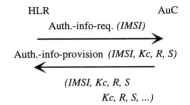

Figure 5.20 Request for authentication set provision.

The algorithms A3 and A8 may be used at each GSM network operator's discretion. The visited public land mobile network (*PLMN*) does not need to know these algorithms to authenticate a GSM mobile or to compute the ciphering key (Figure 5.21). Only the format of their inputs and outputs must be specified. Further, the maximum run time for A3 is set to 500 ms.

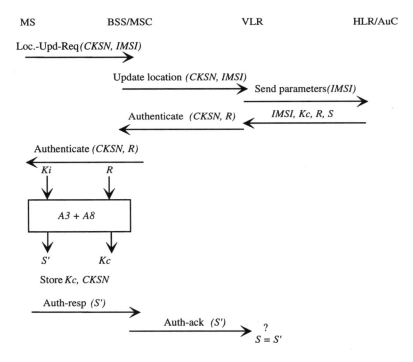

Figure 5.21 Example of authentication with IMSI (location updating).

Note: The CKSN is used to ensure authentication information K_c consistency between the MS and the VLR. It consists of one octet.

Authentication with IMSI occurs when the first connection to the network is performed (e.g., terminal powered on) or whenever either the mobile or network has lost the TMSI.

Actually, the subscriber's identity transmitted on the radio interface may represent a risk, particularly when a mobile does move a lot, because it can lead to the determining of the subscriber's location. With the GSM system, the temporary identity TMSI is used and the fixed identity of the subscriber (the IMSI) is transmitted during initial access only (Figure 5.22). During the first call, the network provides the subscriber, in a ciphered mode, this temporary identity. The TMSI could then be granted again at the discretion of the network serving the terminal.

In addition to authentication and ciphering mechanisms, most of the GSM network operators use an international mobile equipment identity (IMEI) color list, stored in the equipment identity register (EIR). This database includes the identity of terminals that have been reported stolen.

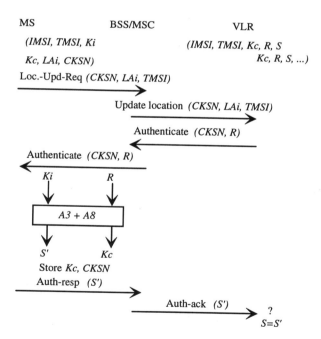

Figure 5.22 Example of authentication with TMSI (location updating).

Ciphering in GSM. The *A5* algorithm used in GSM for ciphering is implemented at the lowest level (Figure 5.23). It uses an "exclusive-or" function, which combines the data bit by bit using modulo 2 addition with a pseudorandom 114-bit ciphering sequence. At each frame, this algorithm produces a 114-bit ciphered/deciphered block (one for the mobile and another one for the base station). The text sequences to be ciphered are also made of 114 bits, marked $e_o, e_1, \ldots, e_{115}$; the bits e_{57} and e_{58} are ignored.

The first bit produced by the *A5* algorithm is modulo 2 added to e_o, the second to e_1, and so on. The blocks produced depend on the TDMA frame number (which determines the *COUNT* parameter used for determining the ciphered/deciphered blocks) and K_c.

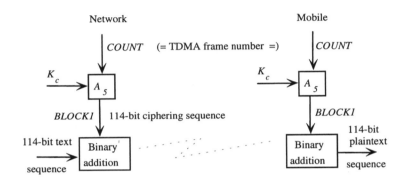

Figure 5.23 Ciphering mechanism in the GSM system.

5.4.2.2 IS-54 and IS-95 Systems

With the digital cellular systems using the IS-41 network standard, such as IS-54 (TDMA Digital-AMPS system) or the IS-95 (CDMA system), specific user information is electronically registered in the subscriber's terminal by the service provider. With IS-54 and IS-95, the user inputs a 64-bit security key called the "A-Key" into his or her terminal via the keyboard. This parameter is confidentially supplied to the user by the service provider usually by mail. The correct entry of the key by the user is checked by a security software program resident inside the terminal. The service provider also saves the subscriber's key in the home network.

In the IS-41 authentication technique [16], the original authentication key is concealed from the visited network by creating temporary authentication data called secret shared data (SSD) for user authentication. Because the SSD are transmitted to the visited network without being ciphered, it is possible for the SSD to be tapped by an eavesdropper. To overcome this problem, a synchronized call counter is used by both the mobile and the network. This counter is updated and transmitted to the network on each new call and is used to detect clone terminals whose counter value does not match that of the network.

Systems based on IS-41 also use a method similar to the one used by GSM for authentication and determination of the ciphering key. The security data is transmitted from the home network to the visited network on request during a location update of the mobile. The authentication process is similar in the visited network to that in the home network.

In an IS-41 network, a unique and global 32-bit key is regularly generated and transmitted in the network service frame on the transmission channel. The terminal attempts to access the system by calculating an authentication response 18 bits in length using an authentication algorithm involving the individual security data and the current global key. During a location update the result of the key calculation is transmitted to the user's home network to be checked. If the mobile is authenticated, the security data is sent to the visited network. The protection functions for the subscriber identity is the same as those used by GSM (TMSI).

Call privacy offers cryptographic protection against eavesdropping between radio base stations and the mobile. The same principles as those implemented in GSM are used in IS-41. Figure 5.24 shows the message flow for authentication from call origination with generation of the two ciphering keys VPMASK and SMEKEY. VPMASK is applied to voice transmission over the air interface between the handset and the serving network, and SMEKEY is used to encrypt certain fields within the signaling message sent by the MS to the serving PLMN.

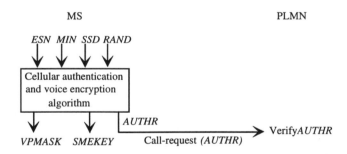

Figure 5.24 Authentication principle in IS-41.

The MS must obtain the RAND from the PLMN at registration (or location updating). It uses this number with the ESN, MIN, and SSD to generate VPMASK, SMEKEY, and the authentication result AUTHR.

In Figure 5.25, the modes for saving confidential data in the terminals and in the network for both GSM and IS-54 are shown.

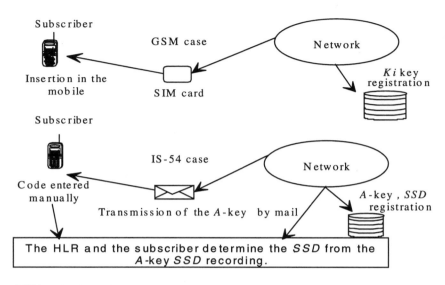

Figure 5.25 Security-data saving in the GSM and IS-54.

5.5 CONCLUSIONS—FUTURE SYSTEMS SECURITY FEATURES

Within the UMTS group for the development of a third-generation system, some trends for the security features have been identified [8]. The most significant characteristics of third-generation system security functions include:

- Use of public-key cryptography,
- Use of more powerful smart card technologies,
- Availability of trusted third-generation parties acting as certification authorities for public keys (see KDC),
- Secret-key- and public-key-based mechanisms that provide mutual authentication and cipher-key agreement for confidentiality, anonymity, and nonrepudiation,
- End-to-end confidentiality of communications,
- Secure billing protocols.

APPENDIX 5A

CRYPTOGRAPHIC KEY DISTRIBUTION METHODS

Distribution of cryptographic keys based on symmetrical algorithms

With this method, several cryptographic keys are used for different purposes. This method uses a hierarchy of keys with primary keys that are used to cipher secondary keys that in their turn cipher user data. Some primary cryptographic keys help decipher secondary ones that are used to cipher the data. For this, a *key distribution center* (KDC) is set up. It contains a key cipher system. An example of the use of a KDC is shown in [17]. Another such example is that of Needham and Shroeder (Figure 5A.1).

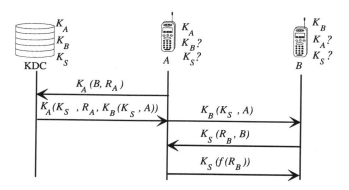

Figure 5A.1 Key management based on symmetrical algorithms.

Assuming that A and B would like to communicate with each other; A sends a ciphered message to the KDC by means of a K_A cryptographic key (the public key of A) which contains the address of B as well as a random number R_A. The KDC, after deciphering the message from A using known K_A, sends a message back to A that is still ciphered with K_A, and comprises three parts: K_S (session key), R_A, and K_s with the address of A ciphered with K_B (B public key).

A then sends the cryptographic key K_S, with A's address enciphered with K_B, to B. B sends back an insignificant number R_B with B's address enciphered with K_S. Finally, A is authenticated by sending back a slightly modified R_B previously agreed by both A and B using the session key K_S.

The disadvantage with such a method is that the system depends on the KDC. To avoid any likely KDC failures, several centers of this kind could be set up but with one major constraint, *viz* the need to secure them.

Distribution of keys based on asymmetrical algorithms

This method could be achieved either by using secret-session keys or public-session keys.

Use of secret-session keys:

The exchange of messages is shown in Figure 5A.2, which reproduces the same conventions as in the previous figures where K_A is the public cryptographic key for A, K_B is the public cryptographic key for B, and K_{KDC} is the KDC public key.

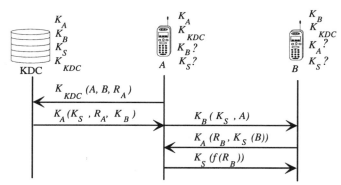

Figure 5A.2 Key management based on asymmetrical algorithms with session secret keys.

Use of public-session cryptographic keys:

The public keys A and B are used with the proviso that they are authenticated by the KDC in advance.

In Figure 5A.3, K_{SE} is the KDC secret key, K_{SP} is the KDC public key, K_A is the A public key, and K_B is the B public key.

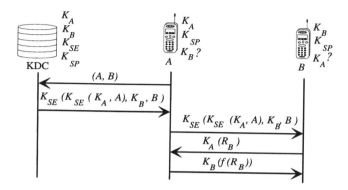

Figure 5A.3 Key management based on an asymmetrical algorithm with public-session keys.

REFERENCES

[1]   McGrath-Hadwen, E., "Fraud After GSM," *Mobile Europe*, pp. 27–28, Jan. 1997.

[2]   Tanenbaum, A. S., *Computer Networks*, Prentice Hall, 1995.

[3]   Hibberd, M., "Tough on Fraud, Tough on the Causes of Fraud," *Mobile Communications International*, pp. 37–41, July/Aug. 1997.

[4]   Mann, N., "You Can't Fool Me ...," *GSM World Focus 1998*, pp. 89–94, Jan. 1998.

[5]   Nilsen, L., "Confidentiality Services in the Telecommunications Area," *Proceedings of the XV International Switching Symposium*, Berlin, Germany, pp. 221–225, Apr. 1995.

[6]   Wilkes, J. E., "Privacy and Authentication Needs of PCS," *IEEE Personal Communications*, pp. 11–15, Aug. 1995.

[7]   Federrath, H., et al., "Minimizing the Average Cost of Paging on the Air Interface - An Approach Considering Privacy," *Proceedings of the Vehicular Technology Conference'97 (VTC'97)*, Phoenix, AZ, pp. 1253–1257, May 4-7, 1997.

[8]   Francis, J. C., H. Herbrig, and N. Jefferies, "Secure Provision of UMTS Services Over Diverse Access Networks," *IEEE Communications Magazine*, pp. 128–136, Feb. 1998.

[9]   Simmons, G. J. (ed), "Contemporary Cryptology, The Science of Information Integrity," *IEEE*, New York, 1992.

[10]   Diffie, W. and M. E. Hellman, "New Directions in Cryptography," *IEEE Transactions on Information Theory*, Vol. IT-22, pp. 644–654, Nov. 1976.

[11]   Rivest, R. L., A. Shamir, and L. Adleman, "A Method for Obtaining Digital Signatures and Public-Key Cryptosystems," *Communications of ACM 21 (2)*, pp. 120–126, 1978.

[12]   Jianwei, L., and W. Yumin, "Authentication of Mobile Users in Personal Communication Sytems," *IEEE International Symposium on Personal, Indoor, and Mobile Radio Communications PIMRC'96*, Taipei, Taiwan, Oct. 15-18, 1996.

[13]   Noerpel, A. R., Y-B. Lin, and H. Sherry, "PACS: Personal Access Communications System - A Tutorial," *IEEE Personal Communications*, pp. 32–43, June 1996.

[14]   Helme, S., "There's Life in the Old Dog Yet," *Mobile Communications International*, Apr. 1998.

[15]   Shimizy, A., and S. Miyaguchi, "FEAL - Fast Data Encipherment Algorithm," *System Computers Japan*, Vol. 19, No. 7, 1988.

[16]   TIA/EIA IS-41, "Cellular Radio Telecommunications Intersystem Operations," *Telecommunications Industry Association*, Dec. 1991.

[17]   Lamère J.-M., Y. Leroux Tourly, *La sécurité des réseaux*, Dunod Informatique, 1989.

SELECTED BIBLIOGRAPHY

[1]   "New Fraud Solutions," *Mobile Communications International*, pp. 23, Sept./Oct. 1995.

[2]   Beller, M. J., L.-F. Chang, and Y. Yacobi, "Privacy and Authentication on a Portable Communications System," *IEEE Journal on Selected Areas in Communications*, Vol. 11, No. 6, pp. 821–829, Aug. 1993.

[3]   Brown, D., "Techniques for Privacy and Anthentification in Personal Communication Systems," *IEEE Personal Communications*, pp. 6–10, Aug. 1995.

[4]   Clissmann, C., and A. Patel, "Security for Mobile Users of Telecommunication Services," IEEE 0-7803-1823-4/94, pp. 350–353, 1994.

[5]   Coléou, G., "Procédés de Chiffrement," *Techniques de l'Ingénieur*, E 6-480, 1989.

[6]   Diffie, W., and M. E. Hellman, "A Critique of the Proposed Data Encryption Standard," *Communications ACM*, Vol. 19, pp. 164–165, Mar. 1976.

[7]   Donegan, P., "Building Defences Against Fraud," *Mobile Communications International*, pp. 27–32, Feb. 1996.

[8]   Federrath H., A. Jerichow, D. Kesdogan, and A. Pfitzmann, "Security in Public Mobile Communication Networks," *Proceedings of the IFIP TC6 International Workshop on Personal Wireless Communications*, Prague, Czech Republic, pp. 105–115, Apr. 24-25, 1995.

[9]   Gong, L., "Increasing Availability and Security of an Authentication Service," *IEEE Journal on Selected Areas in Communications*, Vol. 11, No. 5, pp. 657–662, Aug. 1993.

[10]   ISO 7498-2 Information Processing Systems - Open Systems Interconnection - Basic Reference Model - Part 2: Security Architecture, 1989.

[11]   Jianwei, L., and W. Yumin, "Authentication of Mobile Users in Personal Communication" *IEEE International Symposium on Personal, Indoor and Mobile Radio Communications PIMRC'96*, Taipei, Taiwan, Oct. 15–18, 1996.

[12]   Miyaguchi, S., "The FEAL Cipher Family," *Proceedings CRYPTO'90*, Santa Barbara, CA, Aug. 1990.

[13]   National Bureau of Standards, FIPS PUB 46-1, Data Encryption Standard, Jan. 22, 1988.

[14]   Suzuki, S., and K. Nakada, "An Authentication Technique Based on Distributed Security Management for the Global Mobility Network," *IEEE Journal on Selected Areas in Communications*, Vol. 15, No. 8, pp. 1,608–1,617, Oct. 1997.

[15]   Ford W., *Computer Communications Security*, Prentice Hall PTR, Englewood Cliffs, NJ, 1994.

SELECTED BIBLIOGRAPHY ON CRYPTOGRAPHY

[16]   Stinson, D.R., *Cryptography: Theory and Practice*, CRC Press on Discrete Mathematics and Its Applications, Mar. 1995.

[17]   Menezes, A. J., P. C. van Oorschot, and A. Vanstone, *Handbook of Applied Cryptography*, CRC Press on Discrete Mathematics and Its Applications, Boca Raton, FL, 1997.

[18]   Van der Lubbe, J.C.A., *Basic Methods of Cryptography*, Cambridge University Press, May 1998.

[19]   Rhee, M. Y., *Cryptography and Secure Communications*, McGraw-Hill Series on Computer Communications, Feb. 1994.

[20]   Van Tilborg, H., *An Introduction to Cryptology*, Kluwer Academic Publishers, 1988.

[21]   Schneir, B., *Applied Cryptography*, 2nd Ed., Wiley, New York, 1996.

CHAPTER 6

RESOURCE MANAGEMENT IN CELLULAR SYSTEMS

The concept of cellular radio relies on the fundamental property of radio waves increasing attenuation with distance. Because of this property, a radio frequency used on one site can be reused on another provided that the second is far enough from the first. The implementation of this principle allows radio communication systems to cover very large areas and to serve very high traffic densities.

For some time, cellular radio has been reserved for full duplex public mobile systems where the service area covers regions or entire countries. In the last few years, the growth in demand, combined with technical advances, is such that present and future high-density mobile communications systems rely on cellular architecture. The widespread deployment of cellular networks and the increasing range of services offered (see GSM Phase II+ services, Chapter 12) have driven the increasing importance of this technique in mobile networks [1].

The two main functions encountered only in cellular systems are *handover* (see Chapter 8) and *location management* (see Chapter 8). Nevertheless, DECT, PHS, and PACS systems, for example, which are classified as "cordless systems" (see Chapter 10), as well as some private/professional radiocommunication systems (see TETRA in Chapter 9), include these functions. Handover and location management are, together with frequency reuse, the main characteristics that make cellular systems the most complicated to design, plan, and operate compared with other mobile radio communications systems.

6.1 HISTORY

Historically, the cellular concept was conceived at the end of the 1940s, when it was implemented in some experimental cellular systems constructed by Bell Labs. Radio engineers noted that it was possible to avoid the problem of radio spectrum congestion by "redesigning" the radio system coverage [2]. At the time, mobile radio systems were considered as particular cases of radio or TV broadcasting. Powerful transmitters were placed at high locations and covered areas extending over tens of kilometers (from 60 to 80 km).

It was in the 1970s, however, with the technical advances and a growth in demand, that this technique was to be implemented efficiently in commercial mobile radio communications systems. Early mobile systems, noncellular in layout, were generally not able to maintain more than a few simultaneous calls (typically five). At peak hours, subscribers were obliged to redial for long periods until a channel became available.

Because the first public radio telephone systems did not employ frequency reuse and because coverage was obtained with a few high power base stations, allocated frequencies were all used in the limited coverage area. They could only be reused in very remote areas. Additional coverage to extend existing zones was not possible.

Semicellular networks then appeared that covered wide geographical areas with several base stations, each of which was allocated only a fraction of the spectrum. The cell structure allowed reuse of frequencies and coverage of a wide area with large radius cells (radius between 10 and 15 km). In these networks, handover management (communication continuity during the transition from a cell to another cell, see Chapter 8 for more details) was not implemented. It was then mandatory that cell borders were designed to correspond to low transit areas to reduce the call-dropping rate. Because of the low frequency reuse, these semicellular systems had limited capacity. Indeed, the main objective was to provide a telephone service for subscribers driving in towns or on highways. These systems were, for this reason, called "car phone systems" or "zero-generation systems."

With increasing demand, the operators' objective was to increase the capacity of existing networks. This increase in capacity required the intensive use of a cellular structure and the introduction of handover mechanisms. The systems deployed used an analog air interface. Cell radius in such systems was generally larger than one kilometer. They are called "first-generation cellular systems." One of the first commercial systems of this type was the American system, Advanced Mobile Phone Service (AMPS), developed by AT&T in the 1980s.

The continuous and quasi-exponential growth in demand, combined with improvements in signal processing and component integration, was the main factor that motivated the decision to develop systems based on digital transmission techniques. The evolution of network management (introduction of a signaling network separate from the user traffic network) and distributed processing concepts are also considered in the design of these new systems called "second generation cellular." The most important representative system is the Global System for Mobile Communications (GSM), which offers a range of services much greater than those provided by first-generation cellular systems (typically only a telephone service between two parties). The improved security offered to subscribers and operators is also a substantial benefit brought by the second generation compared with previous systems.

Current work centers around the design of a system that will be able to provide a universal mobile telecommunications service to anybody, anywhere, anytime. Cellular networks will be of very different types (picocellular for indoor coverage, microcellular for outdoor coverage in high traffic density areas, macrocellular for medium or low-density traffic areas). These networks will coexist and subscribers will be able-access any of them according to their location (indoors/outdoors), the services required (voice, data, Internet access) and/or the speed of the user. These systems are called "third-generation," Universal Mobile Telecommunication System (UMTS), or Future Public Land Mobile Telecommunications System (FPLMTS), which became International Mobile Telecommunications 2000 (IMT 2000).

6.2 THE CELLULAR CONCEPT

The cellular architecture was originally designed as a means of providing a region of substantial geographical size (a country or even a continent) with a communications network using a limited frequency allocation and servicing an increasing traffic demand.

Frequency reuse theoretically allows unlimited user densities and coverage areas. The need to operate a single system and increase its capacity indefinitely with a finite amount of spectrum is thus fulfilled.

6.2.1 Frequency Reuse

This mechanism is based on the pathloss property of radio waves, which means that a frequency used on one site can be reused on another site provided that the two sites are sufficiently far from each other. Each site covers an area called a *cell*, the size of which usually depends on user density. Cells that use the same frequencies are called co-cells. These co-cells must be sufficiently far from each other in order that the level of co-channel interference is sufficiently low so it does not degrade the quality of service.

To implement frequency reuse, the frequency band allocated to the system is split into several subbands. Each subband is allocated to a particular base station and reused in its co-cells.

Example 1: The cellular concept advantage
Assume a coverage area A with a radius R and N, the number of frequencies allocated to the system. If $N = 7$ and it is assumed that one frequency supports one communication channel, without implementing frequency reuse, at most, seven communications can be served in area A.

By distributing these frequencies (one frequency per cell) in smaller cells with radius $r (< R)$, the maximum number of simultaneous communications is equal to $n\_com = \dfrac{\pi.R^2}{\pi.r^2} = \left(\dfrac{R}{r}\right)^2$ (see Figure 6.1).

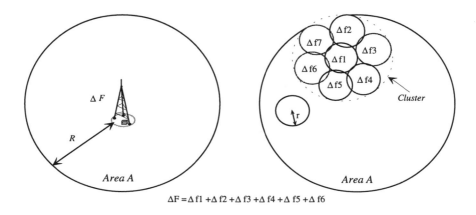

$$\Delta F = \Delta f1 + \Delta f2 + \Delta f3 + \Delta f4 + \Delta f5 + \Delta f6$$

Figure 6.1 Advantage of the frequency reuse mechanism.

If we take $R = 10$ km and $r = 500$m, $n\_com = (10/0.5)^2 = 400$ communications can therefore be served in area A.

A group of cells using all the channels allocated to the system is called a *cluster*. Clusters are placed side by side and can, by this arrangement, cover any given region. They can be juxtaposed indefinitely and cover unlimited geographical areas.

Present cellular systems use fixed frequency allocation; that is, frequencies are allocated to a particular site and this allocation does not change or only changes rarely. To satisfy traffic load and fluctuating C/I (carrier-to-interferer ratio) conditions during the day, dynamic channel allocation can be used as an alternative to the present techniques [3]. This is discussed in Section 6.3.

Comment: The efficiency of frequency reuse depends on the radio environment where the network is implemented and on the techniques used against interference [4]. This efficiency can be represented by the ratio N_C/N_T (≤ 1) where N_T is the total number of available user channels (e.g., frequencies, timeslots, codes) in a unicellular system (a theoretical situation where no co-channel interference is observed) and N_C is the number of available user channels in a multicellular system (the case where co-channel interference is taken into account).

Frequency reuse is more efficient in microcellular networks implemented in urban environments than in macrocellular networks implemented in suburban or rural areas for the following reasons:

- In a microcellular network, the propagation pathloss exponent in "nonline-of-sight" conditions gives rise to better protection against interference (*interference shielding*).
- The standard deviation of shadow fades and interferers experienced in a cell is lower for microcells and leads to lower fluctuations in interference.
- The two-slope propagation phenomenon (with a break point, see Chapter 2) in microcells allows better protection against interference.

Finally, the influence of operating frequency on reuse efficiency should be mentioned. Simulation results [4] show that the reuse factors at 2-GHz range from 40% to 70% greater than those obtained at 900 MHz.

6.2.2 Reuse Distance and Number of Cells per Cluster

The importance of the cellular concept in terms of efficiency is much more obvious when it is observed that co-channel interference does not depend on the absolute distance but on the ratio of the distance between the cells (D) and the cell radius (R).

With a hexagonal structure, with xy coordinates where x and y subtend an angle of 60 deg (Figure 6.2), and with the unit distance equal to the distance between two cells' centers, (i.e., $U = \sqrt{3}R$), the distance D between two points (x_1, y_1) and (x_2, y_2) is such that: $D^2 = i^2 + ij + j^2$ with $i = x_2 - x_1$ and $j = y_2 - y_1$.

Example 2: Optimum cluster size calculation

Consider two cells with coordinates (x_1, y_1) and (x_2, y_2) in a hexagonal structure xy (see Figure 6.2) and with the coordinates (x_1', y_1') and (x_2', y_2') in an orthogonal structure $x'y'$. With the $x'y'$ coordinates, the distance D between the two cells is such that:

$$D^2 = (x_2' - x_1')^2 + (y_2' - y_1')^2$$

where $x_2' = x_2.\cos(2\pi/3) = x_2\sqrt{3}/2$,

$x_1' = x_1.\cos(2\pi/3) = x_1\sqrt{3}/2$,

$y_2' = y_2 + x_2.\sin(2\pi/3) = y_2 + x_2/2$,

$y_1' = y_1 + x_1.\sin(2\pi/3) = y_1 + x_1/2$.

Then $D^2 = \dfrac{3}{4}\left(\dfrac{x_2}{2} - \dfrac{x_1}{2}\right)^2 + \left(y_2 + \dfrac{x_2}{2} - y_1 - \dfrac{x_1}{2}\right)^2$

$= (x_2 - x_1)^2 + (y_2 - y_1)^2 + (y_2 - y_1).(x_2 - x_1)$.

Assuming $i = x_2 - x_1$ and $j = y_2 - y_1$,   then: $D^2 = i^2 + j^2 + ij$.

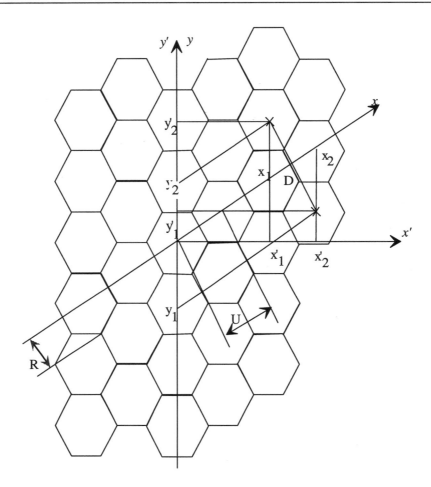

Figure 6.2 Cellular network and associated coordinates.

Note: Cellular planning and hexagonal structure

During the planning phase of the network, cells are identified as hexagons. Indeed, the hexagonal shape is considered the optimum structure for this purpose. It allows coverage of the surface without cell overlap and without holes. The use of the hexagonal shape is dictated by the need to simplify planning and design of a cellular system, at least during the theoretical planning phase.

If radio waves are propagated uniformally in all directions and there are no obstacles between transmitter and receiver, ideal coverage would be provided by means of circular cells. However, in practice, obstacles and phenomena encountered in mobile radio result in cells having highly varied shapes, quite different from the hexagonal or circular. The most individual forms are encountered in the microcellular environment (Figure 6.3).

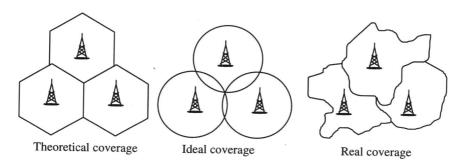

Theoretical coverage   Ideal coverage   Real coverage

Figure 6.3 Differences between theoretical and real coverage.

Nevertheless, the concept of cells remains valid and for a given base station the corresponding cell represents the territorial area where it is best utilized (from the point of view of the received signal level), compared with the other base stations, to serve the mobiles located in the cell area.

6.2.2.1 Reuse Distance

As the reuse distance is the same in all directions, the number of hexagons that are equidistant from the initial reference point is 6. Locating hexagons at the coordinates (p, q), $(p + q, -p)$, $(-q, p + q)$, $(-p, -q)$, $(-p - q, p)$, and $(q, -p - q)$, the distances between these hexagons and the reference hexagon (located at the point $(0,0)$) are then equal to $p^2 + pq + q^2$. This distance determines the positions of the co-cells and the reuse distance has the following formula:

$$D^2 = i^2 + ij + j^2 \tag{6.1}$$

where i and j are positive integers.

6.2.2.2 Number of Cells per Cluster

Assuming that a cluster has a hexagonal shape, the number N of cells per cluster can be determined as follows. Let a and A be the areas of the cell and of the cluster, respectively (Figure 6.4).

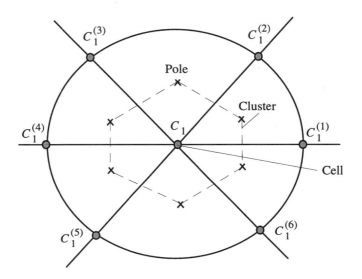

Figure 6.4 Determination of a cluster shape.

Then $a = 3\sqrt{3}\,\dfrac{R^2}{2}$

Assuming that $U = \sqrt{3}R$, the distance unit, then $a = \dfrac{\sqrt{3}}{2}$.

The distance between two co-cells being equal to D, A has the value:

$$A = 3\sqrt{3}\left(\frac{D/2}{\cos 30\deg}\right)^2 \frac{1}{2} = \sqrt{3}\,\frac{D^2}{2}$$

Thus the number of cells per cluster is equal to:

$$N = \frac{A}{a} = D^2$$

where $N = i^2 + ij + j^2$ with i and j integers and where D is expressed in terms of distance unit U.

Introducing U, it becomes $N = \dfrac{D^2}{3.R^2}$, and thus the relation between D and R is:

$$\frac{D}{R} = \sqrt{3N} \qquad\qquad (6.2)$$

Example 3: Determination of the number of cells per cluster

Let c_i be the cell using the frequency f_i and located at the center of the cluster. The interfering cells (6) are in the center of the clusters that are adjacent to the first cluster. The poles of the hexagon are common to the six adjacent hexagons, which, by symmetry, are equidistant from the central cell and its interfering cells.

Cell c_i cluster is thus constituted and has a hexagonal shape with a radius

$$R = \frac{D}{2} \cdot \left[\frac{1}{\cos\frac{\pi}{6}} \right], \text{ where the area of the cluster } A \text{ is}$$

$$A = 3 \sqrt{3 \left[\frac{D}{2} \cdot \frac{1}{\cos\frac{\pi}{6}} \right]^2 \cdot \frac{1}{2}} \cdot$$

With $U = \sqrt{3}\,R$, the previously defined distance unit, then $a = \sqrt{3}/2$ and $A = \dfrac{\sqrt{3}}{6} \dfrac{D^2}{R^2}$.

The number of cells per cluster is thus equal to $N = A/a = D^2/3.R^2$, which is the same as in the relation (6.2).

Integer size clusters can therefore contain 1, 3, 4, 7, 9, 12, 13, 16, 19, 21, ... cells. In practice, 3-, 4-, 7- (Figure 6.5), 12-, 21-, or 27-cell patterns are the most common.

Analog networks are often based on a 7/21 structure, which indicates that in a group of seven contiguous sites with three cells per site, each frequency is used only once. In a digital system such as GSM, a 4/12 pattern can be used.

A single site to cover three cells with sectored antennas (Figure 6.6) is implemented by operators to minimize the number of sites to be installed and managed.

Common configurations for a group of channel allocations are as follows:

- For a 9-cell cluster: 3 BSs with three sectors per BS.
- For a 12-cell cluster: 4 BSs with three sectors per BS.
- For a 21-cell cluster: 7 BSs with three sectors per BS.

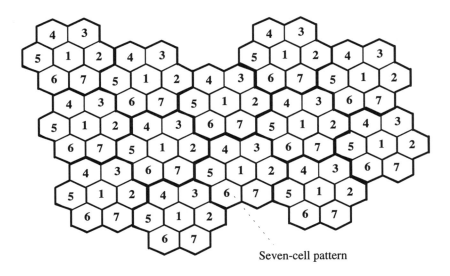

Seven-cell pattern

Figure 6.5 Example of coverage with a seven-cell pattern.

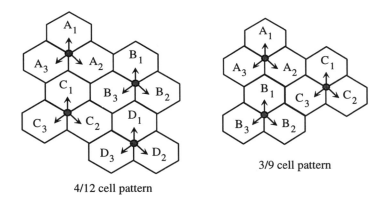

4/12 cell pattern

3/9 cell pattern

Figure 6.6 Examples of clusters with trisectored cells.

Example 4: Nine-cell cluster and 27 frequencies (see Table 6.1).

Table 6.1

Example of channel allocation in a nine-cell pattern

Frequency groups	A_1	B_1	C_1	A_2	B_2	C_2	A_3	B_3	C_3
Channel	1	2	3	4	5	6	7	8	9
	10	11	12	13	14	15	16	17	18
	19	20	21	21	23	24	25	26	27

For each cell of the cluster, there corresponds a group of frequencies L_i such that it is adjacent only to the cells containing groups M_j ($M = A$, B, or C) with $j \neq i$.

6.2.2.3 Stockholm Pattern

Irregular cellular patterns can also be designed, such as the Stockholm pattern represented in Figure 6.7.

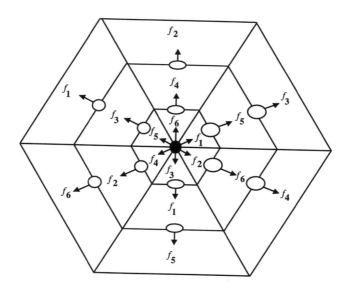

Figure 6.7 An irregular pattern: the Stockholm pattern.

This kind of coverage is produced by use of directional antennas that allow the cells to have this unusual shape. The use of this pattern is justified by coverage that needs to be extended progressively. The operator starts to build the network by covering the center of a town (central hexagon) and gradually extends outward progressively by providing additional cells around the initial site.

6.2.2.4 Reuse Partitioning

The idea of *reuse partitioning* is to provide coverage that involves different cluster dimensions [5]. It is a concept used in high traffic density areas. In this method, each cell is divided into several concentric subcells (see Figure 6.8).

Inner cells are those closest to the base station (which is located in the center of the cell). The required power to reach the desired C/I in the inner cells is much lower than that required in outer cells or rings. Thus, the reuse distance is lower for the inner cells than for the outer cells. This method offers high spectral efficiency. The least severe conditions are those for terminals located near the base station; mobiles located at the center of the cell need a less powerful signal than mobiles located at the periphery [6]. Therefore center cell channels can be reused with a smaller separation. Each cell is divided into several areas, where each one follows a particular cluster size.

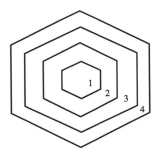

Figure 6.8 Concentric subcells.

In practice, usually only two patterns with different sizes are combined. A cluster pattern with a low number of cells is obtained by using low radius cells. This is realized by authorizing access to the corresponding base station only to mobiles that receive a strong signal (i.e., they are close to the base station). This first pattern is combined with a "classical" pattern with a greater number of cells. In Figure 6.9, a cellular structure combining a seven-cell cluster (for the outer areas) and a four-cell cluster (for the inner areas) is represented.

1, 2, 3, 4, 5, 6, 7: Seven cell cluster
1', 2', 3', 4': Four cell cluster

Figure 6.9 Two-pattern coverage.

Example 5: Radius of cells for the two cluster patterns:
  In Figure 6.9, large size pattern parameters, outer (1), are the following:
  $N_1 = 7$ (no. of cells per cluster), $R_1 = R$, and $D_1 = R_1\sqrt{3N_1} \cong 4.6.R$
  and the parameters of the small size pattern, inner (2), are the following:
  $N_2 = 4$ (no. of cells per cluster), $R_2 = R/2$, and $D_2 = R_2\sqrt{3N_2} \cong \sqrt{3}.R$.

  For pattern (2) base stations to take advantage of pattern (1) sites, D_2 is taken equal to $2.R$, which respects the minimum reuse distance.
  Inner cells serve low power mobiles (i.e., mobiles close to the base station). They can also be used to serve indoor sites. The large size cells cover the outdoor environment. Some GSM operators use this kind of pattern, especially in urban environments.

  When a mobile moves to another area in the same cell, it must move to another channel and make an intracellcell handover (see Chapter 8).

Example 6: Comparison between the capacity of a "classical" network and the capacity of a network using the reuse-partitioning scheme

Let $n = n_1 + n_2$ the number of channels allocated to the system with n_1, the number of channels allocated to the large-size pattern (1), and n_2, the number of channels allocated to the small-size pattern (2).

Thus, $n'_1 = n_1/N_1$: number of channels in each cell belonging to a cluster of type (1) and $n'_2 = n_2/N_2$: number of channels in each cell belonging to a type (2) cluster.

The mean number of channels per unit area S (taken as equal to that of a large size cell) will then be: $C' = \dfrac{n'_1 + n'_2}{S}$ for the network combining the two kinds of clusters and

$C = \dfrac{n}{S.N_1}$ for the "classical" network.

The values of the parameters used in this example are:
$n = 70$, $N_1 = 7$, and $N_2 = 4$.

Table 6.2

Capacities of a "classical" network and that of a network with the reuse partitioning scheme

N_1	70	50	35	20
N_2	0	20	35	50
C.S	10			
C'.S	10	12.14	13.75	15.36

The comparison of capacities with this very simple model shows (see Table 6.2) that implementing reuse partitioning gives a network higher capacity than a "classical" network.

Example 7: Calculation of the optimum capacity of the *partitioning* scheme

Let D_1 (D_2) and R_1 (R_2) be the reuse distance and the cell radius for the inner (outer) patterns respectively. N_1 (N_2) is the size of the outer (inner) pattern, A_1 is the outer area covered by the larger cell, corresponding to the anulus, and A_2 is the area covered by the smaller cell corresponding to the inner pattern. The total area of a cell is thus $A = A_1 + A_2$.

The relation (6.2) becomes: $\dfrac{R_1}{R_2}\sqrt{3N_2} = \sqrt{3N_1}$. Rearranging gives $\rho = \dfrac{R_2}{R_1} = \sqrt{\dfrac{N_2}{N_1}}$.

If $F = F_1 + F_2$, the network frequency band, with F_1, the frequencies used by the outer cell pattern (Q_1 carriers) and F_2 the frequencies used by the inner cell pattern (Q_2 carriers), a cell of the outer pattern has on average $q_1 = \dfrac{Q_1}{N_1}$ carriers and a cell of the inner pattern uses

on average $q_2 = \dfrac{Q_2}{N_2}$ carriers. The total number of carriers used in a cell is therefore $q = q_1 + q_2$.

Assuming that q_1 (q_2) is proportional to the number of users in the area A_1 (A_2), and that the mobile subscribers are uniformly distributed all over the network, we obtain:

$$\frac{q_2}{q} = \frac{R_2^2}{R_1^2} = \rho^2 \text{ and } \frac{q_1}{q} = \frac{R_1^2 - R_2^2}{R_1^2} = 1 - \rho^2.$$

Thus $Q = Q_1 + Q_2 = q_1.N_1 + (q - q_2).N_1 = N_1.q(\rho^4 + 1 - \rho^2)$.

The number of carriers per cell q' in a classical cellular network (i.e., without reuse partitioning) is equal to $q' = Q/N_1$. The maximum increase of capacity obtained by reuse partitioning is achieved when $B(\rho) = q'/q = \rho^4 + 1 - \rho^2$ reaches its minimum value; that is, when the value of its derivative becomes zero, which is obtained for $\rho = \sqrt{\dfrac{1}{2}}$. The network

capacity increase ratio is thus equal to: $B^{-1}\left(\sqrt{\dfrac{1}{2}}\right) = 1.33$, which signifies a 33% additional

capacity availability. The optimum partitioning of a cell is thus obtained when $A_1 = A_2 = A/2$ [7].

6.2.2.5 Different Types of Cell

The size of a cell is directly related to the propagation conditions and to the traffic density in the area covered by the cell (see Chapter 7).

A cell can also have different shapes according to the type of coverage required. In Figure 6.10, various kinds of cell shape are represented.

6.2.2.6 Relation Between C/I and D/R

The radius R of a cell is a function of the transmitter powers of the base station (BS) and mobiles. To decrease the reuse distance D it is sufficient to reduce the transmission power of the equipment. The value of the C/I ratio is kept constant by adjusting the power of the mobiles and base stations. However, the minimum value of the D/R ratio depends on the lowest value of C/I accepted by the system.

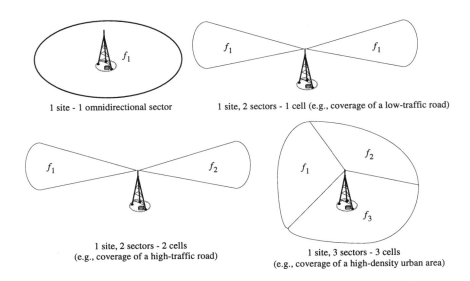

Figure 6.10 Examples of different cell shapes.

The ratio $q = D/R$ is called *co-channel reuse ratio*. It is an indication of transmission quality and traffic capacity. From the transmission point of view, this ratio dictates the co-channel interference level in the system; the higher this ratio, the lower the potential interference level. Finally, from the traffic point of view, the ratio D/R is an indicator of system performance.

Co-channel interference is a function of the parameter q, also defined as the *co-channel interference reduction factor*. As q increases, the co-channel interference decreases.

D is a function of K_I (the number of interfering cells in the first ring) and of C/I. We have: $\dfrac{C}{I} = \dfrac{C}{\sum\limits_{k=1}^{K_I} I_k}$ where C is proportional to $R^{-\gamma}$ and I is similarly proportional to $D^{-\gamma}$. Thus:

$$\frac{C}{I} = \frac{R^{-\gamma}}{\sum\limits_{k=1}^{K_I} D_k^{-\gamma}} = \frac{1}{\sum\limits_{k=1}^{K_I} (q_k)^{-\gamma}} \tag{6.3}$$

The aim of any cellular system must therefore be to operate with a level of C/I that is as low as possible to minimize the reuse distance by reducing the number of cells per cluster.

For example, a system operating with a minimal *C/I* value of 9 dB (GSM case) will have a reuse cluster pattern of seven or nine cells. For a system operating with a *C/I* of 3 dB, the corresponding pattern will be three or four cells.

Comment 1: Relation between *C/I* and signal quality

In practice, consistent quality in a GSM environment requires a worst case *C/I* of better than 12 to 15 dB to be achieved across at least 98% of the target area. This is because at the theoretical level of 9 dB, the degradation of quality against further reduction of *C/I* is very rapid and some practical margin for this has to be provided.

The acceptable *C/I* threshold, and thus the minimum cluster size, depends on the system. For instance, in an FDMA system, the accepted value of *C/I* is around 18 dB. Using mechanisms such as discontinuous transmission, slow frequency hopping, sophisticated coding, and interleaving techniques or power control, the system can accept *C/I* thresholds even lower.

Comment 2: Relation between *C/I* and SNR

The *C/I* ratio can be linked to the minimum SNR (E_b/N_0) for audio quality (in the case of GSM using the standard full rate codec). The following relation can be established between the two ratios:

$$SNR = 0.9 \times C/I + 10 \ dB \tag{6.4}$$

Relation (6.4) is valid for 12 dB $\leq$ C/I $\leq$ 35 dB and without fading. With Rayleigh fading effects included, the use of a fading margin between 10 and 15 dB is generally necessary [8]. The formula in this case becomes:

$$S/N = 1.0 \times C/I - 9 \ dB \tag{6.5}$$

For example, a 21-dB ratio will be obtained for a *C/I* of 30 dB.

6.2.2.7 Adjacent Channel Interference Protection Ratio

The World Administrative Radio Conference held in Geneva in 1979 defined the protection ratio as "the minimum value of the wanted to unwanted signal ratio, usually expressed in decibels, at the receiver input determined under specified conditions such that a specified quality is achieved at the receiver output." This value will have different values according to the type of modulation used and the quality of service required. Mathematically, a protection ratio can be defined as $a = min\{C/(I+N)\}$ (where C is the user signal, I is the interfering signal, and N is the noise) for a specified signal quality.

The focus here is on interference generated by adjacent channels (see Chapter 2). The degradation in quality of service caused by this is quantified by the Adjacent Channel Interference Protection Ratio (ACIPR) and is evaluated with the ratio *C/A* (*A* indicates the adjacent channel interferers). It indicates the threshold below which interference from transmissions on adjacent channels is too high.

Mathematical derivation, objective measurement, or subjective assessment can determine ACIPR. The mathematical derivation of a relationship between the C/I ratio at the receiver input, the receiver transfer function, and the quality of the signal at the receiver output can be established. However, objective measurement is an easier method. The value for the protection ratio can be measured at the input of the receiver by comparison with a predefined signal parameter at the receiver output, which reflects the signal quality. The simplest procedure is to implement subjective tests. It consists of taking a group of people who listen to speech communication over the radio channel. The interferer signal power is adjusted to determine the different interference levels that correspond for instance to "just audible," "intelligibility difficulty," and "minimal acceptable quality." The measurements generally show that the adjacent channel interferer (A) must be at least 9 dB greater than the wanted signal (C).

6.2.3 Capacities

The intrinsic capacity of cellular systems is defined by the following formula:

$$K = \frac{n}{N.B} \tag{6.6}$$

where n is the number of channels per carrier, N is the frequency reuse pattern size, and B is the duplex frequency bandwidth occupied by a channel (in $MHz_{(duplex)}$).

Example 8: Intrinsic capacities of an analog system and a digital system

Let an analog system (e.g., the Radiocom 2000 cellular system) with the following characteristics: $B = 2 \times 12.5$ kHz $= 0.025$ $MHz_{(duplex)}$, $n = 1$, and $N = 21$ (for a C/I between 17 and 18 dB). The intrinsic capacity of the system is thus

$K = 1.9$ channels/cell/$MHz_{(duplex)}$

Let a digital system such as GSM where $B = 2 \times 200$ kHz $= 0.4$ MHz $_{(duplex)}$, $n = 8$, and $N = 9$. Its intrinsic capacity is thus equal to:

$K = 2.2$ channels/cell/$MHz_{(duplex)}$

Note: The relatively small difference in capacity between the two systems is due to the omission of intrinsic capacity improvement techniques such as power control, frequency hopping, and DTX (discontinuous transmission). These mechanisms will be introduced in Section 6.3.

As a better measure, the capacity of a cellular network can be estimated in terms of the subscriber numbers (in erlangs) per hertz per km^2. Two more parameters have to be added to those previously defined. These are R, the cell radius, and η, the throughput of a given channel. This latter parameter is defined as the ratio between the number of traffic channels and the total number of channels (traffic and signaling). The capacity of a cellular network, in this case, is given by the formula:

$$\frac{n.\eta}{N.B.\pi.R^2} \text{ Erl/Hz/km}^2 \tag{6.7}$$

Example 9: Capacity of a GSM network

Consider a GSM network with the following characteristics: $B = 0.4$ MHz$_{(duplex)}$; $n = 8$, and $N = 9$, $\eta = 0.9$, for example, and $R = 1$ km. The capacity of this network is equal to 0.64 Erlang/MHz$_{(duplex)}$/km^2.

6.2.4 Link Budget

In a cellular system, the border of a cell (and thus its area) is defined as the set of points where a mobile, with the maximum authorized power (for instance, 2 W in an urban area and 8 W in a rural area), is able to communicate with the base station serving that cell. That is, it can decode the transmission from the base station and the base station can also receive signals from the mobile. Thus, the size of each cell is determined by balancing the transmitted power against propagation and other losses and the receiver sensitivities of the BS and MS.

The coverage of a cell (and thus the design of a cell) is limited by the transmission link (uplink or downlink) on which the signal power is weakest. This is generally the uplink channel. To minimize interference in other cell sites, uplinks and downlinks must be balanced (Figure 6.11). If a base station transmits at too high a power, it will cause interference in its co-cells.

Balancing the propagation loss on the uplink and on the downlink is done by using a *link budget*. It is important for bidirectional communication with mobiles located near the borders of a cell. Indeed, the transmissions must have the same quality in both directions. However, a good balance can only be obtained for some mobiles because of such things as fluctuating propagation conditions, differing receiver sensitivities and antenna gains that differ from one handset to another. In practice, the values of radio parameters are determined in such a way that the power difference between the uplink and the downlink is less than 10 dB. A difference of between 3 and 5 dB is generally considered satisfactory.

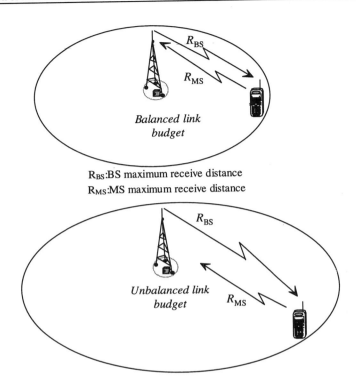

R_{BS}:BS maximum receive distance
R_{MS}:MS maximum receive distance

Figure 6.11 General principle of link balancing.

The link budget should take into account all losses and gains that may be encountered on the communication link, including the fade margin (generally between 6 and 15 dB) and an environmental margin (between 5 dB for outdoor communications, and 25 dB for indoor communications). Transmitter/receiver powers are adjusted for the signal to fulfill these conditions.

Link budgets (measured in dB or in dBm) are calculated with reference to the technical specifications of the equipment (BSs, MSs). The worst-case values are considered for determining the size of the cell. Several factors affect link budget calculation, some which "add" dBs to the radio link (e.g., transmitter and receiver antenna gain), and some which "subtract" dB from the radio link (obstructions or fading). The link budget specifies the available received power for a pair of transceivers (Figure 6.12). The corresponding equation can be written as

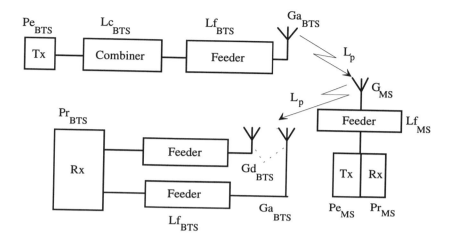

Figure 6.12 Link budget diagram.

$$P_r = P_t + G_t + G_r - L_p - M_f \, (dB) \tag{6.8}$$

where P_t is the transmitter power in dBm, G_t and G_r are the gains of the antennas in dB (they represent a fraction of the power radiated or received by the antenna in any direction and elevation), L_P is the pathloss in dB, and M_f is the fade margin in dB.

The following parameters are considered for the determination of the link budget (where BTS is base transceiver station and MS is mobile station):

- Pr_{MS}: Power received at the MS (in dBm),
- Pr_{BTS}: Power received at the BTS (in dBm),
- Pe_{MS}: Power transmitted at the MS (in dBm),
- Pe_{BTS}: Power transmitted at the BTS (in dBm),
- Lf_{MS}: Loss due to the MS feeder link (in dB),
- Lf_{BTS}: Loss due to the BTS feeder link (in dB),
- Lc_{BTS}: BTS combiner loss (in dB),
- Lp: MS-BTS propagation loss (in dB),
- G_{MS}: MS antenna gain (in dBi),
- Ga_{BTS}: BTS antenna gain (in dBi),
- Gd_{BTS}: BTS antenna diversity gain (in dB).

The link budget balancing formula is determined as shown in the following:

The received signal level at the mobile from the BTS is given by the following formula:

$$Pr_{MS} = Pe_{BTS} - Lc_{BTS} \cdot Lf_{BTS} + Ga_{BTS} - L_p + G_{MS} - Lf_{MS} \tag{6.9}$$

and the equivalent for the BTS receiver:

$$Pr_{BTS} = Pe_{MS} - Lf_{MS} + G_{MS} - L_p + Gd_{BTS} + Ga_{BTS} - Lf_{BTS} \tag{6.10}$$

Expressing Lp as a function of the other parameters in (6.8) and (6.9), gives

$$L_p = Pe_{BTS} - Lc_{BTS} - Lf_{BTS} + Ga_{BTS} + G_{MS} - Lf_{MS} - Pr_{MS} \tag{6.11}$$

$$L_p = Pe_{MS} - Lc_{BTS} - Lf_{MS} + G_{MS} + Gd_{BTS} + Ga_{BTS} - Lf_{BTS} - Pr_{BTS} \tag{6.12}$$

Combining (6.10) and (6.11), it becomes

$$Pe_{BTS} - Lc_{BTS} - Pr_{MS} = Pe_{MS} + Gd_{BTS} - Pr_{BTS} \tag{6.13}$$

The link budget is therefore given by the following equation:

$$Pe_{BTS} = Pr_{MS} + Gd_{BTS} + Lc_{BTS} + Pe_{MS} - Pr_{BTS} \tag{6.14}$$

Thus, the BTS transmit power (Pe_{BTS}) must be higher than that of the mobile (Pe_{MS}) with a value that corresponds to the sum of the diversity gain (Gd_{BTS}), the combiner loss of the BS (Lc_{BTS}), and the difference in the sensitivities ($Pr_{MS} - Pr_{BTS}$).

Note: Definition of EIRP
Effective Isotropic Radiated Power (EIRP) is defined as the radiated power from the antenna referenced to a theoretical point source. It can be calculated by the following expression (in the case of the BTS antenna):

$$EIRP = Pe_{BTS} - Lc_{BTS} + Ga_{BTS} \text{ (dB)} \tag{6.14}$$

In practice, to obtain as large as possible coverage, it is recommended that antennas be designed to give the highest gain possible consistent with the coverage pattern required. In order to balance the link it is usual for a receiving antenna gain to be higher (e.g., 3 dB higher) than or equal to the transmit antenna gain (typical value of 12 dB) and a receiver sensitivity better than -104 dBm (GSM base station specification) [9].

The propagation loss L_p includes three factors:

- P_{Los}, propagation pathloss,
- M_{Fad}, fast fade margin (which classically follows a Rayleigh law),
- M_{Shad}, shadow margin (which classically follows a log-normal law).

Comment: Determination of the parameter M_{Fad}

M_{Fad} takes into account the fast variations of the received radio signal. It can be obtained from the curves that give the BER versus the E_b/N_0 ratio by taking the difference between the level obtained with a Gaussian channel and the level obtained for a Rayleigh channel. These measurements are realized for a BER value of 10^{-3}. Alternatively, M_{Shad} can be determined by assuming that the mean local field in a given area and at a fixed distance from the transmitter follows a log normal law. In this case, the parameter depends on three variables: γ, the propagation pathloss exponent; σ, standard deviation of the model, and *Disp*, the desired availability of the signal (in percentage of the coverage; e.g., 95%). Accurate determination of M_{Shad} requires very long processing times. Consider a simple method that takes into account a linear variation of M_{Shad} as a function of *Disp*:

- If $F < F_0$, then $M_{Shad} = 0$,
- If $F > F_0$, then $M_{Shad} = a(F - F_0)/(100 - F_0)$.

Where $Sens_{Threshold}$ defines the sensitivity threshold of the receiver, communication will be possible if the following condition is fulfilled:

$$Pr_{MS} \geq (Sens_{Threshold} + M_{Fad} + M_{Shad}) \tag{6.15}$$

In the case of GSM (see Chapter 12), for example:

$Sens_{Threshold} + M_{Fad} = -104$ dBm

and in the case of DECT (see Chapter 10): $Sens_{Threshold} = -86$ dBm.

Example 10: Calculation of link budget for GSM

Consider a BTS with a transmit power of 20W (i.e., 43 dBm), a vehicle MS with a power of 8W (i.e., 39 dBm), and a portable MS with a power of 2W (i.e., 33 dBm). The combiner loss form coupling several transmitters to the same antenna is 4 dB.

The gains (compared to an isotropic antenna) of the different transmitters are equal to 8 dB for the omnidirectional antennas, 2 dB for the 8W mobile station (assumed to be equipped with a $\lambda/4$ antenna), and −3 dB for the 2W mobile station.

The loss at the BTS side (with a 7/8-in diameter and 50m long cable) is equal to 2 dB for the feeder cable and 1 dB for the connectors (i.e., a total of 3 dB). The equivalent loss is equal to 2 dB for the vehicle mobile.

Assuming fade margin of 7 dB for the fast moving mobile and only 5 dB for the portable, the signal power receiver threshold on the uplink is equal to −104 dBm, −102 dBm on the downlink for a 2W mobile station and −104 dBm for the 8W mobile station. The other parameters are given in Tables 6.3 and 6.4.

Table 6.3

Pathloss propagation on the uplink

	8W MS ↗ BTS	Handheld 2W MS ↗ BTS
Transmit power (dBm): Pe_{MS}	39	33
MS connector loss (dB): $Lc_{MS} + Lf_{MS}$	2	0
Transmit antenna gain (dB): G_{MS}	2	-3
EIRP (dBm):	39	30
$EIRP = Pe - (Lc_{MS} + Lf_{MS}) + G_{MS}$		
Receiver sensitivity (dBm): Pr_min_{BTS}	−104	−104
BTS cable and connector losses (dB): $Lc_{BTS} + Lf_{BTS}$	3	3
Receiving antenna gain (dB): Ga_{BTS}	8	8
Diversity gain (dB): Gd_{BTS}	5	5
Fade margin (dB): M_{Fad}	7	5
Equivalent signal strength (dBm):	−107	−109
$E = Pr\_min_{BTS} + M_{Fad} + (Lc_{BTS} + Lf_{BTS}) - Ga_{BTS} - Gd_{BTS}$		
Allowable propagation pathloss (dB): $EIRP - E$	*146*	*139*

Table 6.4

Pathloss propagation on the downlink

	BTS ↘ 8W MS	BTS ↘ Handheld 2W MS
Transmit power (dBm): Pr_{BTS}	43	43
Coupling losses (dB): Lc_{BTS}	4	4
BTS cable and connector losses (dB): Lf_{BTS}	3	3
Transmit antenna gain (dB): Ga_{BTS}	8	8
EIRP (dBm):	44	44
$EIRP = Pr - Lc_{BTS} - Lf_{BTS} + Ga_{BTS}$		
Receiver sensitivity (dBm): Pr_min_{MS}	−104	−102
MS connector loss (dB): $Lf_{MS} + Lc_{MS}$	2	0
Receive antenna gain (dB): G_{MS}	2	-3
Fade margin (dB): M_{Fad}	7	5
Equivalent signal strength (dBm):	−97	- 94
$E = Pr\_min_{MS} + M_{Fad} + (Lc_{MS} + Lf_{MS}) - G_{MS}$		
Allowable propagation pathloss (dB): $EIRP - E$	*141*	*138*

Note: With reference to formulas (6.10) and (6.11), inclusion of the terms M_{Fad} and Lc_{MS} provide the final formula for L_p, the propagation pathloss.

The fade margin is introduced to increase coverage from 70% to 90% in the cell. Finally, to improve the sensitivity by 5 dB, a receive diversity scheme is implemented by adding a second antenna at the BTS. As the difference between the uplink and the downlink received signal levels is less than 5 dB, the link is considered reasonably balanced.

6.2.5 Conclusions

The technique of frequency reuse theoretically permits cellular operators to cover unlimited area and support infinite traffic density with a fixed and limited frequency resource. The C/I of the system must be determined (affected by the modulation scheme chosen and the techniques used to design the system) and hence reuse distance and cluster pattern. The hexagonal model is the most often used because of its simplicity at the system planning stage.

Cellular network capacity is dependent on techniques implemented in the system. In the following section the main techniques used to improve communications quality and increase network capacity are discussed. Certain techniques are not reserved solely for cellular systems. This is the case for DCA algorithms, which are currently only implemented in cordless systems.

6.3 SYSTEM CAPACITY EXPANSION TECHNIQUES AND NETWORK QUALITY IMPROVEMENT

Interference problems arising from the implementation of frequency reuse in cellular applications can result in a reduction in the quality of service or reduce system capacity (these two elements are interrelated). Contrary to other types of system limited by noise (known as *noise limited* systems), cellular systems are limited by interference (and are thus called *interference limited* systems). System designers must therefore find mechanisms to reduce or avoid interference. For this purpose, several mechanisms have been defined and are currently implemented in digital cellular.

Error correction, message retransmission, and signal processing are used at different levels (physical layer or data link layer) as introduced in Chapter 4. These techniques can be used on the uplink or downlink and do not contribute directly to the global interference level. They are present for the control and management of errors during communication (e.g., error correction and detection codes, LAPDm protocol, and equalization).

6.3.1 Frequency Hopping

Frequency-hopping can be used to reduce interference on a given channel by moving other transmitters around in frequency and can also allow a transceiver to move away from a channel with interference or a channel experiencing a severe fade. Its use in military systems has, for many years, been driven by the ability of such a system to operate undetected and to be able to continue communication even in the presence of deliberate jamming. The technique provides interferer diversity by using frequency diversity for each communication segment.

There are two types of frequency hopping, known as *slow frequency hopping* and *fast frequency hopping*. As suggested by the names, the difference between them is the rate of the hop. Slow frequency hopping, which is the subject of this section, hops between radio channels at intervals of at most one message or burst; that is of the order of a millisecond. In GSM for instance, the hops can occur up to 217 times per second. In fast frequency hopping, the transmitter hops at each symbol or several times during the transmission of a message or a burst. These hops occur with delays that are of the order of a microsecond. This last technique is employed in the FDMA *spread spectrum* multiple access method (see Chapter 3).

Slow frequency hopping (Figure 6.13) provides interferer diversity. This is particularly important in high traffic density urban areas, where the interference levels encountered (i.e., the number of interferers) can be significant. As interference is one of the main limitations of system capacity, it is important to reduce it as much as possible. In nonfrequency hopping systems, the interference level can be concentrated on particular carriers. Hopping from one frequency to another makes it easier to avoid jamming on discrete frequencies, or at least minimize the number of messages lost because of the interference. This mechanism operates best if the frequency hopping patterns are pseudorandom; that is, decorrelated one from another. In this way, a transmitter can avoid using the same channel simultaneously with others and will therefore cause a minimum of disruption.

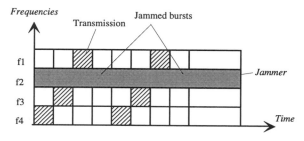

Figure 6.13 Frequency hopping on four carriers with one jammed channel (f_2).

The main drawback of this mechanism is that it requires a number of frequencies at each base station site. To obtain an appreciable gain, at least four frequencies are required, even though improvement can be had with only two.

Frequency hopping can be either *fixed* (the hopping sequence follows a predefined list of frequencies), or *random* (the hopping sequence varies). In the fixed scheme, the radio link quality can be degraded if several concatenated bursts are subject to co-channel interference from other transmitters on the same frequencies and timeslots. In the random scheme, co-channel interference can vary from one sequence to another, which provides better interference diversity.

6.3.2 Discontinuous Transmission and Packet Transmission Mode

Speech communications are characterized by bursty traffic, which means, among other things, that speech bursts are interspersed with periods that contain no user information. Statistics show that a speech channel carries user information on average for only 44% of the time. This characteristic can be used to reduce the mean system interference level (by not transmitting during the quiet periods) or increase system capacity by splitting the channel between several users by using packet transmission techniques.

6.3.2.1 Discontinuous Transmission

The first solution involves transmitting very few or no bursts during quiet user periods. This is called discontinuous transmission (denoted by DTX), employed to reduce the amount of energy transmitted on the radio link, thereby reducing the mean interference level as seen by other users and also increasing the MS battery life.

In Figure 6.14, discontinuous transmission for a TDMA system is represented. Slot i is reserved for the user. When a message is to be transmitted, it is split into several packets, each of which can be transmitted in a slot. When no message is produced by the user the slot is left free or a message (allowing continued supervision of the physical link) is transmitted in the slot periodically (one slot in every four for example). Use of discontinuous transmission supposes that a speech detector is available to differentiate background noise from user speech.

6.3.2.2 Packet Transmission Mode

The two switching mechanisms used in telecommunications systems are *circuit switched* and *packet switched* (also called *datagram*) transmissions. Circuit switched transmission is interesting from the point of view of modem design. In effect, a circuit-switched system allows the use of constant synchronization and equalization [10].

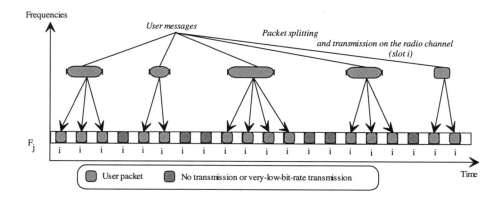

Figure 6.14 Example of discontinuous transmission.

The packet-switched mode supports variable rate transmission but with the drawback of including significant overhead in the packets for equalization and synchronization purposes. Packet access is a generalized form of discontinuous transmission where the channel is shared by several mobiles. This technique is often presented as an alternative to the circuit-switched mode where a terminal keeps a channel from the beginning to end of the communication (even though this channel may change because of a handover: see Chapter 8). The advantages of packet mode are offset by contention problems (see Chapter 3). The technique is applicable mainly in TDMA systems, although it can be implemented with CDMA.

As with discontinuous transmission a mobile terminal uses a channel only when it has information (speech, data) to transmit. As speech generally occupies the channel 44% of the time, a channel could, theoretically be shared by 2.25 users.

Example 11: Capacity increase by using packet access

A channel is occupied on average x % of the time by users. Let y be the maximum access rate obtained with the random access protocol implemented. The number of simultaneous communications on the same channel will therefore be $100.y/x$.

By taking $x = 44$ and $y = 0.8$, a channel will be able to sustain on average 1.8 simultaneous communications. It has been shown that a protocol such as PRMA [11] is able to reach this performance. The increase in network capacity is in this case 80 %.

Compared with discontinuous transmission, packet access allows increased use of the radio resources. However, the benefit to global interference level is lost (see Table 6.5).

Table 6.5
Comparison of the DTX and packet access techniques

Technique	Discontinuous transmission	Packet access
System capacity	Unchanged	Increased
Global interference level	Reduced	Unchanged
Terminal power consumption	Reduced	Reduced

In the two cases, the mobile only transmits when user or signaling messages must be transmitted. This scheme allows a reduction in MS power consumption. This is a very important benefit of these two mechanisms. It is especially significant for portable handsets for which size and talk time between battery charges are very important constraints. Some studies have shown that discontinuous transmission can easily provide more than twice the battery life for the terminal.

6.3.3 Power Control

The signal received by a mobile or by a base station, because of propagation variations and movement of the mobile, fluctuates significantly. It is therefore necessary to define a range of transmit powers that will allow the establishment and maintenance of a radio link with a mobile located anywhere in the service area of a cell. A single defined power level may be higher than required if the mobile is experiencing favorable propagation conditions (base station in line-of-sight or in close proximity) and will produce a level of interference higher than necessary to other mobiles. Similarly, it could be below the required level for a mobile suffering unfavorable conditions (e.g., obstructions, base station far from the mobile). In this case, the quality of service offered to the user can be degraded and breaks in communication can occur. This defined power level would be a mean level required to satisfy the needs of the majority of mobiles located in the service area (e.g., communications with a base level of quality of service for 95% of the cell coverage). The drawback in mobile radio is that the distribution of mobiles in a cell is often very spread. This means that a fixed power value would be satisfactory (in terms of optimum C/I value) for only a very small number of mobiles. It is clear, therefore, that dynamic adjustment of the transmit power for each mobile is advantageous.

Power control from the BS aims to adjust the transmit power of all mobiles in its cell to maintain an approximately equal receive level for them all at the BS, while keeping the quality of service at an acceptable level (i.e., at a C/I value above the design threshold). In this way global co-channel interference level can be minimized and the battery life for mobiles can be optimized.

The transmitter in the mobile will adjust its power within a predefined margin. The upper value can be either the physical capacity of the transmitter (1W, 2W, or 4W), or the service area of the cell (e.g., microcell with a power of less than 1 watt or a macrocell with a power of several watts).

Figure 6.15 represents a simple case where two levels of power are used: a high power (*P_max*), used in the case where the mobile is far from the base station, and a low power (*P_min*), used when the mobile is closer.

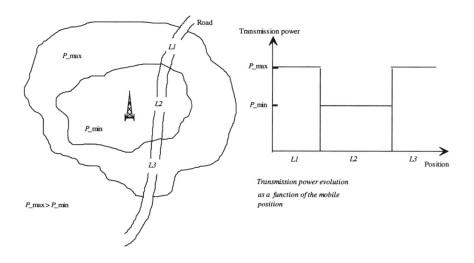

Figure 6.15 Power control of the mobile during transit.

6.3.3.1 Open-Loop or Closed-Loop Power

Power control can be implemented either by use of open or closed loop mechanisms [12].

Open-loop power control is achieved by using channel estimation in accordance with characteristics of the received signal. The transmitter power is adjusted but the controller does not attempt to get feedback information about the efficiency of the adjustment. The technique is obviously not very accurate, but because it does not rely on feedback information, it is quick. This characteristic is useful to compensate for sudden channel fluctuations when a mobile experiences shadow fading (e.g., when it passes behind a building).

A closed-loop power control scheme bases its decisions on real performance measurements of the communication link in the form of the received signal strength, the SNR, the BER, or the Frame Erasure Rate (FER). In the case when power control is implemented on the uplink, quality measurements made of the downlink at the mobile can be transmitted to the base station (as in GSM) and together with uplink data evaluated at the base station determines the optimum power level. Only then need a power adjustment command be transmitted to the mobile.

6.3.3.2 Control of the Mobile Station Power

A mobile moving closer to the base station must reduce is transmit power to:

• Reduce the probability of generating intermodulation products in the base station receiver front end. In fact if the mobile is very close to the base station, high mobile transmitter power can cause saturation in the receiver amplifier if the level received at the base station exceeds a given threshold (e.g., -55 dBm).

• Reduce the global power level to reduce interference probability with other co-channel base stations.

• Reduce adjacent channel interference which may lead to corruption of data or even result in capture effects (a powerful signal transmitted on an adjacent channel can "capture" a receiver attempting to decode less powerful signals on channel).

6.3.3.3 Control of the Base Station Transmit Power

In spite of the various advantages of implementing power control, in most first-generation cellular systems it was not possible to alter the base station transmit power even on a single channel. This was mainly due to the technical limitations of the radio equipment. Power control in these systems tended to be limited to mobiles which were sent, power control commands based on the RSSI measured at the base station. In second-generation systems, dynamic power control of both mobile and base station have been introduced.

By a simple analytical method, it is possible to show that system gains (better quality of service and increased traffic) can be significant when a control mechanism is implemented. Also, the benefits for terminal battery life are significant and the quality of service perceived by the user is consequently improved.

Finally, power control allows the implementation of cells of different sizes inside the same cellular network. Indeed, as the transmitted power in a small size cell is lower than that required in larger cells, the network adjusts the signal power of the mobiles entering a given cell, thus reducing the interference produced by mobiles located in this cell.

6.3.4 Dynamic Channel Allocation

Different assignment policies can be used to allocate a channel to a call. First- and second-generation cellular systems use a fixed channel allocation (FCA) method. Each base station transmits only on its allocated channels and cannot use other channels unless a new frequency plan is processed. Main advantages of this method are first to allow an allocation of channels with an optimal reuse distance, and second, simplified channel management in the system. Fixed-frequency allocation techniques are introduced in Chapter 7.

The FCA technique performs best when the traffic pattern is statistically unchanging. It operates quite well in macrocellular systems with high traffic loads. Dynamic channel allocation (DCA) and random channel allocation (RCA) techniques follow traffic pattern changes and have shown better performance than FCA in terms of local throughput when traffic patterns were variable.

In the FCA method, allocating a channel to a base station is accomplished relative to the estimated traffic in the service area of the base station (i.e., in its cell). Large cells in current systems smooth local traffic peaks. With cell size reduction it is no longer possible to satisfy all traffic conditions. Implementing DCA is necessary in microcellular environments where traffic fluctuations in each cell are much more significant than in large cells.

6.3.4.1 General Processing of DCA Algorithms

The DCA mechanism is processed at establishment of each new communication. At this time, a radio channel is chosen among available channels (all the channels or a subset of the channels allocated to the system) and allocated to the mobile and base station.

The choice of a channel is based on radio channel measurements. One parameter considered is interference level. The location of the mobile station and the propagation conditions (RSSI) on the idle channels when allocating the channel are also important. The choice of a new channel must be realized such that the interference level will not be significantly affected. The channel chosen can be one that presents the minimum interference or one presenting an interference level below a certain threshold. If access to this channel fails, another is chosen.

6.3.4.2 Example of DCA Algorithms

Several categories of DCA algorithm have been defined in the literature. The most popularly cited ones are given here. They can be classified into two categories: *centralized* algorithms and *decentralized* algorithms.

In a centralized algorithm [13, 14], the allocation is achieved by a central controller (mobile switching center). The measurement of channels transmitted by mobiles and base stations to a central controller is computed to make the decision. This type of algorithm uses a compatibility matrix that determines channel availability.

When the algorithm is *decentralized* [15, 16], it is the mobile and base station that choose the channel according to their own knowledge of the channel state. Some examples of decentralized algorithms are as follows:

- *Least interference algorithm*: The selected channel is the one whose instantaneous received interference power is the weakest. This algorithm tries to minimize global system interference.

- *Least interference below threshold algorithm*: the selected channel is the one upon which interference is the weakest among a set of channels with interference below a threshold. The objective of this algorithm is to minimize the global system interference while maintaining the quality of service.
- *Marginal interference algorithm*: the chosen channel is either the one on which the instantaneous interference is highest and is also below a threshold, or is the channel with the least interference. This algorithm provides a good compromise between signal quality and efficient use of the spectrum.

DCA algorithms are currently implemented in second-generation cordless systems (see Chapter 10). Present cellular systems do not yet use this kind of technique but they may be considered for third-generation cellular systems.

6.4 BASIC CELLULAR SYSTEM ARCHITECTURE

In this section, GSM terminology is used, which is also used by most digital systems. The general architecture of a cellular system relies on two subsystems:

- The network subsystem (NSS),
- The base station subsystem (BSS).

The NSS is composed of switches (which manage the call setup and routing) and of databases (which store subscriber profiles). The BSS includes all the equipment required for radio interface management in the form of base stations and their controllers. Actually, this functional distinction in two subsystems appears as a clear difference in physical equipment.

The NSS and the BSS are controlled and supervised by an *operating and maintenance center-network* (*OMC-N*) and an *operating and maintenance center-radio* (*OMC-R*), respectively. These functional entities allow access to statistics, operations, or resource management. The last component of a cellular system is the mobile terminal, which, as in other types of radio systems, provides user communications over the radio interface.

The NSS can be integrated into the PSTN. In fact, many cellular operators, with deregulation happening in an increasing number of countries, try to avoid using the fixed public network by installing their own lines (microwave links were installed initially) [17]. The aim is to use the PSTN only in cases of overload of the private links and for local access to fixed subscribers.

In GSM networks, the Integrated Service Data Network, like services offered to subscribers and the complexity of certain functions, have required the use of *Signaling System Number 7*. This is functionally separated from the transmission bearer network and is mainly used for call control and call processing.

On the radio side, the BSS manages network coverage, which can be as large as a whole country. Management of roaming is achieved by definition of location areas that generally consist of several cells (location management is introduced in Chapter 8).

First-generation cellular architecture is rather simple by comparison with more modern systems because capacity and functionality demands were lower. System design was based around hardware (switches and base stations) rather than the extensive software functionality associated with modern networks.

With an increase in system complexity, a functional approach has been used instead of the classical approach outlined above. ISDN has been one of the first telecommunications networks to be designed according to this new approach. In second-generation cellular systems, the design is partly defined by functions or group of functions that have to be provided by the system. These functions are then implemented in a combination of hardware and software. This growth in functions has extended the range of services provided to the user. As a consequence, new cellular networks are much more complex and software has become increasingly important. The basic architecture presented in this chapter covers the main parts of first- and second- generation cellular systems.

As represented in Figure 6.16, a cellular network contains the three principal functional blocks, NSS, BSS, and OMC. These blocks can correspond to physically distinct equipment as in GSM or they can be totally or partly integrated in the same unit, such as is the case for most first-generation systems.

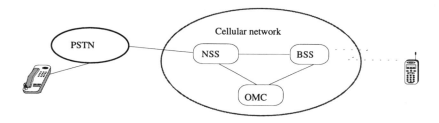

Figure 6.16 General cellular system architecture.

6.4.1 The Network Subsystem

The NSS contains switches and databases. The mobile switching center (MSC) is generally composed of a standard fixed network switch to which some specific functions have been added, such as those related to radio resource management.

The MSC constitutes the interface between the mobile subscribers and the PSTN. It manages calls (e.g., establishment, routing, and transfer). An MSC also ensures functions such as signaling, switching, and analog/digital conversion in analog radio transmission systems. In second-generation cellular systems, some functions such as location are performed by an entity that is different from the MSC. Examples of this are the HLR and VLR databases in the GSM system.

BSS-MSC communications are generally conducted on fixed lines that allow speech and data. The MSC can also be connected to the PSTN. In this case, it is called a gateway-MSC (G-MSC).

The NSS includes databases whose main function is to allow mobile roaming management. In all cellular systems, there is a center in which all information about network subscribers is held. This center can be located in one of the mobile switching centers. It contains the main database of the system, also called *home database* (e.g., the HLR in GSM). The home database contains the subscriber data and location information in the form of the address of the roaming area. The subscriber information is recorded when the user signs up to the service and then generally remains unaltered. The only time-sensitive data that varies according to the user's movement is the *location pointer*. The visitor database (VLR in GSM) contains a copy of the subscriber data of users who are located in the area under its control. These data is deleted when the system detects that the mobile has left the area (home and visitor databases are discussed in greater depth in Chapter 8).

In GSM, a distinction is made between the databases (home and visitor) and the mobile switching centers, even though physically MSC and VLR are presently integrated in the same equipment. In analog systems, MSCs and visitor databases are viewed as integrated.

6.4.2 The BSS

The BSS governs actions controlling radio transmission and radio channel management. The base station, which includes the transceiver (TR/RX) equipment, and the base station controller, which manages the radio resource (e.g., channel allocation, handover) are two distinct entities. The mobile, which constitutes the interface between the user and the network, is sometimes thought of as integrated with the BSS.

6.4.2.1 The Base Station and the Base Station Controller

The base station controller (BSC) may not exist as a separate entity from the MSC, as is the case in most analog cellular systems. The BSC manages radio resources. It supervises access to and communication over the radio channel (power control, handover, etc.). The BSC can be linked to base station sites in different ways: either by bus structure (drop-insert scheme) or by a star configuration (Figure 6.17). The BSC routes the BS traffic to the MSC.

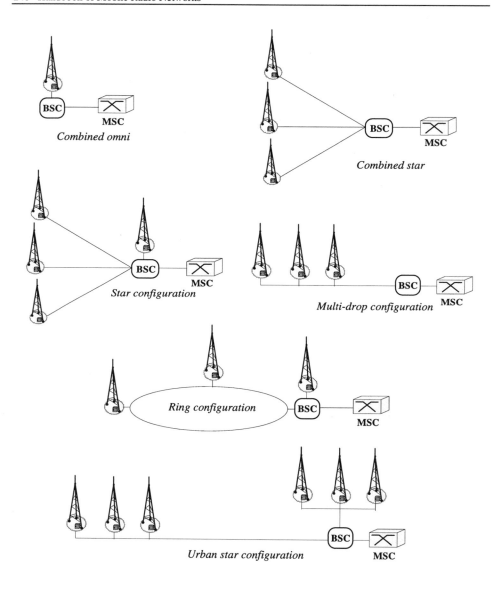

Figure 6.17 Types of links between the BSC and base station sites.

The base station manages the calls coming to or from mobiles located in its service area (i.e., the cell). It is connected to the MSC through the BSC. It includes the TX/RX with the speech coders/decoders and the modems for the digital systems, antenna masts, and antennas. The number of TX/RXs depends on the traffic density within the area managed by the base station. A base station provides functions at the physical level (e.g., modulation/demodulation, equalization) and collects radio channel measurements used in call/network management. These parameters are reported to the BSC.

6.4.2.2 The Mobile Station

The last part of the radio subsystem, sometimes not included, is the *mobile station*. There are several types of mobile station defined, depending on the system in question. Broadly speaking, they can be split into *vehicle mobiles, (trans)portable,* and *handheld.* Vehicle mobile stations were mainly used in first generation networks where the aim was a car telephone service. In this case, the terminal is equipment that is fixed inside the vehicle and has an antenna on the roof of the car. The transmit power of these terminals can reach several tens of watts.

Portable (sometimes known as transportable) mobile terminals also have a fairly high peak power that allows use inside a vehicle and in large cells (which are generally implemented in rural areas). Size is that of a small case with a weight of 4 to 6 pounds.

Handheld mobiles, sometimes just called *portables*, were the most recent ones to appear and are the most common type in GSM networks. Because of their small size, such mobiles are easily carried for use by pedestrians in urban areas (where the cells have small size and low power). They can also be used in environments planned for portable or fixed terminals but, if coverage is poor, it may be necessary to connect to a *booster* amplifier fitted in the vehicle which supplies increased transmitter power. Handheld terminals have become more numerous and portables are becoming obsolete. At present, operators are designing more and more coverage to serve handheld units inside and outside urban areas as it is now estimated that over 90% are of this type.

A mobile integrates with communication management (speech and data), link control functions, and signaling within the system. It is able to synchronize on every channel allocated to the system. Each channel consists of a pair of frequencies that permit bidirectional conversations (*full duplex*). To minimize radio channel usage, digits to be dialed and entered from the keyboard or from an address book memory are transmitted only when the whole number has been entered by the calling party.

A mobile station can be separated into two parts, a user interface and a radio and processing part (Figure 6.18). The interface between the user and the system (the MMI or man-machine interface) includes the speaker/microphone, the display, and the alert signal (sound or visual). It is typically controlled by software. The radio and processing part includes three distinct modules: the radio module, the signal processing and logic module, and the power module.

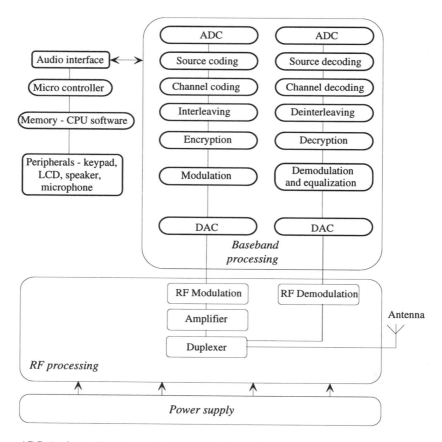

ADC: Analog-to-digital converter; DAC: digital- to-analog converter

Figure 6.18 Block diagram of a generic transceiver.

The radio module provides the transmission, reception, modulation, demodulation, and frequency synthesis functions. It manages all the frequencies of the system in *full-duplex* mode. The transmitter has a power of a few watts. Its power can be controlled by the system to minimize the interference level or to improve the link quality. The receiver demodulates speech and data.

The processing and logic module controls the radio module. It also incorporates functions such as signal processing, information coding and decoding, ciphering, error protection, and correction mechanisms. This unit includes a microprocessor that manages the control tasks of the mobile. For example, the following control messages are managed by the processing module: *transmission of an access demand to the radio channel*; *location updating registration*; and *reception of a channel allocation from the base station to the mobile, reception of a paging message sent to the mobile to inform about an incoming call.*

Finally, the power supply module includes all the required interfaces with the different kinds of battery. The mobile can thus be fed directly by the battery of a car using a cigar lighter, a rechargeable battery, or mains fed equipment. The antenna can be either internal or external to the handset.

6.4.2.3 From the First- to Second-Generation Cellular Systems

The evolution of cellular systems has seen the transition from systems with very simple infrastructures (a few switching centers where user subscriber data is held and base stations provide the interface between the switch and the mobiles) to more complex system (consisting of switches, databases, base stations, and base stations controllers). This increased system complexity is principally due to the need to manage much greater user densities and the introduction of new services that are richer in functionality. This leads to an increase in the number of functions to implement in the system, requiring some decentralization of control.

Access control to the traffic channel, power control, and handover, as examples, are all managed by the network. The only function that can be initiated by the mobile is location update. The decentralization decision (for handover or power control, in particular) has begun with second-generation systems and should continue with third-generation systems. These points are reviewed again in Chapters 12 and 14.

REFERENCES

[1] Cox, D., "Wireless Personal Communications: What Is It," *IEEE Personal Communications*, pp. 20–35, Apr. 1995.

[2] MacDonald, V. H., "The Cellular Concept," *Bell System Technical Journal*, Vol. 58, Jan. 1979.

[3] Yacoub, M. D., *Foundations of Mobile Radio Engineering*, CRC Press, 1993.

[4] Clark, M. V., V. Erceg, and L. J. Greenstein, "Reuse Efficiency in Urban Microcellular Networks," *Proceedings of the IEEE Vehicular Technology Conference '97*, Phoenix, AZ, pp. 279–285, May 4–7, 1997.

[5] Katzela, I., and M. Naghshineh, "Channel Assignment Schemes for Cellular Mobile Telecommunications Systems: A Comprehensive Survey," *IEEE Personal Communications*, pp. 10–31, June 1996.

[6] Halpern, S. W., "Reuse Partitioning in Cellular Systems," *Proceeding of the IEEE Vehicular Technology Conference '87*, Tampa, FL, pp. 322–327, June 1987.

[7]   Lucatti, D., A. Pattavina, and V. Trecordi, "Bounds and Performance of Reuse Partitioning in Cellular Networks," *International Journal of Wireless Information Networks*, Vol. 4, No. 2, pp. 125–134, 1997.

[8]   Garrett, J.M., "Cellular Mobile Radio," *in Land Mobile Radio Systems*, edited by R. J. Holbeche, Peregrinus, 1985.

[9]   Kamanou, P., "Outil de planification. Radio et méthodes associées," *Séminaire régional sur les systèmes cellulaires radiotéléphoniques mobiles*, UIT-BDT, Tunis, June 9–13, 1997.

[10]   Ayanoglu, E., K. Y. Eng, and M. J. Karol, "Wireless ATM: Limits, Challenges, and Proposals," *IEEE Personal Communications*, pp. 18–34, Aug. 1996.

[11]   Goodman, D. J., R. A. Valenzuela, K. T. Gaylard, and B. Ramamurthi, "Packet Reservation Multiple Access for Local Wireless Communications," *IEEE Transactions on Vehicular Technology*, Vol. 37, No. 8, pp. 885–890, Aug. 1989.

[12]   Pichna, R., and Q. Wang, "Power Control," *The Mobile Communications Handbook*, Editor-in-Chief Jerry D. Gibson, CRC Press, 1996.

[13]   Tan, H.-C., M. Gurcan, and Z. Ioannou, "A Radio Local Area Network With Efficient Resource Allocation," *Proceedings of the Second WINLAB Workshop on Third-Generation Wireless Information Networks*, East Brunswick, NJ, Oct. 1990.

[14]   Dimitrijevic, D. D., and J. F. Vucetuc, "Design and Performance Analysis of Algorithms for Channel Allocation in Cellular Networks," *Proceedings of the Fourth WINLAB Workshop on Third-Generation Wireless Information Networks*, East Brunswick, NJ, pp. 285–303, Oct. 1993.

[15]   Nanda, S., and D. J. Goodman, "Dynamic Resource Acquisition: Distributed Carrier Allocation for TDMA Cellular Systems," *Proceedings of the IEEE Globecom '91*, Phoenix, AZ, Dec. 1991.

[16]   Cimini, L. J., G. J. Foschini, and C. M. I, "Distributed Algorithms for Dynamic Channel Access in Microcellular Systems," *Proceedings of the Third WINLAB Workshop on Third-Generation Wireless Information Networks*, East Brunswick, NJ, Apr. 1992.

[17]   Chow, D., "Building out your own network infrastructure," *Mobile Communications International*, Nov. 1995, pp. 43–46.

SELECTED BIBLIOGRAPHY

[1]   Abdul-Haleem, M., K. F. Cheung, and J. C.-I. Chuang, "Aggressive Fuzzy Distributed Dynamic Channel Assignment for PCS," *Proceedings of the IEEE International Conference on Personal Communications*, Tokyo, Japan, pp. 76–80, Nov. 6–10, 1995.

[2]   Calhoun, G., *Digital Cellular Radio*, Artech House, 1988.

[3]   Cheng, M. M.-L., and J. C.-I. Chuang, "Performance Evaluation of Distributed Measurement-Based Dynamic Channel Assignment in Local Wireless Communications," *IEEE Journal on Selected Areas in Communications*, Vol. J-SAC 14, No. 4, pp. 698–710, May 1996.

[4]   Chuang, J. C.-I., "Performance Issues and Algorithms for Dynamic Channel Allocation," *IEEE Journal on Selected Areas in Communications*, Vol. 11, No. 6, pp. 955–963, Aug. 1993.

[5]   Clark, M. V., "Reuse Efficiency in Urban Microcellular Networks," *Proceedings of the IEEE Vehicular Technology '96*, Atlanta, GA, pp. 421–425, Apr. 28–May 1, 1996.

[6] Furuya, Y., and Y. Akaïwa, "Channel Segregation, A Distributed Adaptive Channel Allocation Scheme for Mobile Communications Systems," *Nordic Seminar on Digital Land Mobile Radio Communications*, Stokholm, Sweden, pp. 311–315, Oct. 1986.

[7] Ivanov, K., N. Metzner, G. Spring, H. Winkler, and P. Jung, "Frequency-Hopping Spectral Capacity Enhancement of Cellular Networks," *Proceedings of the International Symposium on Spread Spectrum*, Hanover, Germany, pp. 1,267–1,272, Sept. 1996.

[8] Kuhn, A., W. T. Haggerty, U. Grage, M. Keller, C. Reyering, F. Arndt, and W. Scholtholt, "Validation of the Feature Frequency Hopping in a Live GSM Network," *Proceedings of the IEEE Vehicular Technology '96*, Atlanta, GA, pp. 321–325, Apr. 28–May 1, 1996.

[9] Lee, W.C.Y., "Elements of Cellular Mobile Radio Systems," *IEEE Transactions on Vehicular Technology*, Vol. VT-35, No. 2, pp. 51–59, May 1986.

[10] Lee, W.C.Y., "Smaller Cells for Greater Performance," *IEEE Communications Magazine*, pp. 19–23, Nov. 1991.

[11] Madfors, M., K. Wallstedt, S. Magnusson, H. Olofsson, P.-O. Backman, and S. Engström, "High Capacity With Limited Spectrum in Cellular Systems," *IEEE Communications Magazine*, pp. 38–45, Aug. 1997.

[12] Oetting, J., "Cellular Mobile Radio, An Emerging Technology," *IEEE Communications Magazine*, pp. 10–15, Nov. 1983.

[13] Sauquet, E., "Fractional Frequency Reuse in GSM Networks," *Mobile Communications International*, pp. 60–62, June 1996.

[14] Steele, R., J.W. Whitehead, and W.C. Wong, "System Aspects of Cellular Radio," *IEEE Communications Magazine*, pp. 80–86, Jan. 1995.

[15] Hammuda, H., *Cellular Mobile Radio Systems*, Wiley & Sons, 1997.

[16] Goodman, D.J., *Wireless Personal Communications Systems*, Addison-Wesley Wireless Communications Series, 1997.

[17] Jovanovic, V.M., and J. Gazzola, "Capacity of Present Narrowband Cellular Systems: Interference-Limited of Blocking-Limited?" *IEEE Communications Magazine*, pp. 42-51, Dec. 1997.

[18] Faruque, S., "High Capacity Cell Planning Based on Fractional Frequency Reuse With Optimum Trunking Efficiency," *IEEE Vehicular Technology Conference*, Ottawa, Canada, May 18–21, 1998.

[19] Lucatti D., A., Pattavina, and V. Trecordi, "Bounds and Performance of Reuse Partitioning in Cellular Networks," *International Journal of Wireless Information Networks*, pp. 125–134, Vol. 4, No. 2, 1997.

CHAPTER 7

CELLULAR PLANNING AND ENGINEERING

Planning a cellular network represents one of the most complex tasks a network operator has to achieve. Although it is a repetitive process, directly related to network development, it is at the installation phase of the system and during the growing stages that it is most critical. It is during these phases that the network experiences the highest rate of development (tasks related to site finding, base station commissioning, frequency replanning, and so forth).

Cellular networks are the most difficult to design, engineer, plan, and operate of all land mobile radio systems, especially since the advent of microcellular techniques. Installing a professional mobile radio (see Chapter 9) network or a cordless system, for example, even if it implies the use of the same basic tools, does not require such complex analysis. Fewer parameters need to be taken into account, frequency planning is not always necessary, and so on. Cellular network planning and engineering techniques that are introduced in this chapter also help the planning of other kinds of networks.

Actually, planning a cellular network requires coping with a large number of different aspects simultaneously. Many factors are involved in this form of mobile communications system design, such as system performance, system capacity, cell coverage, user and signaling traffic, environmental topography, and propagation characteristics. Cell numbers and locations of base stations, parameter definition of mobile stations and channel assignment aspects all relate to one another. General planning must take into account many different constraints. For instance, cell locations can be based on coverage, traffic distribution, and propagation characteristics. Finally, the cell channel assignment mechanism can only be determined once the cellular network architecture has been specified [1]. This plays a significant part in improving system performance in terms of traffic capacity and interference levels.

7.1 CELLULAR PLANNING ELEMENTS

7.1.1 Importance of the Cellular Planning Process

Planning a cellular network is a very sensitive process, the outcome of which determines the success of its operator. A poorly planned network will yield bad quality of service, significant call interrruptions, and a high blocking rate (unsatisfied call requests, sometimes referred to as *Grade of Service*). It will also entail extra costs and fewer benefits for the operator. For this reason network planning, more than any other design aspect of the system, is firmly related to expected revenues, in other words, to the number of subscribers that the operator intends to serve.

Network cellular planning basically starts with two tasks, that of defining coverage and capacity. These are crucial for the operator because the subscribers expect to be provided with a complete roaming service (in terms of coverage and resource availability). Not meeting their demands results in lost revenue and market share, more so in today's competitive environment. The software planning tools used by the operators, as well as their network plans (frequency plans and radio sites) are jealously guarded and regarded as highly confidential information and know-how.

For a cellular system, the number of parameters to be adjusted (mainly radio parameters) are significantly more than those used in fixed communications systems. Traffic and propagation conditions, user mobility, and services vary constantly to generate operating characteristics specific and unique to each network. For this reason, a cellular network operator has to constantly monitor and react to these changes in order to determine new optimum values. Parameters (or *working parameters*) such as those related to paging, handover, power control, radio resource management, and location updating algorithms may be adjusted to optimize the usage of the radio resource. A cellular network where the parameters have been appropriately optimized will offer a good quality of service and an optimum capacity.

7.1.2 Objectives and Problems in Cellular Planning

The objectives of the cellular planning process can be summarized as follows (see also Figure 7.1):

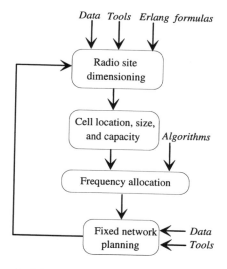

Figure 7.1 General principles of cellular planning.

- Consider the characteristics of an area to be covered (morphology and radio propagation characteristics), the characteristics of the subscribers to be served (density and statistical behavior), and the radio frequency band allocated to the service.
- Minimize the radio and network infrastructure cost according to radio coverage, cell sizes, frequency plans, and fixed network topology.
- Define and observe the quality of service constraints (e.g., blocking rate, C/I level, and so on).

The planning process should lead to

- A base station installation plan (mainly locations, capacities, and transmitter/receiver characteristics),
- An associated frequency plan (frequencies allocated to each base station of the network),
- A fixed network architecture (base station controllers, switching centers, and databases),
- A transmission network to link all these entities.

Albeit simple at first sight, this outcome is actually the end product of a lengthy process involving a wide range of skills and includes various steps that will be introduced in more detail in Section 7.1.3. The main steps characterizing the planning procedure encompass tasks that are as varied as the determination of mobility and traffic models, the definition of radio coverage, the determination of the size of cells, frequency allocation, definition of the switching network, and planning of the signaling network and databases.

For this reason, the planning process is lengthy and costly. With some first-generation systems, the implementation cost for the installation of one base station (overall costs of the planning tasks, equipment, and commissioning) could reach an average of 1 million U.S. dollars [2]. Moreover, the network radio equipment cost can reach about 70% of the total infrastructure investment for a cellular operator [3]. Therefore, it is obvious that precise planning of the network helps optimize the equipment inventory and is of paramount importance to the operator.

7.1.3 Coverage Objectives

In the main, providing good coverage in urban and suburban areas does not entail any investment problems because the traffic load in these areas makes such investment cost effective in the short term. However, rural areas generate more investment problems as the operator must ensure coverage without any guarantee of profit. In practice, in areas characterized by low traffic density, investment in infrastructure (base stations, specifically) is not commensurate with the financial revenues expected. This can be even more complex if there is a lack of appropriate sites and/or transmission links (fixed network leased lines, fibre, or microwave). Consequently, network planning will be determined by various different objectives depending on the type of area to be planned (urban versus rural environments)

- In urban or suburban areas, the objective consists in generating sufficient traffic capacity (in other words, the task consists of meeting the demands of a high number of subscribers).
- In rural areas or ones characterized by a low density of subscribers, the objective consists of ensuring a highest possible coverage rate (cells of several-mile radius, typically) without having to provide a high capacity.

7.1.4 Main Steps

Put simply, the process takes place on two main stages, as follows. First, the engineering stage, per se, that includes the dimensioning of the various network elements, the definition of the cell structure (e.g., the reuse pattern, cell sizes), and radio planning, which consists of selecting the theoretical radio sites, generally by means of a software program. At the end of this stage, the base station sites and number of frequencies per site are determined. Then the task consists of allocating to each site the required number of frequencies so that, on the one hand, the interference level between cells is reduced to the minimum; and on the other, the power level at any point of expected coverage is sufficiently high to provide a satisfactory service quality. This stage is called *frequency allocation*.

The planning process for a cellular network can be illustrated by Figure 7.2, which provides an overview of the various topics that will be dealt with in this chapter. In the first part, both the tools and data used for the dimensioning of the radio sites are introduced. The frequency allocation and radio planning process is dealt with in Section 7.3.1, and the aspects related to the planning of the fixed network are contained in Section 7.3.2.

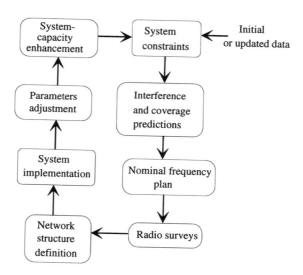

Figure 7.2 Main stages of cellular planning process.

Finally, it is worth mentioning that planning of a network, especially during the growing phase, is a continuous process. In the end of the 1990s, the GSM network operators in Europe were installing new base stations every week and redesigning their networks every month during high-growth phases.

7.2 TRAFFIC DIMENSIONING BASICS

Evaluating the traffic load is a crucial task to be achieved by a cellular network operator. It typically occurs when:

- Planning the network before its implementation,
- Evaluating the impact of introducing a new service,
- Evaluating the impact of new tariffs,

- Studying future demand.

The evaluation of traffic load will take into account a set of parameters, the number of which are determined by the accuracy required from the estimation. This preliminary stage is detailed in the following section.

7.2.1 Traffic Load Prediction

Assessing user traffic load in a given area is not a trivial task, especially in the mobile case where users and therefore the traffic, are not fixed in space. Traffic characteristics of mobile radio systems differ markedly from fixed networks. The mobility of the users entails calls starting at one network access point (a base station), continuing via other access points and being completed at a different one after the mobile has experienced a series of handovers (see Chapter 8 for more details). It is virtually impossible to determine the exact traffic density of any given area as it is quite difficult to determine the range of users who roam into that area. All types of parameters and available means are used to estimate the traffic load that the network is expected to carry at any point of its coverage with as high an accuracy as possible.

7.2.1.1 Parameters Used

Traffic density evaluation process requires the determination of the following parameters.

The most important parameters for evaluating the network load can be separated into two types of traffic:

1. *Call processing traffic*:

    - User call rates (incoming and outgoing calls, per region and for each hour of the day),
    - Average call duration (in the case of a connectionless service, the duration is relative to the bulk of information to be transferred),
    - Resource occupation rate (which is the percentage of time necessary for the information to be transferred),
    - Supplementary services,
    - Voice mail service,
    - Data traffic,
    - Penetration level (density of the population that is likely to subscribe to the service).

2. *Mobility-related signaling*:
    - Location updates,

- Handover,
- International roaming.

Some elements might be ignored for dimensioning, such as

- Signaling and maintenance traffic,
- Intracell handover,
- Data and software program loading,
- Initialization procedures for supplementary service parameters.

In the case where a digital network offers a wide range of different services (e.g., voice, data, short messages), some extra parameters can be added, such as:

- The service symmetry (a weighting factor for the service bit rate according to the bit rate required in each direction, uplink or downlink),
- The coding rate related to the user information bit rate, which gives an overall number of bits to be transferred over the radio interface,
- The throughput (the required bit rate excluding the coding used by the network during information transfer).

The duration of a call varies in accordance with the following parameters:

- Bit rate allocated for the call,
- Reason for making the call,
- Location of the calling and called parties,
- Call charging rate,
- Time of day.

The impact of these factors on the average call duration may be low. Even for services provided by current systems, variations in the average duration of a call with regard to the time of day and environment are generally unknown.

The average call rate per subscriber depends on the following parameters:

- Time of day,
- Location of the calling and called parties,
- Cost of the call,
- Service availability,
- Subscriber penetration level.

The incoming call rate for a particular area can be obtained by multiplying the incoming call rate (per subscriber), the penetration level, and the population density.

Traffic distribution depends on population distribution. The *population density* varies in time and space. It is composed of resident subscribers, visiting subscribers (e.g., for business or entertainment reasons), and an absent population (some of the resident population away from the area). It depends on the following factors:

- Time of day,
- Day of week,
- Season of the year.

For example, in some high-tourist areas, the outgoing-call rate on networks (generated by visiting subscribers) rises during the summer months by several tens of percent compared with the winter [4]. The *penetration rate* is defined as the population percentage subscribing to the service. It depends on

- Service availability,
- Charging rates for the service,
- Value to the subscriber,
- Marketing (for example, a free mailbox service offered with the voice service encourages the subscribers to use it),
- The level of competition between operators.

To estimate the values of these parameters, operators consider the following data (this list is not exhaustive; see Figures 7.3 and 7.4):

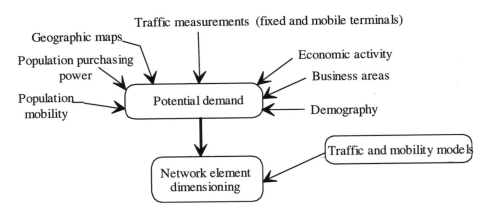

Figure 7.3 Examples of data used for system dimensioning.

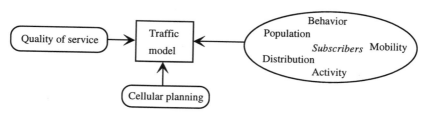

Figure 7.4 Parameters and processes used to define the traffic model.

- Demography and purchasing power of the inhabitants to assess the likely level of service penetration,
- Increases per day, week, or year of economic activity,
- Location of the business areas,
- Road traffic (time and place of traffic jams),
- Distribution of different kinds of terminals (e.g., handheld, fixed),
- Maps and databases to identify main roads, population density, and business areas,
- Volume of traffic measured on the existing (i.e., analog) cellular networks,
- Population mobility (e.g., daily commuting).

Generally, traffic distribution varies in time and space. It can be modeled by a Gaussian distribution. Some authors [5] report that spacial statistics of cellular traffic can be described by a log-normal distribution. Significant concentrations are found in cities during peak hours and particularly at the end of the day, when there are changes of pattern as commuters return home.

Example 1: Traffic growth estimation for a given area
Consider a development area with 1,500 inhabitants. The growth rate is estimated to be 1,000 new inhabitants per year, over 5 years. The average purchasing power leads to a prediction of a penetration rate for cellular services of 5%. Professional, commercial, and leisure activities draw in up to 5,000 people (peak time). Within this visiting population, the penetration rate for cellular service is estimated at 8%. The numbers of these people are predicted to grow by 20% per annum over a 5-year period.

The area is crossed by a main road with a traffic volume of 500 vehicles at any one time, 10% of which are equipped with cellular telephones. This maximum rate is predicted to remain unchanged over the 5-year period. According to prediction, a general growth of 2% in service penetration rate is expected to be achieved per year for 5 years for all categories in the user population.

The traffic rates in the various population ranges are as follows (see the definition of Erlang in Section 7.2.1.):

- Local inhabitants: 0.02 Erlang,
- Commuters: 0.1 Erlang,
- Vehicles: 0.2 Erlang.

Note: The general values used by the European network planners are 30 mErl for mobile users (80 mErl for fixed subscribers) [6].

The traffic to be generated by the system in the area to be considered over a 5-year period is as follows:

Traffic (*year 1*) = $1,500 \times 0.02 \times 0.09 + 5,000 \times 0.08 \times 0.1 + 500 \times 0.10 \times 0.2$ = 52.7 Erl

Traffic (*year 2*) = $2,500 \times 0.04 \times 0.09 + 5,000 \times 1.2 \times 0.10 \times 0.1 + 500 \times 0.12 \times 0.2$ = 81 Erl

Traffic (*year 3*) = $3,500 \times 0.06 \times 0.09 + 6,000 \times 1.2 \times 0.12 \times 0.1 + 500 \times 0.14 \times 0.2$ = 119.3 Erl

Traffic (*year 4*) = $4,500 \times 0.08 \times 0.09 + 7,200 \times 1.2 \times 0.14 \times 0.1 + 500 \times 0.16 \times 0.2$ = 169.36 Erl

Traffic (*year 5*) = $5,500 \times 0.10 \times 0.09 + 8,640 \times 1.2 \times 0.16 \times 0.1 + 500 \times 0.18 \times 0.2$ = 233.388 Erl

For a traffic demand estimation it is useful to have in advance typical user profiles. In other words, a categorization of users into various groups according to their network usage. With the help of the estimated values for the listed parameters, mobility and traffic models can be determined. These help identify traffic for each network area and determine hot spots.

7.2.1.2 Mobility Characteristics

A mobility model describes the probability transfer or flow of users between cells c_i and c_j (once these have been defined) along with transit times per cell (i.e., average time spent by users in each cell). This kind of model is mainly used for the grouping of cells (in other words determining the BSCs and their location), as well as for dimensioning fixed network equipment (typically the switching centers). User movement in a given area is needed to help determine the following two mobility parameters: location updating (LU) and handovers. A knowledge of the visitor rate into various network areas by subscribers from other networks (e.g., business people and tourists) helps determine the internetwork roaming capacity as well as the location management overhead.

Example 2: Determination of mobility model parameters

Consider the theoretical case illustrated by Figure 7.5. A road crosses a given area, and is covered by cells c_i (i = 1...6).

The traffic flow on this road during peak time is 5,000 vehicles per hour west to east, and 3,500 vehicles per hour east to west. This area lies near the borders of network coverage (the neighboring area, covered by another network, lies to the east) and the number of visiting subscriber vehicles is estimated at 15%.

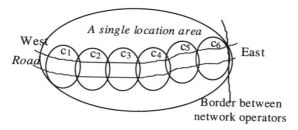

Figure 7.5 Example of an area for the determination of a mobility model.

Table 7.1

Traffic distribution

Traffic direction	West–East	East–West
Home subscribers/hour	$5{,}000 \times 0.85 \times 0.06 = 255$	$3{,}500 \times 0.85 \times 0.06 \cong 178$
Visiting subscribers/hour	$5{,}000 \times 0.15 \times 0.10 = 75$	$3{,}500 \times 0.15 \times 0.10 \cong 52$

If we suppose that a single location area has been designated (this task was achieved during the fixed network planning stage, see Section 7.3.2) and that the mobile service penetration rate is 6% for the home network subscribers and 10% for visitors, the number of subscribers is as indicated in Table 7.1.

Location updating rate:

- For home subscribers, the LU rate is 255/hr for intranetwork locations (e.g., intra-VLR LUs in GSM) as mobiles enter the location area from the west and 178/hr (e.g., inter-VLR LUs using the IMSI in GSM system) as they come in from the east,
- For visiting subscribers the equivalent LU figures are 75/hr, and 52/hr.

If we suppose that a call is continuously established across the whole area, and that the call rate is 50 mErl (see the Erlang definition in Section 7.2.1.2) per subscriber, the number of handovers between the cells c_1 and c_2, c_2 and c_3, c_3 and c_4, c_4 and c_5, c_5 and c_6, is:

$(255 + 75 + 178 + 52) \times 50 \times 10^{-3} = 28$ *handovers/hour/cell.*

The use of a mobility model for cell dimensioning is not an easy task and relies on certain assumptions. It is assumed that sites have already been identified (location and coverage). The model can be used to adjust or validate the planning of cells (Figure 7.6). Several iterations are usually required between cell dimensioning using the traffic model and its validation with the mobility model in order to come up with an optimum result. By contrast, the mobility model is used directly to dimension the fixed network, as can be seen in Section 7.3.2.

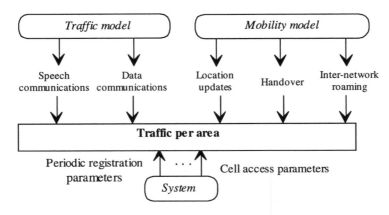

Figure 7.6 Data and procedures considered for traffic-load estimation.

7.2.1.3 Traffic Unit: The Erlang

Telecommunications network dimensioning requires the use of traffic theory. Erlang's work and queueing theory are of paramount importance because they help express in a relatively simple way parameters such as *blocking rates* or *call waiting time*. These tools are all the more useful as they can be applied, within the framework of a slightly restricted hypothesis, to relatively complex services as is the case with a mobile communications system.

Although mobile system traffic characteristics are different from those of fixed systems, cellular system dimensioning is generally based on the same traffic theory used for fixed networks. In fact the *Erlang B formula* is widely used to determine the number of channels required per cell according to traffic in that cell together with an accepted blocking probability. In this case, mobility is not considered and it is generally supposed that no handovers or location updates are performed. To perform accurate evaluations, taking into account the user's mobility, the Erlang formulas can no longer be directly applied (see also Section 8.1.6 in Chapter 8).

Erlang definition: The system load is defined as the number of information units (messages or bits) to be transmitted per unit time. It depends on two parameters: the message or call *average transmission time*, and the message or call *average incoming rate*. The traffic load is obtained by multiplying these two parameters. The common unit in traffic theory is the *Erlang* (Erl), which represents the channel occupation rate. For example, if the channel is occupied for a half-hour over a one hour period, the traffic load during that period is equal to 0.5 Erl.

To calculate any of the following three factors Erlang tables are used (refer to the example in Appendix 7.C):

- Number of traffic channels,
- Traffic load (in Erlang),
- Call blocking rate.

Traffic per subscriber is computed by means of the following formula:

$A = \dfrac{n.T}{3,600}$ Erlang, where n is the number of calls per hour, T the average duration of a call and A is the offered traffic from one or several users. For example, if $n=1$ and $T=90$ sec, the generated traffic is equal to 25 mE.

7.2.1.4 Traffic Modeling by Identifying Several User Groups

The system load varies both in time and space. Traffic measurements in several mobile radio systems have shown that the load can vary within 5 minutes. Moreover, daily traffic peaks hardly reoccur at the same time and with the same amplitude.

In the following model, the user population is subdivided into groups where each group member displays similar traffic and mobility characteristics. The type of traffic generated by a given group of users is generally totally decorrelated from the traffic produced by another group of users. To assess system load, the primary task is to characterize the various groups of users. As a first stage, and from the mobility point of view, users can be categorized into three approximate groups: *significant mobility* (such as business travelers), *medium mobility* (about four trips per day), and *low mobility* (about one or two trips per day). The same kind of categorization can be adopted for traffic.

Consider N as the number of the identified groups of users, P_i the average traffic load of group i, and σ_i the standard deviation of the traffic load corresponding to the same group.

If $\rho_{s,i}$ is noted as the average traffic load of a user belonging to group i, $\sigma_{s,i}$ its standard deviation, and $N_{s,i}$ the number of users in this group, the average traffic load for this group i could be approximated by the following formula $\rho_i = \rho_{s,i}.N_{s,i}$ and the standard deviation by $\sigma_i = \sigma_{s,i}.\sqrt{N_{s,i}}$. In practice, observations of $N_{s,i}$, $\sigma_{s,i}$, and $\rho_{s,i}$ parameters have shown that these generally follow a log-normal distribution.

Finally, to evaluate the optimum traffic load of the system, a supplementary factor, denoted by k, is introduced, which varies according to the system and the average duration of traffic measurements ($k \approx 1.5$ typically). The optimum system load can now be assessed simply using the following formula:

$$\hat{\rho}_{sys,k} = \sum_{i=1}^{N} \rho_i + k \sqrt{\sum_{i=1}^{N} \sigma_i^2} \qquad (7.1)$$

Traffic load evaluation helps to assess the traffic load in each cell and timeslot. The percentage levels of identified groups of users (with their particular mobility and traffic characteristics) are specified for each area. These percentages can vary from one area to another. The prediction model can thus be refined and accurate statistics can, theoretically, help evaluate the traffic and mobility rate in each network cell.

7.2.2 Quality of Service Parameters

During the dimensioning phase, two parameters must be taken into account for indication of quality of service (see Appendix 7A for other quality of service parameters). First, the channel waiting time and blocking probability, and second, the density of radio coverage.

7.2.2.1 Blocking Probability

Assessing the traffic load at peak hours helps dimension system elements (mainly the radio resources, fixed equipment, and links between these pieces of equipment). These elements require substantial investment. As these resources are expensive and have to be kept to a minimum, dimensioning has to assume that the demand for a low proportion of users cannot be met when the system is very heavily loaded (in the busy hour). Those attempts that are not satisfied because of a lack of resources are called *blocked calls*, and the probability of a call being blocked is the *blocking probability*. This probability is a key element and has a direct impact on the quality of service as perceived by subscribers.

As the value of the accepted blocking probability increases, the investment cost and service quality decreases. This parameter determines the amount of resource available to the users all the time, but the main objective is to meet the demands of the highest number of users at peak hours. At other times, a large quantity of the system resources remain under utilized. This is what justifies the idea of allowing the probability of having blocked calls. On the other hand, when a subscriber is paying, it is reasonable to expect a service which is always available so the operator must install sufficient equipment to make sure that there is only a low probability that a call will be blocked. The blocking probability is determined with the help of the Erlang B formula, which relies on queuing theory. The hypotheses on which this model is based are as follows:

- Random call arrivals: call arrivals follow a Poisson process and, therefore, the inter-arrival period of calls follows an exponential distribution,
- Call durations follow an exponential distribution,
- Infinite traffic and homogenous resources,
- The system statistically at equilibrium,
- There is a known system load.

Let c equal the number of services (i.e., the number of channels and lines allocated to users) and P is the system load, then the blocking probability is given by the following formula (Erlang B formula):

Blocking probability = loss probability

$$P_B = \frac{\rho^c}{c!} \left[\sum_{i=0}^{c} \frac{\rho^i}{i!} \right]^{-1} \tag{7.2}$$

The behavior of this function is represented by Figure 7.7, where the blocking probability varies in accordance with system load and the number of available channels.

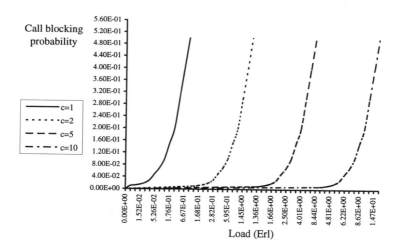

Figure 7.7 Blocking probability as a function of load against the number of channels (c).

The accepted blocking probability varies in accordance with the environment. In a dense urban area where traffic loads are high, the blocking rate can reach 2 to 5%. In suburban and rural areas where traffic conditions are less constrained by available infrastructure blocking rates can be less than 2%.

Example 3: Determination of the number of cells needed to cover an area

Consider an area covering a population of 10,000 subscribers, each of whom generates 25 mE of traffic. There are 24 frequencies available (e.g., with eight channels per carrier, in the case of GSM). These frequencies are allocated to cells in accordance with a 12 cell reuse pattern (12 frequency groups). An acceptable blocking rate is evaluated as 2%.

The number of trisectored sites is calculated as follows. First, the number of frequencies per cell is 24/12 = 2. The number of traffic channels per cell equals 2 × 8 − 2 (control channels), in other words, 14 traffic channels. Thus, the maximum traffic that a cell is able to serve is 14 simultaneous calls. With a blocking rate of 2%, the number of Erlangs will be 8.2. Each cell could, therefore, serve 8.2/0.025 Erl; in other words, 328 subscribers. Therefore the number of cells for the total area will be 10,000/328, or about 30 cells. With each site able to carry 3 cells, the final number of trisectored sites will be 30/3 = 10.

Note: In this example, differences between outdoor and indoor areas have not been considered. The number of radio sites for indoor coverage is significantly higher than that required for outdoors. In practice, and during the preliminary planning phase of a network, only outdoor coverage is usually taken into account. Urban area coverage is usually planned for outdoor handheld portables (2W), and in rural areas vehicle mounted units of 8W. It is estimated [6] that indoor 2W coverage is only cost-effective when subscriber density is greater than 300,000 units per square mile.

7.2.2.2 Channel Waiting Time

To help subscribers avoid resending their call requests (and loading the control channel unnecessarily) if these remain unserved, queueing lists can be managed at the system level. In order to provide a satisfactory service quality for the subscribers, the operator has to ensure that the caller does not have to wait too long before the request is served. This waiting time can be calculated using the *Erlang C formula*. The probability of a call being delayed is given by the following formula:

$$\text{Probability for a call being delayed} = P_D = \frac{\rho^c}{c!}\left[\frac{\rho^c}{c!}+\left(1-\frac{\rho}{c}\right)\sum_{i=0}^{c-1}\frac{\rho^i}{i!}\right]^{-1} \qquad (7.3)$$

Waiting list management strategy (e.g., first come, first served priority) yields a particular delay-distribution. Nevertheless, the average time during which a caller is expected to wait before seeing a call answered does not depend on the way the waiting list is handled. If h is the mean duration of a call and ρ the overall load being offered, then the mean delay $<T>$ is provided by the following formula:

$$<T> = P_D \frac{h}{c-\rho} \tag{7.4}$$

When the first come first served strategy is applied, the delay function distribution T_D follows an exponential law as follows:

$$P(T_D > t) = e^{-(c-\rho)\frac{t}{h}} \tag{7.5}$$

7.2.2.3 Radio Coverage Density

The radio coverage density is considered as an indicator of the quality of service and is particularly important. The radio coverage density indicates the probability of establishing a good-quality communication anywhere in the service area (i.e., a communication link with a C/I level higher than a given threshhold). Typical values of this indicator range between 90 and 95%.

Note: The extension of radio coverage density is generally implemented by the addition of base stations after a network replanning strategy has been undertaken. Coverage holes (e.g., in tunnels) can remain in some cells in spite of network upgrading with additional base stations. These holes can be filled by means of *repeaters*. A repeater is entrusted with the task of boosting the signal received from the base station (downlink) and retransmitting it into the hole and also retransmitting the mobile signals (uplink) located in the hole back to the base station. It is worth noting that the frequencies used by the repeater are identical to those allocated to the base station serving the cell and has the risk of introducing frequency planning problems. This is usually overcome by fitting the repeater with directional antennas.

7.2.3 Cell Dimensioning

Dimensionally, a cell is characterized by two major parameters:. its size (or radius) and its capacity (or the number of channels it provides). Dependent on traffic conditions, the former or the latter is dimensioned first.

As shown in Section 7.1.2, the planning objectives for high density (Figure 7.8) or low density areas (Figure 7.9) are not the same. In an urban area, the first factor to be taken into account is traffic capacity. The process is based on using an estimated traffic density to determine the number of cells to be installed.

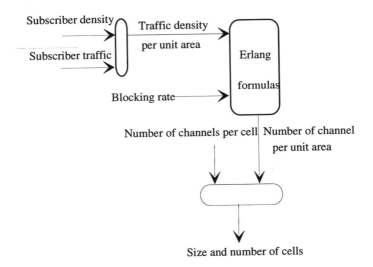

Figure 7.8 Cell dimensioning in a high-traffic load environment.

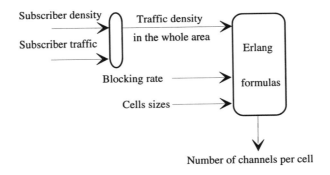

Figure 7.9 Cell dimensioning in a low-traffic environment.

Example 4: Cell dimensioning in a high-traffic environment

Assuming the development area as considered in Example 1, the size of the area is 2 km by 2 km, and the maximum traffic density to be handled during the first three years is 120 Erl.

If each cell is able to carry 20 Erl, then 120/20 = 6 cells are needed, each with a radius of $R = \sqrt{\dfrac{2 \times 2}{\pi.6}} \cong 0.460$ km.

For a low-density traffic area, the primary task consists of ensuring that coverage is as extended as possible using a minimum number of base stations. What is required here is to start from the estimated traffic density, the size and number of cells, and then determine the number of channels per cell.

Example 5: Cell dimensioning in a low-traffic environment

Suppose a rural area of 200 km by 200 km, where a maximum traffic density of 60 Erl is expected. If the maximum size of a cell is 30 km in radius, coverage can be achieved with about 15 cells. If the assumed traffic is uniformly spread over the area to be covered, each cell has to be provided with sufficient channels to carry a 4-Erl traffic load.

7.2.4 Dimensioning Process for a GSM Network

GSM system dimensioning requires sizing both the base station subsystem (BSS) and the network subsystem (NSS). In Figure 7.10, the various dimensioned elements are reproduced along with the sequencing of the various phases. The following example is based around dimensioning of the GSM SDCCH signaling channel (see SDCCH definition in Chapter 12).

Example 6: GSM SDCCH channel.
Traffic parameters are as follows:

- Mean duration of a call (T) = 120 sec
- Mean traffic per user (A) = 0.033 Erl

The mean number of calls per user at peak time is therefore equal to:

$A \times 3{,}600/T = 0.033 \times 3{,}600/120 \cong 1$

The SDCCH is a signaling channel used for *Location Updating, Periodic Location Updating, IMSI Attach/Detach procedures, Call Connection, Short Message Service* (SMS), *Fax Transmission*, and *Supplementary Services*.

The signaling traffic carried on the SDCCH depends on the behavior of the user and the network parameter values defined by the operator (e.g., the value of the periodic location updating parameter). Busy hour traffic generated by a subscriber on SDCCH is 6 mErl on average. This value can change significantly from one network to another. The SDCCH/8 is able to serve $1/0.006 \cong 167$ subscribers. To serve these subscriber traffic channels (TCH), with a capacity of $167 \times 0.033 = 5.5$ Erl are required. This equates to having 11 TCHs if the system is designed to a blocking rate of 2% (see Erlang Table). This is equivalent to having one and a half RF carriers. The first supplies the SDCCH and seven TCHs and the second four TCHs.

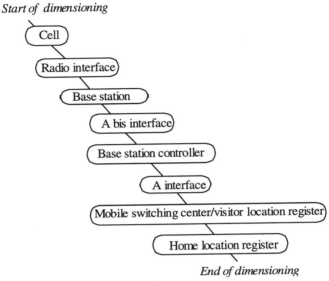

Start of dimensioning

Cell

Radio interface

Base station

A bis interface

Base station controller

A interface

Mobile switching center/visitor location register

Home location register

End of dimensioning

Figure 7.10 Network entities dimensioning process: GSM case.

7.2.5 Conclusion

Dimensioning the various elements of a cellular network (e.g., radio channels, base station controllers, switching centers, databases, transmission links) requires the definition of two main quality of service indicators that are the *blocking rate* and the *waiting time* before obtaining a channel.

A compromise has to be found between either a high value for these parameters (which results in poor service quality but minimal infrastructure costs) or very low values (which give an excellent quality of service but with high infrastructure costs). This dilemma is further complicated by the fact that the equipment is dimensioned for an estimated traffic load at peak times. In other words, for a maximum duration of two hours per day, representing only one-twelfth of the time, or less, if nonworking days are considered. Figure 7.11, shows an example of the network dimensioning process.

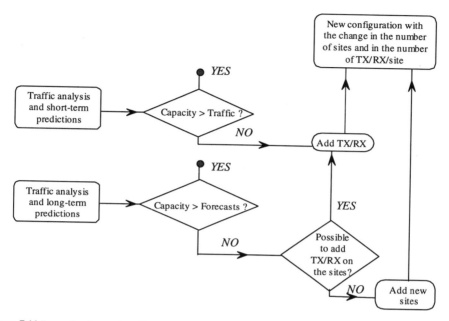

Figure 7.11 Example of capacity adjustment in number of sites and in number of TX/RX per site.

7.3 PLANNING STAGES OF A CELLULAR NETWORK

7.3.1 Radio Planning

One key to success in optimizing network performance is data management. The potential for data overload exists in an operational network [7]. Design plans, configuration data (typically up to 250 parameters per cell), switching center settings and routing analysis, performance loggers for every signaling transaction at each network equipment, customer billing, and call records—all of these add to the data burden. In addition to the technical performance data, operators want to understand how the network performs from the point of view of the customer. To achieve this, operators organize tests by driving thousands of miles to collect large amounts of data for interpretation and benchmarking of performance and optimization. Operators often use vehicle based collection devices for these measurements. These include fast scanning receivers, spectrum analyzers, and test mobile phones fitted with special software. In the following sections the different stages of radio sub-system planning are presented. One of the preliminary steps is to identify potential base station sites.

7.3.1.1 Searching for Practical Sites

In practice, the theoretical location of a base station can sometimes correspond to an inappropriate location (e.g., lack of premises, poor RF performance due to trees or buildings). What is needed is that, having started from a given theoretical configuration, practical sites must identify morphology and propagation characteristics of an area. The main tools and data used by engineers entrusted with network planning are discussed below.

Field data. There are two categories of data required in site finding:

- General data (e.g., morphology, structure, composition), which can be obtained from cartography suppliers. This data is quite often provided in the form of digital maps (see Section 7.3.1.1.3.).
- Specific data leading to the identification of target sites.

The preliminary stage is a *survey.* It mainly consists of:

- Detecting visible sites,
- Choosing base station towers and available space at the correct height,
- Determining feeder lengths (supply cable located between the antenna and the transmitter/receiver),
- Checking off the existing infrastructure (e.g., premises, power supply, access),
- Noting down the accurate address of the site,
- Undertaking propagation tests using back-to-back handheld units.

At this stage, the highest points are surveyed to find the most appropriate sites, including buildings belonging to the operator, grain silos, water towers, and any other tall structures.

All the information gathered during the measurement and survey campaigns are processed by a software program, which uses digital maps. This tool calculates interference levels, draws up cellular coverage, and determines the ideal location for the antenna and radio transmission equipment. Some targeted sites are selected, and for each one some extra pieces of information are gathered (e.g., video reports and pictures, administrative data, engineering task costs, work duration). At this stage these theoretical cellular sites are based on setting up base stations with up to 5/4 radius relative to the ideal. The tolerance range of site locations has much more impact on the transmission quality than on cost or capacity.

Radio measurements. Cellular planning should be accompanied by measurement programs which have as their objective the optimization or upgrading of the propagation models used in the planning software (to determine pathloss attenuation). These models help determine the radio coverage from various sites and establish link budgets (determination of fade and obstruction margins in accordance with the distribution of mean local fading).

A measurement campaign is characterized by the following parameters:

- Types of measurement, including signal power, impulse response, and signal quality,
- Type of survey, including outdoor, indoor coverage and building penetration,
- Type of environment, such as rural, mountainous, urban, and indoor,
- Radio parameters, including frequencies, antenna characteristics, and maximum communications distances.

Measurement surveys are also used to determine base station installation sites (in particular, determine local fading). They are generally carried out on roads lying in areas characterized by a uniform morphology [8] and consist of estimating a mean signal power averaged to eliminate fast fading while maintaining the characteristics of slow fading. The collection of radio data is carried out by taking at least one sample at an interval of $\lambda/4$ (λ is the wavelength) and averaging these samples over at least 40 wavelengths [9]. The mean rates are averaged for a local area to eliminate fast fading. Finally, measurements are filtered to suppress noise.

Propagation prediction tools. Propagation prediction tools are mainly used to:

- Determine sites for the location of base stations,
- Allocate frequencies between the various sites,
- Determine the technical characteristics of the base stations (e.g., antennas).

These software programs can be used to plan different kinds of system, such as cellular, radio broadcasting, paging, microwave links, and radio in the local loop.

A cellular planning software program integrates various modules among the following (some are optional):

- A graphic user interface,
- Radio propagation prediction models (e.g., Okumura-Hata, Walfish-Ikegami, Bullington, Lee, FCC Carey),
- Frequency allocation algorithms (see Section 7.3.1.2),
- Network dimensioning algorithms (e.g., numbers of BTSs and BSCs),
- Software modules for radio spectrum management,
- Data interfaces with the network to recover data collected by the operations and management center for the radio part (OMC-R); this data is used for the

optimization and reconfiguration of the network (adjustment of working parameters),
- Interfaces with field measurement software tools (coupled with position data, usually GPS, to determine location),
- Integration of advanced functions (e.g., frequency hopping, diversity),
- Mapping software,
- Handover simulation program.

Input parameters for these tools are generally as follows (see also Figure 7.12):

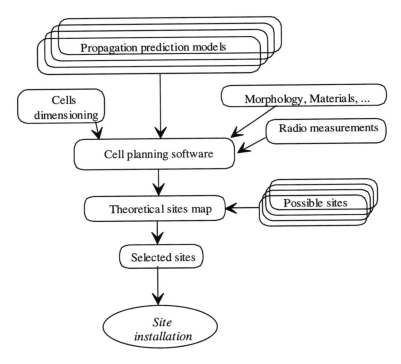

Figure 7.12 Example of data used for cellular planning.

- Limits on certain parameters such as receiver sensitivity (the level below which a receiver cannot demodulate the received signal) and protection ratios (e.g., C/I ratio which is the user signal power relative to a co-channel interfering signal power, and adjacent channel interference, which is the ratio between the users signal power and unwanted adjacent channel signal power).

- Radio measurements (see Section 7.3.1.1.2).
- Digital maps containing data about the area to be covered by the network. They deal with the morphology, nature of the terrain (contours), above ground composition (e.g., kind of vegetation), position of the road network, and streets. In microcellular network planning, a map describing the location and height of buildings is also integrated with the tool. Whenever available (by measurement or simulation), a map of traffic density can be used to assist in dimensioning the cells.

Two types of geographical databases exist, namely, raster and vector data. In the first case, the digital terrain map represents the field data in the form of a bit map. The map resolution depends on the size of the mesh (pixel) and this becomes a factor in the prediction of radio coverage. The typical resolution for an urban area ranges up to 50m whereas for rural areas 100 to 500m resolution can be sufficient. The nature of the ground can also be found in these databases. This can vary considerably and may include several areas such as sea, open or quasi-open space, forests, towns, suburbs, or mountains. In general, between 5 and 10 parameters per pixel can be provided. The main drawback of these databases is that the information is not very precise. For instance, exact details of buildings are not given.

In a vector database, the data is much more accurate in terms of the nature of the ground and especially the heights and nature of obstructions. The main drawbacks of these databases are first their cost, which is very high (such that they are used only for high-density urban areas), and second, their sensitivity to changes in the environment (e.g., new buildings).

Simulation test beds. Theoretical methods obtain an initial configuration for the network by only taking into account some of the aspects (e.g., radio channel allocations, cell sizes). Simulation test beds help complete this work by integrating procedures and mechanisms that are rather difficult to model. They can be relatively simple or very complex. They are most helpful in studying the performance capacity of systems and more particularly, the interactions between various aspects of the system such as power control and handover algorithms.

Microcellular systems case. Planning microcellular systems is very complex. Taking into account the very specific propagation conditions for this kind of environment (importance of the topology, location, and height of buildings), it becomes obvious that the planning methods used for macrocellular systems cannot be used directly. In this case, high-definition maps with structural details of the area are necessary. Antennas are often located below roof level which require different propagation models to be used (e.g., more obstacles).

7.3.1.2 Base station positioning problem

The positioning of the base stations can be modeled as follows:

Given the following set of criteria:

- Traffic demand,
- Potential geographical site position,
- Required coverage,
- Economic evaluation,
- Existing network infrastructure,

and using the input data:

- Installation costs,
- Radio wave propagation laws,
- Traffic demand,
- Network description,

to meet the following objectives:

- Minimize the interference level,
- Maximize the traffic,
- Minimize the investment,
- Minimize the network evolution.

7.3.1.3 Frequency Allocation

At the end of the first planning stage (estimation of traffic demand and network dimensioning), the number of cells, their capacity, and their radii are identified. The propagation prediction tools, along with all the data previously collected, help determine the sites where the base stations are to be installed. The following stage, called *frequency allocation*, consists of allocating frequency carriers to the base stations.

Informal methods that try to identify such assignments have been used since the beginning of the twentieth century, when maritime applications of Marconi's wireless telegraph first appeared [10]. At that time, a simple way to minimize interference was to assign different transmitters to different noninterfering frequencies or to come as close to this as possible. This approach led to heavy spectrum use but remained viable so long as the growth of the usable spectrum kept pace with the growth for its demand.

During the 1950 to 1990 period, the growth of the usable spectrum was relatively slow, while at the same time demand increased dramatically. This has led spectrum managers to design a different approach to frequency assignment. In this approach the amount of spectrum required by an assignment is the function to be minimized, and instead of trying to eliminate unwanted interference completely, acceptable interference levels are included among the criteria that must be satisfied by the frequency assignment (*interference limited assignment*).

In hexagonal regular structure network models, each frequency is reused according to a regular and fixed pattern (see Chapter 6). In practice, it is very difficult to apply theoretical reuse patterns. There are various reasons that cause problems in practical cases [11]:

- In a real system, the actual base station sites are rarely organized in regular patterns,
- Propagation conditions are generally quite irregular and therefore interference levels do not depend simply on the ratios between the reuse distances and the cell radii (i.e., the D/R ratios),
- Spatial variations of traffic density can lead to capacity demands that differ from one cell to another,
- Quite often, environmental constraints impose limits on the use of certain frequencies (e.g., the case of networks along a country border).

In such cases, a more general approach than that of hexagonal cells with regular patterns is required. The analogy between the graph-coloring problem and that of frequency allocation helps identify methods used to address this problem.

Note: The map coloring problem and its analogy with frequency allocation
A map-coloring algorithm should color the nodes in such a way that all nodes connected by an edge have distinct colors and that the overall number of colors used is limited. The analogy with the problem of frequency allocation is immediately obvious. Base stations constitute the graph nodes and the existence of an edge between two nodes indicates that the two base stations cannot use the same frequency. Here is a classical way to solve this problem.

The information describing the relationship between cells can also be represented by a specific matrix called an *interference matrix*. This is the result from a planning tool which has been input with digital maps, propagation models, cellular sites coordinates, and so forth. This tool determines the C/I ratio at each point of coverage (the frequencies considered during this stage are all the same, the sites being considered as co-channel sites) and for a fixed coverage rate (e.g., 95%). A histogram of C_i/I_{ij} values (where C_i is the signal level of the cell i and I_{ij} represents the interference level generated by cell j within the i cell) is plotted for each pair of cells i, j. Each histogram acquires the shape illustrated by Figure 7.13.

The interference matrix determines the *reuse matrix*, the shape of which is reproduced in Table 7.2.

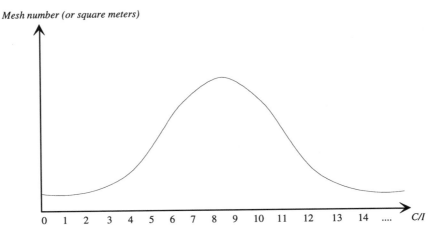

Figure 7.13 C/I_{ij} values histogram.

Table 7.2
Reuse matrix

Cells	c_1	...	c_i	...	c_n
c_1	α_{11}				
...					
c_j			α_{ij}		
...					
c_n					α_{nn}

Where α_{ij} represents the necessary interval in number of carriers between the carriers used in cells i and j, taking into account the interference limits imposed by the system (e.g., $C/I \geq 17$ dB for an analog cellular system) and by the propagation characteristics.

For example, taking a GSM system:

- $\alpha_{ij} = 0$ (the i and j cells could use the same carriers) if at least 95% of the histogram demonstrated that $C/I_{ij} > 5$ dB,

- $\alpha_{ij} = 1$ (the carriers being used in the i and j cells must be separated by at least one carrier) if at least 5% of the histogram demonstrated that $C/I_{ij} \leq 5$ dB,
- $\alpha_{ij} = 2$ (the carriers being used in the i and j cells must be separated of at least two carriers) if at least 5% of the histogram demonstrated that $C/I_{ij} \leq 5$ dB $-$ 18 dB (18 dB is the adjacent channel interference protection ratio).

Taking the frequency allocation problem model as it is presented in [11]:
Let C, a cellular system composed of n cells be marked c_1, ..., c_n, assuming that the following data is known:

- For each cell c_i, the demand in radio channels equals m_i,
- For each pair of cells (c_i, c_j), the necessary frequency difference between channels is determined by α_{ij} (stored in the reuse matrix),
- For each cell c_i, we define a set A_i of preallocated frequencies and a set of forbidden frequencies B_i.

A frequency plan is thus defined as the set $F = (F_i)$, where each F_i represents a set of positive integers (i.e., of frequencies). This set exists if the following conditions are met:

- F_i contains m_i elements,
- If $f \in F_i$ and $f' \in F_j$, then $|f - f'| > \alpha_{ij}$,
- F_i contains A_i,
- F_i does not contain any frequencies belonging to B_i.

The demand in the number of frequencies for a frequency plan F corresponds to the highest integer appearing in the set F_i. In other words, it corresponds to the number of the highest channel frequency (assuming that the channels are numbered from 1).
The problem of frequency allocation, therefore, consists of finding a frequency plan that can be achieved with the minimum of frequencies. A solution to this problem does exist as long as the data is coherent, that is:

- A_i does not contain m frequencies,
- A_i and B_i have no links.

The map-coloring problem is an NP-complete (or NP-hard) problem. Most importantly, this means that an algorithm able to solve this problem accurately for random data input would not complete after a polynominal time-delay (i.e., this time delay increases in, at least, an exponential manner). Consequently, and as is the case for medium sized networks (e.g., beyond a certain number of cells), frequency planning can only be achieved through heuristic methods. It was obvious that with the GSM network such as that covering the *Ile de France* area, where in 1999 the number of cells forecast was greater than 2,000, the impact of frequency allocation constituted a task that is far from trivial.

For the moment cellular network planners use approximate or heuristic algorithms that do not automatically yield optimum solutions. There are two major kinds of heuristic methods. The deterministic method (leading to a single and unique frequency plan) and the nondeterministic method (that can lead to several and various frequency plans, which are determined by random choices being taken during the "search for solution" stage).

A brief survey of the principles of two problem resolution methods for frequency planning is given below. The first is the *Gamst method* [11], and the second is based on the *simulated annealing optimization method*.

The Gamst approach. Gamst's algorithm consists of two main phases: data analysis and iterative processing. The purpose of the first phase is to determine the value of the *largest lower bound* (*LLB*), which represents the number of frequencies required to meet the frequency demands of the cellular network (i.e., the number of radio carriers to be used). The second phase consists of generating a frequency plan that satisfies the interference and channel demand constraints. When the number of frequencies used in the frequency plan is equal to LLB, the algorithm stops.

For inputs the algorithm uses the following data:

- The geographical area,
- Traffic density,
- Frequency band,
- Propagation law,
- Minimum signal level,
- The C/I threshhold,
- Blocking probability, also called outage probability.

Gamst's algorithm principle is as follows:

To each cell i a number of channels n_i are assigned (physical channels: RF carriers or timeslots). The total number of channels is denoted by D ($= \Sigma n_i$). A permutation of D channels (from 1 to D) is denoted by *chan*. To each channel k ($1 \leq k \leq D$) a value *diff(k)* is associated with *diff(k)* > *diff(k+1)*. This function is used to minimize the total number of carriers used in the frequency plan. A channel number is allocated to each cell according to its traffic demand.

Note: Two different channel numbers or cells can correspond to the same frequency. Gamst's algorithm varies *diff* to minimize the number of used channels by recomputing *chan*. The algorithm is stopped when the number of used frequencies is lower or equal to LLB. As an output the algorithm provides the number and location of base stations, their RF output power and the frequency allocation plan. A large number of cellular planning tools are equipped with the Gamst method. It generally offers satisfactory results.

The simulated annealing approach. The simulated annealing (SA) approach is a general method used to find the approximate solution of NP complete combinatorial optimization problems. In its principle, it consists of the generalization of an iterative scheme (local search) using a simulation of the heat treatment annealing process. The general description of an SA algorithm is the following:

In the first step, the problem has to be expressed as a "cost" optimization function. A configuration space S, a cost function C, and a neighborhood structure N are defined. The second phase consists of defining a cooling schedule. The third phase performs the annealing process.

To avoid blocking in local optima, "uphill moves" are authorized by the use of the Metropolis criterion [12].

For the frequency assignment process, the algorithm is defined as follows:
Parameters:

The "cost" function is defined by the expression

$$C(s) = \frac{1}{2} A \sum_{\substack{(i,j),(i',j') \\ (i,j) \neq (i',j') \\ |i-i'| \leq c_{jj'}}} s_{ij}\, s_{i'j'} + \frac{1}{2} B \sum_j \left(\sum_i s_{ij} - traf_j \right)$$

where $traf_j$ is the demand expressed as number of channels in cell j.
Note: The second term of $C(s)$ becomes very small when using the flip-flop method [13], which defines the transition method from one state to another state by replacing a channel i_1 in cell j by a free channel i_2.

The parameter $p = \exp\left(\dfrac{C(s') - C(s)}{T}\right)$ is also defined, where T is the temperature. $T > 0$ and decreases towards 0 during the cooling phase.
Process:
In the initialization phase, an initial solution s_1 is selected in S, the best cost C^* is initialized with the corresponding solution:

$s^*.\ C^* := C(s_1)$ and $s^* := s_1$

Each configuration s corresponds to a (m, n) matrix where $s_{ij} = 0$ if channel i is not used in cell j and otherwise $= 1$.

The following steps are repeated until either the number of iterations becomes greater than a maximum or the improvement in the cost function becomes very small.

s_n denotes the current solution.

Determine s randomly in the neighborhood $V(s_n)$ of s_n.

If $\min[1, p] > 0.92$ then $s^* = s'$, else $s^* = s$; (Metropolis criterion).

One of the most difficult tasks in SA is parameter tuning. The following hypothesis can be assumed:

The initial temperature T_0 is set according to the following method:

Define the ratio r as

$$r = \frac{number \quad of \quad accepted \quad moves}{total \quad number \quad of \quad moves}.$$

T_0 is arbitrarily fixed and the algorithm is computed.
r is calculated and the parameter readjusted:

- If $r < 0.7$, T_0 is multiplied by 3,
- If $r > 0.9$, T_0 is divided by 2,
- Otherwise, T_0 is used for the following phase (cooling phase).

T_{min} stop temperature that corresponds to a value of r is less than 0.01.
    The maximum number of iterations is fixed at 1,000.
    As with the Gamst's algorithm, the simulated annealing method is widely used in frequency planning.

Conclusions. Two types of frequency allocation methods can be used: conventional or advanced methods, the latter being more recent. Conventional methods combine hard limits with threshold comparisons of interference by use of interference and reuse matrices. Advanced techniques take full interference table information as a cost function in an optimization problem which is subject to hard constraints. They calculate frequency assignments while minimizing network wide interference over a given number of radio frequency channels. In addition, new work on optimization is being introduced based on genetic algorithms and neural networks.

7.3.2 Fixed Network Planning

Fixed network planning commences once radio planning has been completed. It consists of determining the capacities and locations of various concentrators (i.e., BSCs and MSCs) and databases (e.g., HLR and VLR) as well as transmission links between equipment (mainly links between switching centers and base stations, links between switching centers, and other switching centers). The fixed network planning phase uses as its inputs the locations and capacities of cells, the traffic matrices between base stations, and those between base stations and various points in the fixed network. In the main, this phase consists of the following stages [14]:

- Determination of the number and location of base stations controllers (BSC) and mobile switching centers (MSC),
- Defining base stations links to the BSCs (allocation of BTSs to their BSCs) and of BSCs to MSCs (allocation of BSCs to their MSCs),

- Defining connections between MSCs,
- Determining the capacity and location of databases (HLR and VLR).

Several parameters are fixed in advance so as not to have a problem with a large set of solutions. For example, the locations of gateway switching centers for connections between gateways and MSCs, and locations of certain MSCs can be fixed beforehand [15]. Similarly, interconnection link capacities can be fixed between MSCs and BSCs and between BSCs and BTSs.

7.3.2.1 Constraints to Be Considered

The main objective is to minimize the investment cost of BSCs together with link transmission costs. In fact, signaling exchange bearers between network sites account for a considerable part of the the investment and management costs of a network. They are estimated to represent between 25 to 35% of the network operational costs. Various alternatives can be used to connect sites to the network but depend, to a large extent, on the designated security level (e.g., whether redundancy is to be used or not) and local cost factors. Finally, it should be noted that the two elements that make the most significant contribution to the costs of the distribution network are the base station controllers and links with the base stations.

During the fixed subnetwork planning phase mobility related aspects must be considered. This task consists of minimizing the inter-BSC handover rates and the location update management load. High signaling loads are due to the interswitch traffic initiated by handovers, location updating, and roaming (e.g., through call routing). The planning objective is to allocate cells to switches evenly to distribute this load while keeping the network overhead at a minimum.

7.3.2.2 Site Location Problem

Given the following data:

- A set of terminal sites $I = \{1, \dots N\}$,
- Possible sites for concentrator $J = \{1, \dots M\}$,
- Cost C_{ij} to connect the terminal concentration site i to the concentration site j,
- Cost of implementing F_i on a concentrator on site j,
- Capacity K to connect a concentrator,

the task consists of minimizing the following function: $\sum_{i,j} \left(c_{ij} \cdot x_{ij} + F_j \cdot y_j \right)$

under the contraints:

$$\forall \, i \in I, \, \Sigma \, x_{ij} = 1$$

$$\forall\, j \in J, \Sigma\, x_{ij} \le K.y_j$$
$$\forall\, i \in I, \forall\, j \in J, x_{ij}, y_{ij} \in \{0, 1\}$$

The location of concentrator sites is a complex issue as the related fixed subnetwork must also be optimized.

Examples of methods used to resolve this problem are the Lagrangian relaxation and simulated annealing methods or the use of neural networks and genetic algorithms.

Example 7: Determination of the location of an MSC

A simple planning method takes into account each important city successively in relation to its hierarchy. For each city, the economic benefit to host a local MSC is estimated. If. the cost is estimate is too high, an alternative consists of using a switch located in a city that is higher up in the hierarchy [15].

What should be highlighted here is that a local MSC helps shorten the links with the BSCs. Most of the traffic is local and occurs between the mobiles and subscribers on the fixed network located in the same area. Thus, with a local MSC (see Figure 7.14 and 7.15), PSTN calls may be at local calls costs, whereas for remote MSCs, the majority of calls would incur an interregional charge.

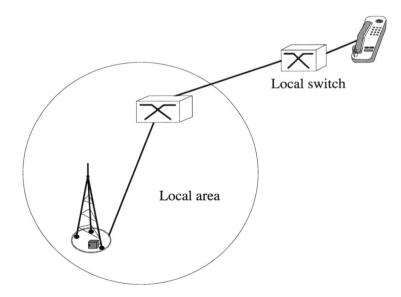

Local switch

Local area

Figure 7.14 Local area connected to a local MSC.

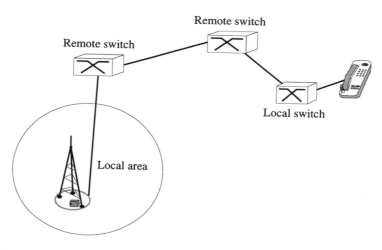

Figure 7.15 Local area connected to a remote MSC.

7.3.2.3 Star or Ring Structure

Consider the case for a GSM system. With a star architecture, base stations are directly linked to the BSC. With a ring architecture, the network is designed as a circular structure and the BSCs are linked around its circumference. In this structure, traffic channels can use two different paths to reach a BSC, thus allowing the automatic rerouting of channels if one of the links is broken. In practice, both approaches are combined to obtain a practical, cost effective, and optimized transmission solution.

The star topology (Figure 7.16) is mainly used in current networks because it requires relatively quick planning and helps, if the links are not protected, to reduce the amount of equipment to be installed. Yet, such configurations are less reliable because a broken link close to a BSC can lead to the isolation of several base stations with resultant loss of service. To avoid such risks one solution is to duplicate the critical links.

The ring topology (Figure 7.17) is more complex to use, especially in areas characterized by a high density of base stations, and requires an extra transmission link for each ring. This structure helps give satisfactory transmission protection without having to duplicate links between individual sites. It is particularly useful when implementing networks where microwave links (operating with frequencies higher than 13 GHz) are used. In fact, in these frequency bands, where heavy rain is likely to disrupt traffic, ring topology offers better protection than the duplication of BTS-BSC direct links.

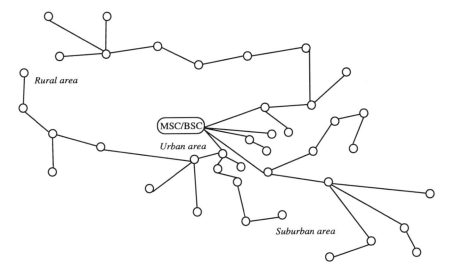

Figure 7.16 Star structure.

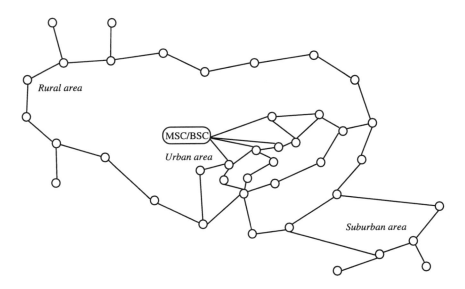

Figure 7.17 Ring structure.

The investment cost for a ring structure is also lower than that for a secure star topology, because fewer radio links and cables are required. The number of cables or radio terminal equipment for a ring network with N sites is $2 \times (N - 1)$ versus $2 \times 2 \times N$ for a protected star topology. The larger the ring, the more significant the impact of minimizing transmission costs and the more important network efficient planning becomes.

Linking cellular sites to networks is very often accomplished by microwave links. The advantage of such a solution is its relatively short installation time, low cost, and the possibility of change to the network topology by redirecting antennas or changing sites. The constraints of cellular network dimensioning require distances in urban areas ranging between 1 and 5 km and in rural area distances ranging between 20 and 35 km. Most of the radio links use frequencies between 13 and 38 GHz (the main bands are 15, 18, 23, and 38 GHz) for hops which can reach up to 25 km. A 2 Mbps capacity is generally enough for a BTS-BSC direct link. With a chain or ring structure, where the capacities of several BTS are combined, the required capacity is likely to reach bit rates up to 4.2 Mbps.

7.3.2.4 Centralized Architecture or Decentralized Architecture

An MSC supervises a group of cells, the number of which is determined by its capacity and the traffic levels in the coverage area. For a centralized architecture (Figure 7.18), an MSC has considerable capacity and controls a substantial number of base station groups (clusters) [2]. This form of architecture is generally representative of first-generation cellular or small-size systems. For a decentralized architecture (Figure 7.19), which is applied to large area coverage systems, the MSCs are smaller and control only a low number of cells.

Calls between mobiles that are located in the same network can be routed through the PSTN if there are no links inside the cellular network between the calling party MSC and that of the called party. If there is link capacity available, then the call is routed within the cellular network independent of the PSTN.

7.3.3 Conclusion

The initial planning of the network is followed by engineering work that is equally important. As the load increases with the growth in the number of subscribers, the network capacity will need to increase. Also, during the network life, quality of service measures are undertaken by teams who regularly make measurements related to radio signal levels and communications quality. The planning process, therefore, is constantly revisited in line with network upgrading. Frequency plans can be changed several times during the life of a network. Finally, and not relevant to fixed systems, the efficiency of the radio resource has to be maximized by fine tuning management procedures (e.g., power control, handover) while maintaining a satisfactory quality of service. Traffic forecasting is a rather sensitive task, the outcome of which can, a posteriori, prove to be inaccurate. Note that commercial activities, such as the use of punitive tariffs for peak hour calls, can have significant impact on system load and are available for smoothing out traffic peak overload.

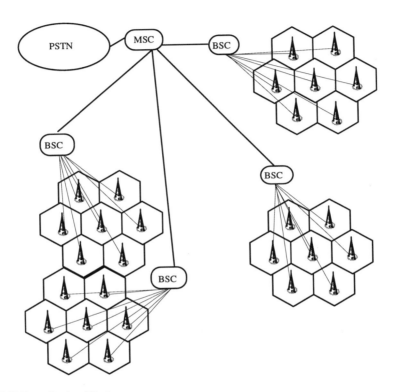

Figure 7.18 Centralized architecture.

In Figure 7.20, the main stages of the planning process are summarized together with the most important parameters taken into account throughout the process.

7.4 SYSTEM TUNING: EXAMPLE OF THE GSM BSS

System tuning is a major task done after planning and implementing a cellular network. Radio interference related parameters are probably the most difficult to adjust and optimize. Furthermore, it is never done just once but requires an iterative process based on observation of the system behavior. The main parameters involved with this task are introduced here. A GSM system is used as an example.

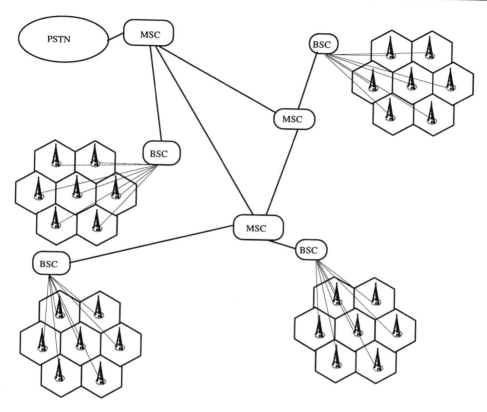

Figure 7.19 Decentralized architecture.

A system like GSM includes a lot of procedures and functions used for radio link management (see Chapters 6 and 8). Their parameters must be adjusted and optimized. The most important items and their corresponding parameters that require tuning or setting are listed below:

- Cell and network identification,
- Selection,
- Radio link control,
- Power control,
- Handover,
- Location management.

The last two procedures are dealt with in detail in Chapter 8.

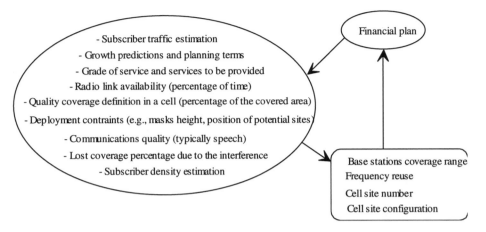

Figure 7.20 Stages and main parameters of the planning process.

7.4.1 Cell and Network Identification Parameters

These parameters allow a mobile to identify the network in which it is located (e.g., to determine if it has right of access to that network and to present the identity of the network to the subscriber) and to identify the cells (see mobility management in Chapter 8). The parameters whose values must be fixed when initializing the system are the following:

- The cell global identity, CGI, composed of the following parameters:
    - mobile country code, MCC, which fixes the country (for instance, MCC = 208 for France),
    - mobile network code, MNC, which identifies the PLMN,
    - location area code, LAC, which identifies the location area (see Chapter 8),
    - cell identity, CI, which is composed of the number of the station as well as the number of the cell sector (in the case of a bi or trisectored BTS).

- The base station identity code, BSIC, composed of the following parameters:
    - network color code, NCC, which gives the network code (3 bits used to avoid problems of PLMNs overlapping at the country borders),
    - base color code, BCC, which identifies the reuse pattern.

The BSIC number and the BCCH frequency identify the cell for handovers (see Chapter 8).

7.4.2 Cell Selection Parameters

This procedure is described in detail in Chapter 8. It uses the following parameters:

- RXLEV_ACCESS_MIN, which indicates the minimum signal level received by the MS below which it is not authorized to access the cell (typical values of - 95 dBm in an urban environment and -99 dBm in rural areas). Increase or decrease of this parameter reduces or enlarges the cell size.

- $C1 = A - MAX(B, 0)$ is the criterion whose parameters are:
  - $A = RXLEV: RXLEV\_ACCESS\_MIN$ where $RXLEV$ is the actual received signal level,
  - $B = MS\_TXPWR\_MAX - P$, where MS_TXPWR_MAX is the maximum authorized transmitter power of the MS in the cell, and P is the transmitter power capability of the MS
- $C2$, which is the second selection criterion. It uses timeouts to take into account the problem of repetitive location updates (when a mobile station is near the border of a location area). Its parameters are $T3212$ (which is the periodicity of location updates) and $MAX\_RETRANS$ (which is the maximum number of access attempts to the RACCH).

7.4.3 Link Control Parameters

These parameters are used to estimate the decoding level of signaling frames. The main ones are:

- $RADIO\_LINK\_TIME\_OUT\_BS$ for the downlink control,
- $RADIO\_LINK\_TIME\_OUT\_MS$ for the uplink control.

The value of each of these parameters is increased by 2 in the case when the SACCH has been decoded, and decreased by 1 if not. The values of these parameters lie between 4 and 64.

7.4.4 Power Control Parameters

The power control procedure mainly uses the following parameters:

- $POW\_RED\_STEP\_SIZE$ indicates the power decrease step,
- $POW\_INC\_STEP\_SIZE$ indicates the power increase step,
- $U\_RXQUAL\_XX\_P$ is the upper threshold for the received signal quality,
- $U\_RXLEV\_XX\_P$ is the upper threshold for the upper received signal level,

- *L_RXQUAL_XX_P* is the lower threshold for the received signal quality,
- *L_RXLEV_XX_P* is the lower threshold for the upper received signal level.

7.5 INCREASING THE CAPACITY OF A CELLULAR NETWORK

One of the many advantages of a cellular architecture is its ability to grow progressively in capacity (both in area and traffic channels). The dimensioning and planning processes that have been described in the previous sections, and that are achieved at the network implementation stage, are followed by increasing the network capacity, which in turn follows an increase in traffic demand. One of the initial goals for an operator is to cover the service area as quickly as possible. To accomplish this, large cells with omnidirectional antennas are often installed, as they ensure maximum coverage density with a quite low traffic density. During this start up phase, network capacity is not a major problem for an operator because the number of users is low. A main priority is to ensure the most expansive radio coverage. A cell size depends on the frequency, antenna gain, environmental characteristics, and the equipment used. At this stage, cells, and consequently clusters, are large. This characteristic of the network is likely to limit the risks of co-channel interference.

After the first phase, the network has either of the following shapes (see Figure 7.21). Either the base stations are located in the middle of the cells (omnidirectional antennas), or they are located at the corners of the cells (sectored antennas).

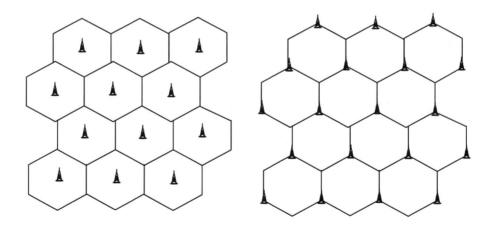

Base stations at the center of the cells   Base stations at the corner of the cells

Figure 7.21 Cellular geometry with base stations in the middle and at the corners of cells.

As traffic demand increases, network capacity is increased in the areas where required. Several techniques can be used. In the sections below, those most commonly used are presented.

Example 8: Calculation of cellular network capacity

For the sake of comparison of various capacity increase techniques the following parameters are commonly used to define cellular network capacity:

- S: surface area of the cell,
- N_o: number of cell sites per km^2,
- B_o: bandwidth allocated to the network,
- A_o: maximum capacity per site (Erl).

The maximum capacity of a cellular network in its first stage configuration will therefore be:

$$C_0 = \frac{A_0 \cdot N_0}{B_0} \ Erl/Hz/km^2$$

Suppose a network covering an area of 10 km by 10 km with cells of radius $R = 0.5$ km and with a bandwidth $B_o = 2 \times 12.5 = 25$ MHz (the case of a GSM network using half the GSM allocation).

To cover the whole area will require $\quad \dfrac{S}{\pi . R^2} \cong 127$ cells.

Assuming that the reuse pattern size is 19, then the number of clusters is approximately 6. The 2×12.5 MHz band is split into 62 duplex channels all of which are used within each pattern. Thus, $62/19 \cong 3$ duplex channels can be allocated to each cell. Assuming that a duplex radio channel is able to carry an average of about 7 duplex calls, one cell will be able to handle a traffic load of 14.036 Erlangs (with a blocking rate of 2%).

The system is capable of a total traffic load over the whole area equivalent to:

$A_0 = 127 \times 14.036 \cong 1,783$ Erlangs.

Finally, the network capacity will be:

$$C_0 = \frac{1783 \times 1/\pi \times R^2}{25 \times 10^6} \cong 0.091 \ mErl/Hz/km^2$$

7.5.1 Adding New Channels

The most immediate and quickest method of increasing the capacity of a network consists of adding new channels to cells (Figure 7.22). This can occur only in the case when all the frequencies allocated to the system have not been fully used. It would mean adding transmitters/receivers (TX/RX) at each base station site.

During the initial implementation phase of a network, not all the channels allocated to the system are used. Indeed, the growth in the number of subscribers is often planned with the idea of bringing on line all the channels progressively. Once all these ones have been allocated, then alternative methods have to be sought.

Example 9: Calculation of the new capacity after doubling the number of channels

Doubling the number of channels (i.e., using the whole 50-MHz GSM band while keeping the same reuse pattern) helps double up the number of duplex channels at each cell site. With 6 RF channels per site (see Example 8) a capacity of 42 logical channels is available which can support a traffic load of 32.836 Erl (2% blocking rate). The total traffic load the network will be able to carry will, therefore, be equivalent to

$A_1 = 127{\times}32.836$ Erl $\cong 4{,}170$ Erl

Therefore, the network capacity will be:

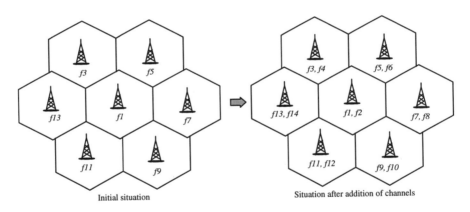

Figure 7.22 Method of adding new channels.

$$C_1 = \frac{4{,}170{\times}1\big/{\pi{\times}R^2}}{50 \times 10^6} \cong 0.106 \; mErl/Hz/km^2$$

The normalized capacity (in $mErl/Hz/km^2$) increase between this network and the first is

$$\Delta C_{01} = \frac{C_1 - C_0}{C_0} = \frac{0.106 - 0.091}{0.091} \cong 16\%$$

The greater advantage offered by this technique is that it does not require any change to the cellular pattern. Its drawbacks are considerable consumption frequencies (for each channel the capacity is increased by only 16%), and the need to add more equipment (transmitters/receivers).

7.5.2 Channel Borrowing

After the planning process, the channels are allocated to cells in accordance with the geographic distribution of the expected traffic (or *traffic relief*). As has been emphasized, this distribution varies in time and space.

When demand in traffic exceeds the local network capacity in a given area and there is spare capacity in neighboring cells, a rebalance between the various areas can occur by a temporary or permanent transfer of resources (see also Figure 7.23). The advantage with such a technique is that it does not require significant changes in equipment.

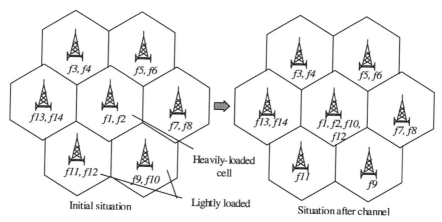

Figure 7.23 Channel borrowing method.

This channel borrowing technique can be achieved automatically and is often included in networks. Dynamic channel allocation (DCA) algorithms, around which several studies have focused, are predicted to be used in third-generation systems (see Chapter 6). The same kind of algorithms are being used in cordless systems. They imply common use of all or part of the band allocated to the system. Channel monitoring can be carried out centrally or in a decentralized way by the base stations. Thus a base station can use a radio channel if local interference conditions allow it to do so.

7.5.3 Modification of the Cell Reuse Pattern

Small cluster sizes help improve traffic density. It is possible, when traffic demand increases considerably, to reduce the number of cells per cluster (Figure 7.24). This reduction is generally accompanied by a significant decrease in the value of the C/I ratio. This phenomenon may have a negative impact on the quality of service below a given threshold.

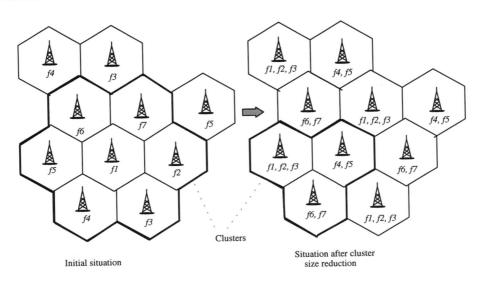

Clusters

Initial situation

Situation after cluster size reduction

Figure 7.24 Reducing the cluster size.

Example 10: Cell pattern reduction

The network uses all of the 50-MHz frequency band. There is a change from a 19-cell reuse pattern to one of 15 cells. The number of available channels per cell now becomes $124/15 \cong 8$. This equates to a traffic load of 44.00 Erl (with a blocking rate of 2%). The traffic load transmitted by the network will, therefore, be

$A_2 = 127 \times 44 = 5{,}588$ Erl.

And the network capacity becomes

$$C_2 = \frac{5,588 \times 1 \Big/ \pi \times R^2}{50 \times 10^6} \cong 0.142 \; mErl/Hz/km^2$$

The increase in the new network capacity, compared with the previous configuration (see example 9), is:

$$\Delta C_{21} = \frac{C_2 - C_1}{C_1} \cong 34\% \; .$$

The advantage with this technique is that it does not require the use of additional frequencies. Its major drawbacks are that it leads to a decrease in the C/I level and that it might require extra transmitter/receiver equipment.

7.5.4 Cell Splitting

The classic solution to increase capacity consists of reducing the cell service area. The cell splitting technique consists (Figure 7.25) of reducing cell sizes with an immediate consequence of increasing network capacity. Each cell is split up into a number of cells of a smaller size. A splitting by a factor K over the coverage area entails an increase in the number of base stations by a factor K^2. For example, if a frequency plan with 15-km radius cells requires a reuse distance of 45 km, a reduction in cell radius to 1.5 km will help bring down the reuse distance to 4.5 km. Each reduction in cell radius by a factor of 2 allows multiplication by 4 in the number of circuits per MHz/km². For a system based on 1-km radius cells, the number of circuits will be 100 higher than for a system based on 10-km radius cells. The major drawbacks of such a method are

- An increase in the interference level,
- The necessary re-planning of frequencies,
- The costs entailed by the implementation of new cell sites,
- An increase in the number of handovers.

Example 11: Cell splitting
The cell radius that was in the network of the previous example, R = 0.5 km, moves down to R' = 0.3 km. The number of cells in the network will, therefore, be equal to $S/\pi.R'^2$ $\cong 354$. If we consider the size of the new reuse pattern as being 15, each cell can carry a traffic load of 44.00 Erl and the network is able to carry a traffic load of 15,576 Erl.
The network capacity becomes

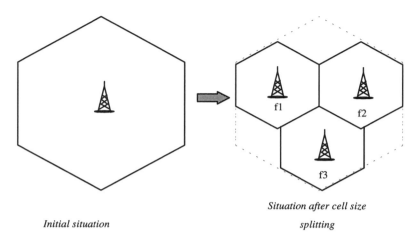

Initial situation

Situation after cell size splitting

Figure 7.25 Cell-splitting method.

$$C_3 = \frac{1{,}5573 \times 1 \Big/ \pi \times R^2}{50 \times 10^6} \cong 1\ 102\ mErl/Hz/km^2$$

The capacity increase of the new network compared with the previous one is:

$$\Delta C_{32} = \frac{C_3 - C_2}{C_2} \cong 676\%$$

Theoretically, cell-splitting can be carried out indefinitely. In practice, some constraints put limits on the use of this technique:

- When the distance between cells reduces, co-channel interference increases. In fact, the characteristics of signal propagation show that the ratio *useful distance/reuse distance* reduces as the cell size reduces. Hence, the size of a cell split by a factor k does not imply that the capacity is multiplied by k.
- Finding cell sites and installing the equipment for new base stations is a rather lengthy and complex issue.
- The overall cost of the system increases with the increase in the number of base stations.
- The handover rate increases with decreases in cell size (which entails extra signaling and processing loads).

7.5.5 Sectorization

One alternative to cell splitting is *sectorization* (Figure 7.26). This technique consists of splitting a cell into several sectors, with each sector using a different set of channels and a directional antenna. Each sector can, therefore, be considered as a new cell. The most commonplace configurations comprise three or six sector cells. Typical configurations are trisectored sites in urban areas, omnidirectional in rural areas, and bisectored sites for road coverage. Base stations can be situated either in the middle or at corners of the cell.

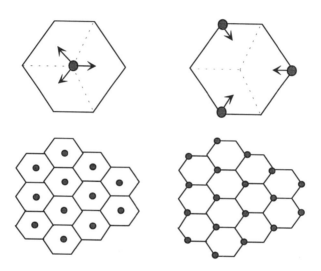

Figure 7.26 Implementation of sites for cell sectorization.

Omnidirectional antennas can be replaced by directional sectored antennas to help multiply the number of cells without having to increase the number of radio sites (Figure 7.27). Another advantage is that sectorization helps increase the C/I ratio which, in turn, can enhance the quality of service.

7.5.6 Down-Tilting

Down-tilting offers two major advantages. The first consists of reducing interference levels to a minimum. This is obtained by reducing the effective transmitted power in the direction of a co-channel cell because radiation is concentrated in the served cell. It also helps in obtaining better power control usage, allowing the set up of lower antenna towers, and in helping obtain higher reuse factors. The second advantage of down-tilting assists in avoiding coverage gaps in areas located close to the transmission site by directing the antenna beam into those areas.

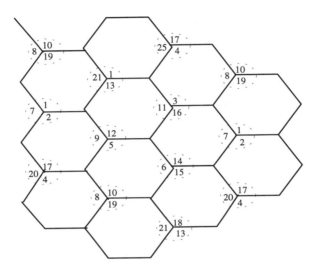

Figure 7.27 Example of trisectored cell implementation.

The principle of down-tilting consists of directing the base station antenna in such a way as to help it transmit in one direction so that its transmissions cause the least amount of interference to other cells while maintaining a satisfactory service quality in the served cell (Figure 7.28). For this it is necessary to use a high-gain antenna. Two kinds of down-tilting are available: *electrical* and *mechanical*.

The antenna transmission lobe is directed towards the area to be covered with a decrease in the transmission of the signal into neighboring cells. Typically, angle θ ranges from 10 to 20deg. For example, experiments have shown that when an antenna is tilted by 10deg, power in the horizontal direction is reduced by 4 dB. At the co-channel cell, the interference level reduces by only 0.25 dB. When the antenna is tilted by 20deg, the decrease in interference is up to 1 dB. In electrical down-tilting, an incremental phase shift is applied in the antenna array feed network. This kind of down-tilting is also commonly used in omnidirectional cells.

The main drawbacks of this technique are:

- The reuse pattern has to be modified because of interference levels changes,
- The area covered by the cell is reduced and signal levels can decrease at the border of the cell.

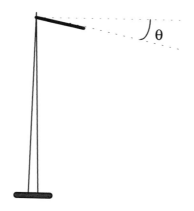

Figure 7.28 The down-tilting technique.

7.5.7 Cell Layering

The increase in number of subscribers combined with the large number of services and size of environments covered (especially in urban areas) has further complicated the task of providing universal access from only one type of network. The network has to provide very high user densities for low mobility users within an indoor environment, medium densities for pedestrians, and low densities for fast moving users (vehicles) out of doors. The same network cannot ensure optimum access to these three groups of users with very distinct, different characteristics. To provide coverage adapt to each group of people and the environment, the concept of superimposed cells (also called *overlaid, layered, hierarchical,* or *multitiered cells*) has been introduced (Figure 7.29). Three types of cells can be identified:

- *Macrocells* have a radius ranging between about 1 to about 30 km (a distinction between small and large macrocells can also be made). These are used to serve the fast moving subscribers using mobiles with power ranges between 1 and 10W. Base station antennas are usually mounted above 30m. Coverage from these cells is economical because it does not require the implementation of a large number of base stations. These cells can be used for filling coverage holes between microcells. Also, whenever a microcell is overloaded, a mobile can be directed towards a macrocell. Therefore, macrocells act as backup for radio problems and carry traffic overflow to assist the microcells' capacity problems. Cells covering highways are a specific case of macrocells where their capacity is much higher than that of other types of macrocell [16].

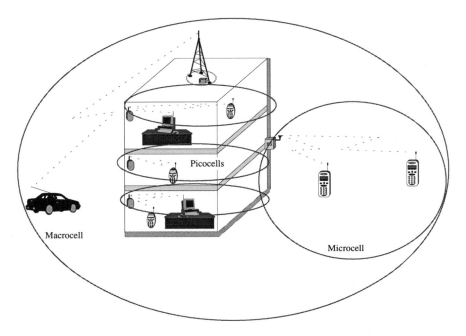

Figure 7.29 Multilayer cell structure.

- *Microcells* are served by small low power base stations located in streets or large indoor spaces (e.g., airports, stations, shopping malls). A microcell radius usually ranges from 100 to 300m. They mainly serve pedestrian subscribers and slow moving users (especially vehicles stuck in traffic jams). Coverage of a microcell caters the needs of mobiles with power ranges of between 10 and 100 mW. The antennas of microcell base stations are usually located beneath roof level. Their two main drawbacks are, on the one hand, the number of significant obstacles which limit blanket coverage, and on the other, the difficulties in forecasting traffic volume. These difficulties increase as the cell radius decreases because it is rather difficult to smooth or average the traffic load over the whole cell surface. Several kinds of microcells can be identified, among which are highlighted the *cigar-shaped*, *four-leaf-shaped* cells, generally located at crossroads, those serving pedestrians, and finally those serving fast vehicles.
- *Picocells* are used to cover the indoor environment (offices, mainly). For this purpose, their coverage can be three dimensional so that they are able to cover several floors of a building. The terminal power is generally below 10mW. The radius of these cells ranges from 10 to 100m. Their main drawback is the considerable variation in fields that leads to the virtual impossibility of predicting

either the traffic or the radio coverage. Systems using picocells generally belong to the cordless category (see Chapter 10).

7.5.8 Trends Toward the Microcellular Techniques

In helping to increase the density of their radio networks, operators have several methods at their disposal. The most common method is the reduction of cell size, used notably in cellular systems which operate in the high frequency bands with propagation characteristics which lead to the efficient use of small cells. Since the end of the 1990s, most European GSM operators have been progressively introducing microcells into large cities.

These microcellular systems are used to augment service in high-traffic areas. The microcell layer is deployed in buildings, malls, and corridors to supply local capacity in specific areas. The network is capable of operating at two layers. One, by allocating distinct frequencies inside a band (in other words, by partitioning the spectrum so as to supply sub-bands for macro and microcells), or by using two distinct frequency bands (using terminals which operate in two bands), or by using two distinct transmission modes in one or two frequency bands (in other words, dual-mode (e.g. DECT/GSM) and dual-band (e.g., 900/1800 MHz) terminals). In the first case, coverage determines distinct terminal groups by the signal levels received by the users. In the second and third cases, users can opt to have a specific subscription that will allow them to have access to one or both levels of the hierarchy.

7.5.9 Capacity Solutions Comparison

In the previous and present chapters various solutions have been presented that can increase the capacity of a mobile network (Table 7.3). These can be divided into three types:

- Core solutions, which include techniques such as power control, frequency hopping, and discontinuous transmission. Their main benefits are increased capacity and enhanced service quality through interference reduction and extended mobile battery life by minimization of battery consumption.
- Network solutions, which include techniques such as cell splitting, reuse partitioning, hierarchical cells, smart antennas, and dynamic channel assignment.
- Mobile solutions, which include two categories: those requiring extra spectrum (e.g., dual band) and those requiring new codecs (half-rate, enhanced full rate, and adaptive multirate).

Table 7.3

Capacity solutions comparison

	Implementation time	Capacity improvement	Coverage improvement
Core solutions	Fast (about one month)	20–70%	Low
Cell splitting	Slow (six to eighteen months)	300–500%	Significant
Reuse partitioning	Fast (about one month)	20–40%	Low
Layered cells (macro and micro)	Medium (three months to one year)	200–400%	Significant
Layered cells (macro, micro, and pico)	Medium (three months to one year)	1,000%	Very significant
Smart antennas	Medium (three to six months)	200–300 %	Significant

7.6 CONCLUSIONS

Cellular network planning is of paramount importance, quite often reiterated several times during the life of the network (even weekly during high growth phases). Operators may have to set up, during this stage, about a hundred new sites every month [17]. One of the main objectives consists of minimizing the number of radio sites to be installed while trying to satisfy the criteria of density and service quality. The implementation cost of one site ranges between U.S.$100,000 to U.S.$1 million. The costs related to the leasing of lines, site rentals, various taxes, maintenance, and to the supply of power amount to sums ranging between U.S.$0.5 and 1.5 million over 10 years of a site life. The same has been observed for PCS networks where the capital costs can be as high as U.S.$1 million per cell site before the first dollar of revenue can be realized by a network operator [7]. To make these investments worthwhile, the operators are more and more tempted to opt for site-sharing, especially in less dense areas where the cost effectiveness of a site is low, or sometimes even negative.

APPENDIX 7A

QUALITY OF SERVICE

The first and main quality-of-service measure is subscriber satisfaction. In telecommunications networks, quality-of-service (QoS) is a broad subject that groups together a wide range of topics. It is defined by the ITU-T in recommendation E.800 as "The collective effect of service performance which determines the degree of satisfaction of a user of the service." There are four main factors on which QoS depends: support (through secondary services such as information, provisioning, and billing), ease of use, integrity of transmission, and serviceability. The latter is the network's ability to set up calls on demand and maintain them for the desired duration.

For mobile networks, the QoS parameters are mainly related to call procedures and speech criteria:

- No service available (the phone is in continuous "network search" mode and is unable to find a network on which it can register). This parameter is related to the coverage quality and traffic capacity.
- Call attempts fail even if the network is available. On a mobile originated call, after dialing, the mobile falls back into idle or network search mode, while a mobile terminated call receives a busy tone. This parameter is also related to the coverage.
- Call drops after successful setup. Speech connection terminated followed by nothing or a busy tone. This parameter is usually related to handover and to the capacity in each cell (which must be sufficient to accept handover calls).
- Completed calls with bad transmission quality. Communication is degraded even if control on call setup and hold to termination is maintained.

In digital systems, QoS can be described by the mean BER, which can be obtained from dividing the number of bits received in error by the total number of bits transmitted during a predefined period of time. For data communications, QoS can be described by the average BER (e.g., $BER < 10^{-5}$), packet-error rate (e.g., $PER < 10^{-2}$), signal-processing delay (e.g., 1–10 ms), multiple access collision probability (e.g., $< 20\%$), the probability of a false call (e.g., false alarm), the probability of a missed call, and the probability of a false call (synchronization loss).

For voice communications, QoS is generally expressed in terms of *mean opinion score* (MOS), which is made after subjective evaluation by the service users. The different scores can be: *0* = bad, *1* = poor, *2* = fair, *3* = good, and *4* = excellent. For example, in pulse coded modulation (PCM) technology the C/I ratio for a speech channel has a value of 35 dB and has an excellent quality. In the case of mobile services, the subscribers give a value of QoS = 1 for a C/I = 15 dB and a value of QoS = 4 for a C/I > 25 dB. For more objective speech-quality evaluation in real conditions (e.g., with fading, BER = 10^{-2}, Doppler, background noise, ignition noise) supplementary tests can be used such as the diagnostic acceptability measure (DAM), and the diagnostic rhyne test (DRT).

QoS has a major impact on network cost. For instance, the definition of two different levels of radio coverage lead to very different investment costs. As an example, if the marketing decision is to design a new network with a 97% coverage probability, then a cell area of 39.4 km^2 with a 95% coverage probability reduces to 31.0 km^2. If it is decided to go for indoor locations near windows (which requires an allowance for penetration loss of about 15 dB) with 90% coverage, then the cell area shrinks to 8.0 km^2 and the number of cells has to increase.

APPENDIX 7B

ERLANG FORMULAS

In a mobile radio system, it is generally considered that call arrivals follow a Poisson law. It is also considered that the waiting time for a channel release follows a negative exponential distribution and that blocked calls are lost (i.e., they are not processed).

Call Arrival Process

The call arrival process follows a Poisson law and the probability of having exactly k arrivals during a time interval of t is given by

$$p_k = \frac{(\lambda t)^k}{k!} e^{-\lambda t}$$

(7B.1)

Mean Interarrival Time

The probability for the interarrival time τ to be lower or equal to t is denoted by

$$A(t) = 1 - exp(-\lambda t)$$

(7B.2)

The interarrival time is given by

$$E[t] = a/\lambda$$

(7B.3)

The Number of Waiting Blocked Calls

The blocking probability is given by

$$B = \sum_{k=N}^{\infty} \frac{A^k}{k!} exp(-A)$$

(7B.4)

with $A = \lambda/\mu$ (or traffic in Erlang) and N is the number of servers (number of channels).

Blocked Calls Cleared

In an N channel system where the arriving calls are lost if all the channels are busy, the probability p_k becomes

$$p_k = \frac{A^k / k!}{\sum_{i=0}^{N} A^i / i!} \qquad (7B.5)$$

Blocking occurs when all N channels are busy. This case occurs with the probability

$$E(A,N) = p_N = \frac{A^N / N!}{\sum_{i=0}^{N} A^i / i!} = \frac{E(A, N-1)}{N / A + E(A, N-1)} \qquad (7B.6)$$

This formula is called *Erlang-B formula*.

For high values of N, we approximate $E(A, N)$ by

$$E(A,N) \approx \frac{A^N}{N!} \exp(-A) \qquad (7B.7)$$

Note that because of the user's mobility, handover and roaming reduce the channel usage time in the cell where the call was set up; whereas traffic increases in cells where mobiles are sent by handover. The Erlang-B formula is no longer valid in this case.

Blocked Calls Delayed

In this model, the blocked calls are queued before getting a resource:

$$\text{If } p_k = \begin{cases} \dfrac{A^k}{k!} p_0 & \text{for } 0 \le k \le N \\[2ex] \dfrac{A^k}{N!} N^{N-k} p_0 & \text{for } k \ge N \end{cases}$$

where $p_0^{-1} = \sum_{k=0}^{N-1} \dfrac{A^k}{k!} + \dfrac{A^N}{N!} \dfrac{1}{1 - A/N}$

The probability for a cell to be delayed (i.e. to enter the queue) when no channel is available is:

$$C(N,A) = \frac{A^N}{N!} \frac{1}{1 - A/N} p_0 \qquad \text{for } K \ge N \qquad (7B.8)$$

Erlang-C Formula

If s is the mean communication duration, the following formulas are obtained (see Table 7B.1).

Table 7B.1

Parameters derived from Erlang formulas

Mean waiting time for all calls	$T_m = \dfrac{s}{N-A} C(N,A)$
Mean waiting time for calls in the queue	$T_m = \dfrac{s}{N-A}$
Probability for the calls to be delayed for more than t seconds	$p(\tau \geq t) = C(N,A) \cdot e^{\frac{-t}{T\mu}}$
Probability for all the calls in the queue to be delayed of more than t second	$p(\tau \geq t) = e^{\frac{-t}{T\mu}}$
Probability for a call in p (or higher) position in the queue to be delayed	$p = C(N,A) \cdot (\dfrac{A}{N})^p$

APPENDIX 7C

ERLANG TABLE EXAMPLE

Table 7C.1

Grade of service (maximum blocking rate)

Channel	1%	2%	3%	5%	10%	20%	40%	Channel
1	.01010	.02041	.03093	.05263	.11111	.25000	.66667	1
2	.15259	.22347	.28155	.38132	.59543	1.0000	2.0000	2
3	.45549	.60221	.71513	.89940	1.2708	1.9299	3.4798	3
4	.86942	1.0923	1.2589	1.5246	2.0454	2.9452	5.0210	4
5	1.3608	1.6571	1.8752	2.2185	2.8811	4.0104	6.5955	5
6	1.9090	2.2759	2.5431	2.9603	3.7584	5.1086	8.1907	6
7	2.5009	2.9354	3.2497	3.7378	4.6662	6.2302	9.7998	7
8	3.1276	3.6271	3.9865	4.5430	5.5971	7.3692	11.419	8
9	3.7825	4.3447	4.7479	5.3702	6.5464	8.5217	13.045	9
10	4.4612	5.0840	5.5294	6.2157	7.5106	9.6850	14.677	10
11	5.1599	5.8415	6.3280	7.0764	8.4871	10.857	16.314	11
12	5.8760	6.6147	7.1410	7.9501	9.4740	12.036	17.954	12
13	6.6072	7.4015	7.9667	8.8349	10.470	13.222	19.598	13
14	7.3517	8.2003	8.8035	9.7295	11.473	14.413	21.243	14
15	8.1080	9.0096	9.6500	10.633	12.484	15.608	22.891	15
16	8.8750	9.8284	10.505	11.544	13.500	16.807	24.541	16
17	9.6516	10.656	11.368	12.461	14.522	18.010	26.192	17
18	10.437	11.491	12.238	13.385	15.548	19.216	27.498	18
19	11.230	12.333	13.115	14.315	16.579	20.424	29.498	19
20	12.031	13.182	13.997	15.249	17.613	21.635	31.152	20
Channel	1 %	2 %	3 %	5 %	10 %	20 %	40 %	Channel

Table 7C.1

Grade of service (maximum blocking rate)

Channel	1%	2%	3%	5%	10%	20%	40%	Channel
21	12.838	14.036	14.885	16.189	18.651	22.848	32.808	21
22	13.651	14.896	15.778	17.132	19.692	24.064	34.464	22
23	14.470	15.761	16.675	18.080	20.737	25.281	36.121	23
24	15.295	16.631	17.577	19.031	21.784	26.499	37.779	24
25	16.125	17.505	18.483	19.985	22.833	27.720	39.437	25
26	16.959	18.383	19.392	20.943	23.885	28.941	41.096	26
27	17.797	19.265	20.305	21.904	24.939	30.164	42.755	27
28	18.640	20.150	21.221	22.867	25.995	31.388	44.414	28
29	19.487	21.039	22.140	23.833	27.053	32.614	46.074	29
30	20.337	21.932	23.062	24.802	28.113	33.840	47.735	30
31	21.191	22.827	23.987	25.773	29.174	35.067	49.395	31
32	22.048	23.725	24.914	26.746	30.237	36.295	51.056	32
33	22.909	24.626	25.844	27.721	31.301	37.524	52.718	33
34	23.772	25.529	26.776	28.698	32.367	38.754	54.379	34
35	24.638	26.435	27.711	29.677	33.434	39.985	56.041	35
36	25.507	27.343	28.647	30.657	34.503	41.216	57.703	36
37	26.378	28.254	29.585	31.640	35.572	42.448	59.365	37
38	27.252	29.166	30.526	32.624	36.643	43.680	61.028	38
39	28.129	30.081	31.468	33.609	37.715	44.913	62.690	39
40	29.007	30.997	32.412	34.596	38.787	46.147	64.353	40
41	29.888	31.916	33.357	35.584	39.861	47.381	66.016	41
42	30.771	32.836	34.305	36.574	40.936	48.616	67.679	42
43	31.656	33.758	35.253	37.565	42.011	49.851	69.342	43
44	32.543	34.682	36.203	38.557	43.088	51.086	71.006	44
45	33.432	35.607	37.155	39.550	44.165	52.322	72.669	45
46	34.322	36.534	38.108	40.545	45.243	53.559	74.333	46
47	35.215	37.462	39.062	41.540	46.322	54.796	75.997	47
48	36.109	38.392	40.018	42.537	47.401	56.033	77.660	48
49	37.004	39.323	40.975	43.534	48.481	57.270	79.324	49
50	37.901	40.255	41.933	44.533	49.562	58.508	80.988	50
51	38.800	41.189	42.892	45.533	50.644	59.746	82.652	51
Channel	1 %	2 %	3 %	5 %	10 %	20 %	40 %	Channel

REFERENCES

[1]   Hao, Q., B.-H. Soong, J.-T. Ong, C.-B. Soh, and Z. Li, "A Low-Cost Cellular Mobile Communication System: A Hierarchical Optimization Network Resource Planning Approach," *IEEE Journal on Selected Areas in Communications*, Vol. 15, No. 7, pp. 1,315–1,326, Sept. 1997.

[2]   Yacoub, M. D., Foundations of Mobile Radio Engineering, *CRC Press*, 1993.

[3]   Rhodes, T., "There Are Many Ways to Roll-out DCS 1800," *GSM Quarterly*, pp. 22–25, July 1996.

[4]   Dance, S., "Planned Migration," *Mobile Europe*, pp. 63, Mar. 1997.

[5]   Gotzner, U., A. Gamst, and R. Rathgeber, "Spatial Traffic Distribution in Cellular Networks," *Proceedings of the IEEE Vehicular Technology Conference '98*, Ottawa, Canada, May 18–21, 1998.

[6]   Kamanou, P., "Outil de planification. Radio et méthodes associées," *Séminaire régional sur les systèmes cellulaires radiotéléphoniques mobiles*, ITU-BDT, Tunis, Tunisia, June 9–13, 1997.

[7]   Owens, D., "The Big Picture," *CDMA Spectrum*, pp. 36–39, Dec. 1997.

[8]   Evans, G., B. Joslin, L. Vinson, and B. Foose, "The Optimization and Applications of the W.C.Y. Lee Propagation Model in the 1,900-MHz Frequency Band," *Proceedings of the IEEE Vehicular Technology Conference '97*, Phoenix, AZ, May 4–7, 1997, pp. 87–91.

[9]   Lee, W. C. Y., *Mobile Cellular Telecommunications Systems*, McGraw-Hill Book Co., 1989.

[10] Hale, W.K., "Frequency Assignment: Theory and Applications," *Proceedings of the IEEE*, Vol. 68, No. 12, pp. 1497–1514, Dec. 1980.

[11] Gamst, A., "A Resource Allocation Technique for FDMA Systems," *Alta Frequenza*, Vol. LVII, No. 2, Feb.–Mar. 1988.

[12] Metropolis, N., A. Rosenbluth, M. Rosenbluth, A. Teller, and F. Teller, "Equation of State Calculations by Fast Computing Machines," *J. Chem. Phys.*, Vol. 21, pp. 1,087–1,092, 1953.

[13] Duque-Anton, M., D. Kunz, and B. Rüber, "Channel Assignment for Cellular Radio Using Simulated Annealing," *IEEE Trans. on Veh. Techn.*, Vol. 42, No. 1, pp. 14–21, Feb. 1993.

[14] Dreyfuss, F., P. Eskenazi, M. Ribeyron, and B. Liau, "Aspects of GSM Itinéris Mobile Network Planning," *Networks'94*, Sept. 1994.

[15] Zwinkels, A.M.E., and H. F. van Oortmarssen, "Planning the Fixed Network for GSM," *Networks'94*, Sept. 1994.

[16] Afonso, R., L. Cupido, N. Anjos, and N. Vidal, "UMTS Cell Design," *RACE Mobile Telecommunications Workshop*, Metz, France, pp. 306–310, 16-18 June 1993.

[17] Sayer, P., "Plan the Planet," *Mobile Europe*, pp. 35–36, Feb. 1997.

SELECTED BIBLIOGRAPHY

[1] Bajiva, A., "Cellular Radio Planning Tools," *Cellular Radio Systems*, edited by D. M. Balston and R.C.V. Macario, Artech House, 1993.

[2] Faruque, S., and M. Maragoudakis, "A Cost-Effective PCS Deployment Methodology," *Proceedings of the International Conference on Universal and Personal Communications'95*, ICUPC'95, Tokyo, Japan, pp. 868–872, Nov. 6-10, 1995.

[3] Jabbari, B., G. Colombo, A. Nakajima, and J. Kulkarni, "Network Issues for Wireless Communications," *IEEE Communications Magazine*, pp. 88–98, Jan. 1995.

[4] Lee, W.C.Y., *Mobile Communications Engineering*, McGraw-Hill Book Company, New York, 1982.

[5] Madhavepeddy, S., and K. Basu, "The Design of Self Engineering Mobile Telephone Systems," *XV International Switching Symposium*, pp. 426–430, Apr. 1995.

[6] Pottie, G. J., "System Design Choices in Personal Communications," *IEEE Personal Communications*, pp. 50–67, Oct. 1995.

[7] Subramanian, S., and S. Madhavapeddy, "System Partitioning in a Cellular Network," *Proceedings of the IEEE Vehicular Technology Conference '96*, Atlanta, GA, pp. 106–110, Apr. 29-May 1, 1996.

[8] Tcha, D.-W., Y.-J. Chung, and T.-J. Choi, "A New Lower Bound for the Frequency Assignment Problem", *IEEE/ACM Transactions on Networking*, Vol. 5, No. 1, pp. 34–39, Feb. 1997.

CHAPTER 8

MOBILITY MANAGEMENT

The main advantage of mobile radio communications systems compared with fixed networks is *mobility*. The "mobility service" groups together several functions that allow users to have access to telecommunications services over a wide area and maintain communication while moving. Land mobile radio systems able to provide this service most comprehensively are cellular networks, and more specifically, those such as GSM that cover a whole country or even a continent. They allow a user to maintain his or her communication while moving by switching from one cell to another. Therefore, the major focus of this chapter is mobility management as provided by cellular systems.

Note, that wide area professional radiocommunications systems (see TETRA, in Chapter 9) and some cordless systems (see DECT, in Chapter 10) include some mobility functions similar to those defined in cellular systems.

Two mobility management levels should be highlighted: first, at a *microscopic* level, mobility management that allows a subscriber to move from one cell to another while keeping communication with the network, then at *macroscopic* level, management of network mobility that allows a user to benefit from services for which a subscription has been paid (and more precisely, the one that allows him or her to receive and transmit calls) within the coverage of the home network and also in some other (visited) networks (Figure 8.1).

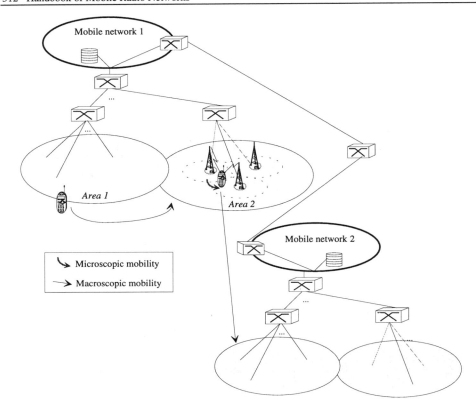

Figure 8.1 Different kinds of mobility in different networks.

The first of these two processes is called *handover*. The second includes the *cell selection/reselection* algorithm and the *location management* process. They both constitute two major functions that are specific to cellular mobile radio systems.

8.1 MANAGEMENT OF RADIO MOBILITY: THE HANDOVER PROCEDURE

The cellular structure that is capable of serving high-subscriber densities has a major disadvantage in the complexity required to ensure communication transfer between cells. This process is called *handover* and occurs only during a call (i.e., when the terminal is not transmitting, no handover is needed).

Note: In some countries, the United States in particular, this mechanism is called *handoff.* The same term is used by radio systems engineers in the civil aviation field where airplanes are in communication with one of the air traffic control stations on the ground. Handoff means, in this context, the transfer of communications from one control station to another, and is employed when the airplane flies away from one control station and comes closer to a new one. In cellular communications this term is particularly appropriate as it corresponds to the transfer of communication from one base station to another because of the motion of the cellular terminal.

The handover mechanism aims at preserving an acceptable quality of communication between a mobile and the network while reducing the level of global interference. This objective can be realized if the mobile's channel frequency or cell is changed. The movement of the mobile or nearby elements is the main reasons behind the triggering of this mechanism. These generate variations in signal behavior caused by fading, obstructions, and interference. Handover can be considered as one of the most complex and important processes in cellular communications (Figure 8.2). To cater for the fast growth in demand, small cells have been introduced and this increases the number of handovers required in a network.

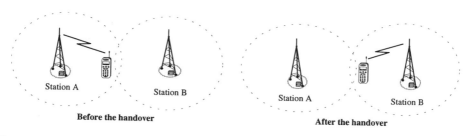

Before the handover   After the handover

Figure 8.2 The progress of an intercell handover.

The handover is without doubt the most critical procedure that is processed during a call (and of any other kind of communication link established between the mobile and the network) because it is used to ensure continuity at the very time that the radio resources change either in the same cell (*intracell handover*), or between two cells (*intercell handover*) either inside or outside the same switching area. Handover performance is of paramount importance as far as quality of service provided to the subscriber is concerned. Two of the quality-of-service factors are the *probability of call rejection* (i.e., a connection request not met) and the *probability of a call interruption* (i.e., a call interrupted while under way). A call interruption, the probability of which increases during a handover, is obviously less appealing to a user than a call rejection. The success rate of a handover is all the more important because it is considered a sensitive indicator in the measurement of the quality of service.

In a cellular network handover has to perform the following tasks:

- Allow the users to move while the call is under way,
- Allow the call to resume and avoid disconnection of the mobile to network link because of bad radio transmission conditions (the phenomenon of *"rescue handover"* [1]),
- Balance the traffic load between cells (the phenomenon of *"traffic handover"* [1]),
- Preserve a quality of service acceptable to the user in case of interference,
- Optimize the use of radio resources,
- Minimize the mobile power consumption and global interference level (*"confinement handover"* [1]).

The handover procedure is a mechanism that affects both the radio and the network. More than any other in call procedure, handover involves several levels: physical and data link layers with complete change of the radio link characteristic and network level access point (the mobile goes from one base station to another and eventually from one switch to another) which is performed by link switching inside the MSC.

In this section the handover procedure is illustrated by taking examples from the GSM standard. It is one of the most exhaustive procedures in terms of parameters used and triggering criteria.

8.1.1 Basic Handover Principle

During a call (transfer of user data or signaling), the radio link is regularly measured and evaluated. Detection of an abnormality causes the base station to alert its mobile switching center or the base station controller. Upon the receipt of this alarm, the switch or the controller searches for a new cell and/or a new channel. If it finds one suitable it then triggers a handover, otherwise communication continues on the same channel and the handover attempt procedure is periodically repeated. Following the success of a handover, the old channel is released.

8.1.2 Growing Importance of the Handover Procedure

In first-generation systems (analog), the size of cells is large (with radii of some tens of kilometers), which makes the handover procedure (Figure 8.3) a phenomenon that is unlikely to occur (it wasn't even available in some of the very first systems).

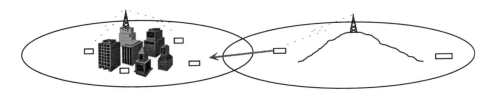

Figure 8.3 Handover in first generation systems.

With second-generation systems (digital) the sizes of the cells are smaller (their radii can go below 500m within high traffic environments), which consequently leads to a higher probability of crossing a cell border before a call is finished (Figure 8.4). Handovers are frequent (several handovers can happen during a call in an urban area) and are now commonplace, whereas they used to be quite rare in first-generation systems.

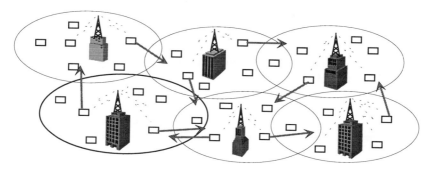

Figure 8.4 Handover in second-generation systems.

With second-generation systems the constraints are much more important. The procedure has to be quick and accurate, and the choice of the target cell has to be the optimum. Further decreases in cell size to meet the high capacities and data rates per user is to be expected in future third-generation systems (Figure 8.5). In addition, different systems (for example, cordless systems for indoor high density environments and outdoor cellular systems to fill holes in coverage) will require totally transparent roaming between the telecommunications services in the various environments. Handover procedures therefore will acquire a crucial role because they will not only have to maintain the quality of service, but they will also be used to ensure the correct service is offered to the user.

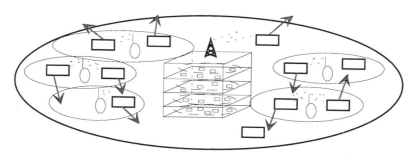

Figure 8.5 Handover in third-generation systems.

8.1.3 The Various Handover Phases

Quite often, the handover procedure is subdivided into three main phases (see Figure 8.6):

- Monitoring and link measurement,
- Target cell determination and handover triggering,
- Handover execution (effective link transfer).

These three phases follow each other chronologically. Target cell determination occurs during the measurement phase.

Each handover phase has to meet certain constraints. Obviously measurement delay has to be less than the duration of a cell crossing. Its duration therefore, corresponds to the size of the cell and hence high constraints are put on it within pico- and microcellular environments. In addition, the duration of the handover processing decision together with the determination of the target cell should be short enough to facilitate a timely handover. Finally, the execution phase should be achieved as quickly as possible to minimize the probability of radio link loss and any degradation in the quality of service generated by the transfer of the link.

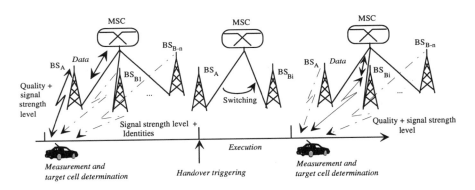

Figure 8.6 Various handover phases.

During the measurement phase, the network and mobile terminal (only the latter in the case of second-generation systems) evaluate the link quality while regularly measuring other channels and cells required for the establishment of a new physical link if required. A handover is triggered as soon as the quality of the current link falls below a preset threshold. The decision to trigger a handover is then taken by relying on these measurements. A handover can be initiated and totally controlled either by the network (as is the case for first-generation systems) or assisted by the mobile (in second-generation systems).

8.1.3.1 Measurement and Collected System Parameters

The handover procedure requires the measurement of a number (of varying degrees of importance, depending on the system) of indicators as well as a collection of system parameters. This data is processed and used by the handover triggering and target cell determination algorithm. Among the measured parameters, the principal ones are

- The *received signal strength indication* (RSSI), which is an indicator of the quality of the radio link,
- The *bit error rate* (BER) in digital systems, which helps deduce the value of the C/I ratio,
- The *distance between the mobile and the base station* (this distance is calculated from the value of the timing advance),
- The base station identity,
- The frequencies of the neighboring base stations broadcast channels (e.g., BCCH in GSM),
- The position (in frequency and time) of the various channels.

The number of measurements required for handover triggering should not be excessive to avoid overload of the processing equipment (base stations and eventually mobiles). On the other hand, measurement intervals should be small enough so that the system can react in time to a decrease in signal. For example, in GSM the number of measurements to be averaged is between 3 and 10 and the interval between measurement reports is about half a second (reports from the MS to the BSC are every 480 ms). The number of samples included varies according to the channel type (or the bit rate). Finally, the mobile reports to the network measurements it has made of its six most powerful neighboring base stations.

8.1.3.2 Candidate Cell Selection

During the measurement and data collection phases, a list of candidate cells is generated. Take the example of the IS-95 CDMA system [2]. In this system, it is the mobile that processes the list of candidate base stations. Each mobile maintains several lists of channels specified by pseudo-noise (PN) (see Chapter 12) offset and frequencies. The members of these lists depend on various threshold criteria and include:

- The *active set*, which contains the currently used channels and has more than one member during a soft handover (see Section 8 1.3.4.1).
- The *candidate set*, which contains channels that are almost as good as the ones in the active set (see selection criteria hereafter). It is from the candidate set that new channels are chosen for soft handover.
- The *neighbor set*, which is the set of channels that do not meet the criteria to be included in the active set and candidate set, but is reasonably strong.

- The *remaining set*, which contains every other channel.

As soon as a traffic channel is allocated to the mobile, the latter determines the candidate cells and sets up the *candidate set*. A base station is included or withdrawn if it meets one of the following criteria:

- The pilot channel power goes above a predetermined threshold.
- The mobile, initially, receives a message containing the network identity of the base station, and second, it receives signals from this base station at a power that is higher than a predetermined threshold.
- If the calculated delay (see below) associated with this base station expires it is withdrawn from the list of candidate cells.
- If the mobile adds a base station to the list of candidate cells but the list already contains the maximum number of candidate cells. The base station whose calculated delay expires first will then be withdrawn from the list.

A calculated delay (called the *handoff drop timer*) can be associated with each candidate base station. It is triggered as soon as the power of the pilot channel from a particular base station drops below a predefined threshold. This calculated delay is reset to zero and deactivated as soon as the channel power exceeds the threshold level.

8.1.3.3 Handover Triggering Algorithms

A handover is triggered after evaluation of several conditions. If at least one of these is met, a handover is triggered. The evaluation of handover triggering criteria occurs after each collection phase of measurements from current and neighboring base stations. Within a descriptive framework of handover procedures for first- and second-generation systems details of parameters, measurements, and triggering criteria are presented. Generally these criteria are based on three variables as follows: duration of the measurement averaging window, threshold level (signal power and quality thresholds typically), and the hysteresis margin [3].

Several methods have been proposed in the literature for handover triggering. Some are currently implemented in operational cellular systems. As examples the principles of some of them are shown below.

Signal relative power. A handover is triggered when signals transmitted by a neighboring base station are received with a higher power level than those of the current base station [4]. This criterion does not take into account the absolute power level of the current base station. Thus this method can trigger a handover when the current base station is still being received with sufficient power and is still capable of serving the mobile [5].

If $RXLEV\_DL/UL(BS_n)$, the received power level of the neighboring (current) base station n (BS_n) on the downlink (DL) or uplink (UL), and $RXLEV\_DL/UL(BS_{current})$, the received power level of the current base station ($BS_{current}$) on the downlink (DL) or uplink (UL), this criteria can be expressed by:

$$RXLEV\_DL/UL(BS_N) > RXLEV\_DL/UL(BS_{CURRENT})$$

Relative signal power with threshold. In this case, a threshold level is added to the previous case of difference in signal power [6]. The handover is triggered only if the power of the current signal drops below a given threshold and if the mobile has received a neighboring base station whose signal is more powerful than that of the current base station. The performance of this method is highly dependent on the threshold value.

With the same notation as before and with *RXLEV_THR*, the received signal level threshold, the criteria can be formulated either as:

$$RXLEV\_DL/UL(BS_N) > RXLEV\_DL/UL(BS_{CURRENT})$$

and

$$RXLEV\_DL/UL(BS_N) > RXLEV\_THR$$

or as

$$RXLEV\_DL/UL(BS_N) > RXLEV\_DL/UL(BS_{CURRENT})$$

and

$$RXLEV\_DL/UL(BS_{CURRENT}) < RXLEV\_THR$$

Relative signal power with hysteresis. A handover is triggered only if signals from the new base station are greater than that of the current base station's signals by a value called the *hysteresis margin* [7]. As in the previous case, adjustment of this hysteresis value determines the performance of the handover.

With the notations defined previously, and with *HYST*, the hysteresis margin, a handover will be triggered if the following criterion is met:

$$RXLEV\_DL/UL(BS_N) > RXLEV\_DL/UL(BS_{CURRENT}) + HYST$$

Relative signal power with hysteresis and threshold. With this procedure the two previous criteria are combined. A mobile will only carry out a handover to a new base station if the power of the current base station signal drops below a predetermined threshold and if the power of the target base station signal exceeds that of the current base station by a value that is at least equal to the hysteresis margin. This method is used in the GSM standard.

With the notations defined previously, the handover criteria can be formulated either as follows:

$$RXLEV\_DL/UL(BS_N) > RXLEV\_DL/UL(BS_{CURRENT}) + HYST$$

and

$$RXLEV\_DL/UL(BS_N) > RXLEV\_THR$$

or as:

$$RXLEV\_DL/UL(BS_N) > RXLEV\_DL/UL(BS_{CURRENT}) + HYST$$

and

$$RXLEV\_DL/UL(BS_{CURRENT}) < RXLEV\_THR$$

Prediction techniques. These techniques are based on an expected value of signal strength and use the velocity and direction of the mobile station. The major drawback with this kind of method is the complexity of processing it requires and the difficulty of predicting radio signal fluctuations [8].

8.1.3.4 Execution Procedures

When a handover is triggered, a new channel is set up, the connection is transferred to a new link, and the previous one is released. The handover process can be classified, first, by the way the new link is set up (whether the previous link is released before, during, or after the establishment of the new link), and according to the source of messages for new link establishment as in the case of a mobile initiated handover.

Hard, seamless, and soft handover. Three different cases of handover can be defined which depend on whether the existing link is released before (*hard handover*), during (*seamless handover*), or after (*soft handover*) the establishment of a link with the target base station.

In the case of a hard handover, the mobile uses only one radio channel at a time, a phenomenon that gives rise to an interruption of communication during transfer (Figure 8.7). The new link is established in advance in the network so that the interruption be made as short as possible.

Figure 8.7 Hard handover.

The communication and routing of data via the new link are achieved simultaneously. This supposes that the handover is controlled by the network from beginning to end. One of the major advantages of this method lies in that the mobile uses only one radio channel at a time. Its drawback is that the transfer is accompanied by a somewhat lengthy suspension of the call. Also if the mobile is being used as the source of measurement all the data required for the handover are carried up to the network through the radio interface (mobile toward network), which causes significant loading at this interface. This method is used in the GSM system [9, 10].

In a seamless handover, the new link is set up and used in parallel with the old one and the data flow is transferred by the mobile on both links (Figure 8.8). During handover only the original link is active. At the end of the handover the new flow of data is activated by switching at the network level; the previous flow is stopped and its link is released. This type of handover supposes that the mobile transmits on two channels simultaneously and therefore consumes more resources than a hard handover but with a preserved quality of service and a lower call interruption probability during the transfer phase. *Dynamic channel allocation* (DCA) is more appropriate for this kind of procedure than a fixed channel allocation method. This kind of handover is implemented in the DECT standard (see Chapter 10) [11].

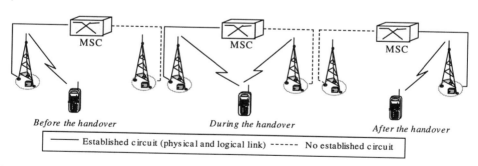

Figure 8.8 Seamless handover.

Soft handover has been introduced with the CDMA systems (IS-95) [2, 12]. With a soft handover, two links and two corresponding flows are activated during a relatively lengthy period of time (Figure 8.9).

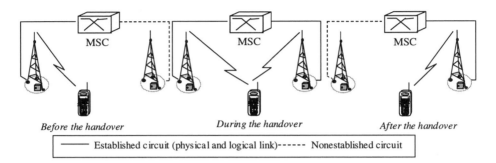

Before the handover   During the handover   After the handover

——— Established circuit (physical and logical link)------ Nonestablished circuit

Figure 8.9 Soft handover.

The mobile terminal is connected simultaneously to two (or more) base stations. The two links, seen from both the mobile and the network, are considered as paths that are carrying only one flow of data. The use of this technique can be extended by allowing the mobile to be permanently connected to more than one base station during a call.

This technique is known as *macrodiversity*. During its travels the mobile sets up and releases connections with the network according to radio propagation conditions. The handover is implicitly carried out with the help of the link's attachment/detachment process. This method has a major advantage in terms of quality of service for the user but suffers from a big drawback in that it generates a heavy load on the network. This is because the system has to process several links at a time instead of just one. This type of handover is being considered in the deliberations for the definition of future systems.

Backward and forward handover. The establishment of a new link is accomplished through signaling exchanges between the mobile and the target base station. The signaling exchanges can take one of two possible paths, either through the fixed network, or through the radio interface [13].

If the exchange occurs through the old link, the handover is called a *backward handover*. In this case, the new links are initiated from the current base station. This solution can be used for handovers between cells of the same type. It is slower than the next solution (forward handover) but in return offers better control of the radio interface resources [14].

When the data are transmitted directly from the mobile station to the target base station over the radio interface the handover is then called a *forward handover*. In this case, it is the new base station that is requested to establish links with the fixed network. Obviously this case is only of interest when the mobile controls the handover procedure. The advantage with control of handover by the mobile is that activation of the new link is according to its own perception of radio propagation conditions. In this way a significant reduction in the transmission of a large number of measurements over the radio interface is achieved. When handover control is carried out at network level, the delays and spectrum usage generated by the transmission of network handover commands to the mobile are difficult to justify. Forward handovers can be used for handovers between cells of different kinds (in other words between different networks). The advantage of this solution lies in its speed. It is, therefore, better suited to handovers between small cells where delay constraints are significant. However, this method suffers by reducing the network capability of controlling its radio resources.

In current cellular systems only the backward handover method is implemented. Forward handover is used, for example, in cordless systems such as DECT.

Synchronous, asynchronous, and pseudosynchronous handovers. In TDMA systems it is possible to temporarily synchronize the target and current cells with each other [15].

The timing advance (see Chapter 3)—determined by the network and used by the mobile to adjust its transmission time in the uplink slots—will be different in the target and current cells (Figure 8.10).

When the cells are not synchronized the handover is called an *asynchronous* or *non-synchronous handover*. The network can only carry out the determination of timing advance for the new cell after the mobile has been successfully connected to the new cell. With a GSM network this requires that the mobile transmits an access packet on a dedicated channel in the new cell before resuming its communication (the packets are generally transmitted on the access channel). To be exact, the mobile station starts by transmitting special bursts on the allocated channel allocated to the target base station. The value of timing advance for the handover access bursts is set to 0. The target base station evaluates the propagation time on receipt of these bursts and sends the value for timing advance to the mobile station before it may start its transmission. This kind of handover introduces a longer interruption time than synchronous handover because of the time required for the timing advance determination [16]. In GSM, depending on network implementation, communication interruption is between 200 and 300 ms.

When the two base stations (current and target) are synchronized, the network can estimate the value of timing advance that the mobile requires for the target cell without any presynchronization phase. In this case, the mobile can immediately transmit its messages in the timeslots of the dedicated channel that has been allocated to it in the target cell. This kind of handover is called a *synchronous handover*. In GSM the BSC informs the new base station of the timing advance value. Assuming an optimized implementation, the interruption, in the mobile to network direction during a synchronous handover is about half what it is with an asynchronous handover. In GSM, for example, the duration of a call interruption during the execution phase of a handover can be reduced from approximately 200 ms (asynchronous handover) to 100 ms (synchronous handover) [1].

Remark: A *pseudsynchronous handover* scheme can be defined if the network is pseudo-synchronized. This type of handover is almost the same as the synchronous case and shows similar performance. The only difference lies in the way in which the mobile station computes the new timing advance before changing channel. The pseudosynchronous handover is performed as follows: a mobile station that has to hand over from a base station I to a base station J. Base station I indicates to the mobile station (in the *handover command* message) the time difference between base stations I and J synchronization times $RTD = local\ GSM\ time\ (BS_I) - local\ GSM\ time\ (BS_J)$. The mobile station can therefore measure the observed time difference (OTD) between the two base stations. Let i_t be the mobile station to base station I propagation time and J_T the mobile station to base station J propagation time. The time difference $OTD = RTD + t_J - t_I$. The mobile station is therefore able to determine the value of timing advance before accessing base station J.

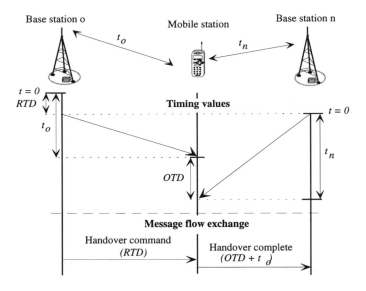

Figure 8.10 Use of the timing values during the handover process.

8.1.4 Various Kinds of Handover Seen by the Network

From the mobile point of view, a handover is the transferring of communication from one or many current radio links to one or several other radio links. Viewed from the network, a handover can be characterized by several levels of complexity. In fact, the network may have to deal with various different types of handover, which are: *intracell*, *intercell*, *inter-BSC*, *inter-MSC*, *subsequent*, and *internetwork* (Figure 8.11).

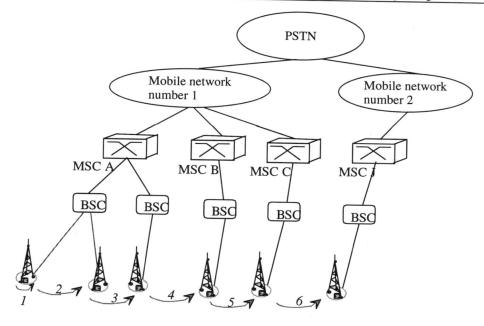

1: *Intra-cell HO*   2: *Intra-BSC HO*   3: *Intra-MSC HO*
4: *Inter-MSC HO*   5: *Subsequent HO (A toward B then B toward C)*
6: *Inter-network HO*

Figure 8.11 Various kinds of handover seen by the network.

8.1.4.1 Intracell Handover

The *intracell handover* (marked 1 in Figure 8.11) happens when the mobile remains connected to the same base station but changes the radio channel. This can occur when the mobile is still located in the service area of its current base station (it is receiving a strong enough signal), but the interference level on its radio channel is too high (i.e., degradation of the C/I ratio).

In the case of sectored cells, an intracell handover occurs when the mobile moves from one sector of a cell to another sector of the same cell. The intracell handover involves only the mobile base station and eventually the BSC. It is not processed by the MSC. From a network point of view, it is considered the simplest form of handover.

8.1.4.2 Intercell and Intra-BSC Handover

During an *intercell handover* (marked 2 in Figure 8.11), which occurs within the same BSC, the mobile changes base station. This type of handover is triggered when the mobile enters a new base station service area.

An intercell handover is triggered because of degradation in quality and signal strength between the mobile and the current base station while a neighboring base station offers better radio link conditions than the current one. The intercell handover criteria are different from those used for triggering an intracell handover. The intercell handover can be triggered for reasons other than radio conditions. Sometimes handovers are triggered to help migrate mobiles from a very loaded base station to less loaded neighbors. A handover that is triggered for reasons of traffic loading and occurs during call setup is called a *directed retry*.

8.1.4.3 Inter-BSC and Intra-MSC Handover

This kind of handover occurs when the mobile changes base station and changes BSC (case 3 in Figure 8.11). It is the MSC that has to manage the handover. The triggering criteria are the same as in the intercell and intra-BSC handovers. Yet in this type of handover the MSC chooses a target cell to take into account capacity criteria that the BSC may be unable to consider. The MSC selects a target BSC from the list of candidate cells sent by the BSC.

8.1.4.4 Inter-MSC Handover

With this kind of handover the mobile changes cell, BSC, and its MSC. Nevertheless, in GSM, for example, [17] an *MSC anchor* (the MSC on which the call had been initialized) remains involved in the communication since the network always links the call through it and then on via the new MSC to the mobile. One of the reasons lying behind this choice is the difficulty of processing several different billing tickets for a single call (which would be the case if the MSC anchor did not control the communication from beginning to end). The successive MSCs visited by the mobile during its communication are called *relay MSCs*.

This type of handover is particularly delicate and is more difficult to achieve than the intra-MSC kind. It also presents more risks (temporary loss of audio and a higher probability of call breakdown).

8.1.4.5 Internetwork Handover

Even though this kind of handover is specified in certain systems such as IS-95 (where the *analog handover* allows the transfer of a call between an IS-95 network and an AMPS analog network), it is unlikely that it will be operational for several years yet. It draws upon a set of technical aspects which are far from being trivial: real-time internetwork switching, authentication while the call is under way, recovering and checking out of the subscribers' rights, and so forth.

8.1.5 Evaluation of the Handover Procedure

The main objectives of a handover procedure are, first, to minimize the number of transfers. This minimizes the probability of call interruptions and reduces the switching load, and second, to minimize the handover processing delay by correct choice of target cell with speedy execution. If the handover is not fast enough, the quality of service may undergo some degradation and drop below the minimum threshold level [18]. A reduction in this delay also helps reduce the global channel interference level.

Received signal strength indication (RSSI) and BER can both be used during the selection of handover triggering criteria. The most simple and most commonly used parameter is the RSSI. To reduce the number of redundant handovers, a hysteresis and averaging window are used. The frequency of redundant handovers is at a maximum near the crossing point of the propagation pathloss curves; in other words on the borders of cells, and decreases as the mobile moves closer to the base station. The number of unwanted handovers can be reduced by means of hysteresis. With an increase in the hysteresis value, the number of redundant handovers decreases but the triggering time is delayed which can cause a handover to fail. Another means of eliminating redundant handovers is to use an averaging window for the signal strength measurements. This is also likely to delay the triggering time of the handover and can also result in failure.

Because of the highly fluctuating behavior at the radio interface, all situations (e.g., obstructions, interference, fading) are possible causes of signal variation (e.g., on those signals transmitted by current and neighboring base stations) for the handover procedure. To evaluate it (generally through software simulations) it is best to select a simple case of a mobile located close to the borders of two cells for verification. After a rather lengthy period of time (the period during which observations are carried out), the simulated mobile is able to identify the output powers of two base stations which are notably stronger than all other surrounding base stations (Figure 8.12). The evaluation of this process, for use in more complex cases, can only have significance if the behavior of the signal is tightly controlled. If it provides accurate answers to the evaluated criteria (with the help of indicators defined later) in this simple case, then it would be expected to react correctly when presented with more complex conditions.

The definition of quantitative values is difficult because they are closely linked to both and the circumstances of the handover context its scenario. Parameters that affect handover-triggering decisions are numerous (they include parameters that have an impact on the signal behavior). This is why the handover procedure is hard to define or evaluate. Therefore, this is why in the following sections an attempt is made to define these parameters only from a qualitative point of view.

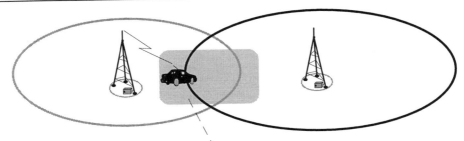

Observation area of the handover procedure process

Figure 8.12 Typical situation for the evaluation of the handover procedure.

8.1.5.1 Microscopic Indicators

First of all, a handover should be evaluated as to its impact on the mobile to network connection. This evaluation is achieved by means of a certain number of indicators called *microscopic indicators* which help identify the behavior and, hence, the performance of this procedure. Those that are most significant and most commonly used are defined below [19].

First of all there is what is called a *reliable link*. A link can be considered as reliable whenever the following two conditions are met. First, the target base station is happy to accept the mobile. Second, no other handover is required before a certain period of time has elapsed (which depends on the environment and behavior of the mobile). The concept of a reliable link in the definition of parameters relating to handover evaluation is used to determine the beginning and end of the link transfer. This concept, for all the reasons given above, can only be defined in a qualitative manner.

- *Number of handover attempts*: in this case the number of connection attempts between the mobile and a new base station before the establishment of a reliable link. The value of such an indicator has to be minimized.
- *Rate of failure*: stands for the handover failure probability rate (in other words, a failure occurs each time the target base station fails to receive access bursts from a mobile or when the connection with this base station is lost after a short period of time) which causes the mobile station to connect back to its old base station. The rate of failure should be minimized so that service quality will not be affected.
- *Ping-pong handovers*: these are handovers during which the mobile is connected alternately to the target base station, goes back to its initial base station, is reconnected to the target base station, and so on several times before establishing a reliable link (Figure 8.13). The ping-pong handover phenomenon should be avoided to minimize the use of resources (radio and network signaling resources) and reduce the chance of connection probability rate loss.

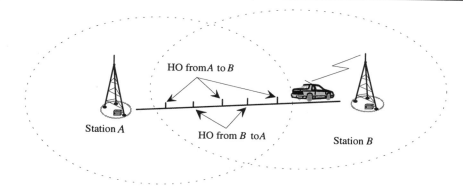

Figure 8.13 Example of ping-pong handover phenomenon.

- *Link transfer duration*: stands for the period of time between the decision to trigger the handover and the establishment of a reliable link with the target BS. The duration of the execution of a handover should be as short as possible in order to minimize the transfer period during which the connection may be lost.
- *Point where the handover is triggered off*: it is the point where the handover is triggered. It should be triggered as close as possible to a cell border. This border is roughly determined during the cellular planning phase.

The use of these few indicators helps evaluate a handover procedure and appraise its performance. Yet, such a procedure can demonstrate very good performance at a microscopic level and give poor performance at a macroscopic level. For the procedure evaluation to be exhaustive, macroscopic indicators must be used as are introduced below.

8.1.5.2 Macroscopic Indicators

The handover procedure should be evaluated from the point of view of its impact on the overall system, especially on the processing capacity (processing and traffic) of the system. The following two kinds of parameter are defined.

- *Number of handovers*: this number is evaluated, on average, for all communications. It can be estimated as a function of the distance covered by a mobile, time of day, and call rates. The result determines the sensitivity of the handover procedure. An excessively high rate indicates that the procedure is over sensitive to signal fluctuations (which causes high rates of radio and network signaling with an increased risk of disconnection). If handovers are too few (for instance, the number is less than cell border crossings), this indicates that the procedure is insensitive to signal variations (which can cause possible connection losses or degradation in the quality of service).

- *Overall quantity of consumed resources*: the execution of a handover entails an increase in the consumption of resources at the transmission level (radio and network interface) as well as in terms of processing (link switching and choice of the candidate and target cells). The resource consumption of the procedure has to be evaluated and interpreted in terms of signaling and processing operations. The value of such an indicator should be minimized.

8.1.5.3 Adjustment of Working Parameters

The behavior of the handover procedure depends on the handover triggering criteria and on the target cell selection method as well as on the values of the parameters used in these criteria. The value of these parameters should, therefore, be adjusted in such a way as to optimize performance. In practice, adjustment of these working parameters is one of the most delicate tasks to be achieved during a network's operating phase [20]. Indeed, contrary to fixed systems, two mobile networks with similar technical characteristics (e.g., capacity, equipment), but implemented in two different areas may well be provided with working parameter values that are likely to be quite different. The task is that with each network implementation or modification these values must be adjusted according to the local propagation environment.

To use these parameters the objectives to be aimed for are directly determined by the indicators introduced above and which are the following:

- Minimization of the number of handovers by distance covered,
- Triggering of the handover as close as possible to the cell border,
- Choice of the target cell in the most accurate way (in terms of quality of service and consumption of resources during and after the handover),
- Connection quality maintained during handover phase.

All these constraints can be directly interpreted in terms of minimization of the following performance indicators:

- Number of handover attempts,
- Handover probability rate failure or communication disconnection,
- Ping-pong impact,
- Handover execution duration (between the decision to trigger off the handover and the establishment of connection with the target base station),
- Quantity of resources consumed.

The adjustment of these parameters, until recently, has been performed manually but has been progressively automated along with increased capacity systems and the resultant increase in numbers of cells. Expert systems have been used to adjust values of these parameters in accordance with collected statistics and help assist or relieve human operators of the complex tasks which are becoming more frequent and more difficult to handle.

8.1.6 Handover Traffic

During network dimensioning, predicted traffic is estimated within each cell. Calls initiated within the cell (*fresh traffic*) must be considered as well as calls that arrive after a handover from another cell (grouped under the term *handover traffic*).

Most of the work dealing with the analysis of cellular mobile systems teletraffic assumes that fresh traffic and handover traffic follow a Poisson process [21, 22]. This hypothesis relies on the memoryless characteristics that have been adopted in telecommunications traffic engineering for some time. When only fresh traffic is considered, such a hypothesis is similar to the Poisson distribution of traffic generated in fixed telephone networks. Theoretically, it has been demonstrated [23] that handover traffic is also of a Poissonian nature if no blocking occurs in the network. This is not the case in a real environment, where the traffic flow characteristics from handover are modified by call blocking. The Erlang formula is, nonetheless, used to determine blocking, assuming that the traffic is Poissonian. This model produces approximate results, but these results are sufficiently accurate. The Poisson handover traffic hypotheses are currently used for planning cellular systems.

8.1.7 Handover Procedures in Analog Systems

Handover procedures were implemented in the analog systems only when problems of density arose with the implementation of medium- and small-size cells. Previously, cells were large enough so a user who had started a call within one cell had time to end it before being leaving that base station's coverage. Handover procedures began to be developed at the end of the 1970s. The main parameter measured and used to initiate triggering and select the target cells is the RSSI. Measurements are only carried out at base stations and at intervals that can reach several seconds [4]. The mobiles are only involved in executing handover commands sent by the controller. These handover procedures are not appropriate for current small sized cell systems, for the following reasons:

- The measured parameter is not accurate enough: RSSI is the sum of the received signal strength, interference, and noise. What is important in a cellular system is the ratio between the received signal and the interference level. (The latter is the sum of the interference and the background noise.)

- Measurements intervals are too long (several seconds), whereas considerable signal variations can occur within a second, especially in an urban environment.

- Processing is computed at switch level which is likely to generate significant delays (this includes measurement time, processing, decision making, and command signaling) and requires processing capacity at the MSC that could be used for other things. This is because the high traffic densities combined with the management of small cells are functions that first generation networks did not have to address.

Second-generation digital systems use handover procedures that require the use of a greater number of parameters to evaluate radio link behavior more comprehensively. These procedures are faster than those of the earlier analog systems thanks to more decentralization of processing operations and decision making. In the following section their major characteristics are introduced.

8.1.8 Handover in Second-Generation Systems

The rapid growth in the number of mobile subscribers added to a reduction in cell size requires speedy and accurate channel measurements [24]. In fact, the handover rate rises whenever the traffic per cell rises and when the size of the cells decrease, which are both characteristics of second-generation systems.

The main innovations related to handover with these systems are first, the measurement of BER for the point-to-point connection—this indicator helps deduce the level of C/I— and, second, the *mobile-assisted handover* (MAHO). Systems based on a TDMA technique offer mobiles the capability of measuring the RSSI of neighboring base station channel signals and then report these measurements to the current base station. These measurements are evaluated in the network, in other words, at the base station controller and mobile switching center (BSC and MSC). These entities indicate to the mobile the base station to which it is to be transferred when it leaves its current cell. In the GSM system, for example, the distance separating the mobile from its current base station can be evaluated with the help of the timing advance to assist in making a handover trigger decision.

The main advantage of MAHO is that the handover is managed from inside the network, which allows more flexibility with regard to future modification of the system and, in particular, the handover algorithms that have to keep pace with traffic and network growth.

In picocell systems, the very fast changes of radio propagation conditions require a reaction that is as fast as the handover procedure. In this case, it is the mobile station that selects the target base station and instigates a forward handover. This procedure is called *mobile controlled handover* (MCHO) with transfer controlled by the mobile and not by the network [25].

8.1.9 Handover in Third-Generation Systems

In third-generation systems handovers will increase considerably, even more than in GSM [26]. First, the characteristics of third generation systems are examined to identify the constraints that have to be met by the handover procedure when implemented in these networks [27].

Third-generation systems are composed of numerous different networks with varying characteristics (e.g., user density, speed of mobiles, services offered, environments served). Simplistically, they are comprised of various types of cells: macrocells to cater for fast vehicles and to cover areas with a low traffic density, microcells for average speed vehicles or pedestrians, used in urban areas and outdoor environments, and picocells for communication inside buildings and environments with traffic high densities. The networks will, therefore, be composed of cell layers. To each cell layer will correspond a specific network that could be provided by a specific operator. Various operators will be able to operate concurrent networks covering the same sites. These operators will technically be able to offer their respective subscribers, handover between networks. (They may be unwilling to do so on commercial grounds as has been seen with current cellular systems, both analog and digital.)

The most compelling requirements for handover are found in urban areas where cells are small (from radii of a few meters to a few hundred meters) in order to offer high subscriber capacity and high transmission bit rates. In addition, these networks are expected to cover a wide range of disparate environments (inside and outside buildings) and meet very different conditions (traffic levels and velocity of mobiles, in particular) that are expected to be quite varied. The internetwork handover is one of the highlighted concerns for third-generation system designers as its technical feasibility is somewhat in doubt in an environment to be characterized by good quality of service conditions.

To offer these various possibilities and functional facilities, the handover procedure is required to meet high-pressure demands, the most important of which are [28]

- Reacting in "real time" (especially for fast handovers in a pico- or microcellular environment),
- Meeting the demands of the subscribers (in accordance with tariffs offered by the various operators, for instance),
- Being able to move from one network to another (interoperator handover),
- Detecting subscriber behavior changes (mobility—transfer a communication from a picocell to a micro or macrocell network in the case of a speed increase by the subscriber, for instance).

8.1.10 Conclusion

The handover (inter or intracell) is one of the most crucial moments during a mobile communication. The parameters of the management procedure for a handover should guarantee an acceptable quality of service for the user and minimize the use of the system resources (network and radio). This process implies a change of the connection between the radio and network within the constraint of maintaining good communication.

The difficulties with this procedure go on increasing commensurate with decreases in cell size and an increase in the number of networks between which handovers are to be processed. The decentralization of handover management that started with second-generation systems will be further increased with third-generation systems, which will impose even more difficult constraints on the handover procedure [29]. This procedure that did not exist at the time of the very first cellular systems, and which with first-generation systems used to be considered rather exceptional, is considered as commonplace in such systems as GSM (the mean number of handovers per call is about 0.5 with a GSM system in a dense environment). It will become even more frequent and complex with future systems.

8.2 NETWORK MOBILITY: CELL SELECTION AND ROAMING

Handover, as described in the previous section, occurs when the mobile is active (that is "in a dedicated state"), in other words when a point to point radio link exists between the network and a mobile. When the mobile is inactive (i.e., not in use), two different processes become relevant: selection/reselection of cells and roaming. The first process allows the mobile to receive data transmitted by the network (especially cell characteristics, access parameters to the network, and location) and to camp (listen) on a specific cell. This cell becomes the access cell if a radio link is to be established between the mobile and network. The second process, which is not used with the small sized cells, allows the system to identify the approximate position of a mobile with some degree of accuracy. These two functions allow the mobile to receive or transmit calls when the subscriber is on the move and within coverage of the network. Their operation is totally transparent to the user but they are, nevertheless, large consumers of radio and network resources [30].

8.2.1 Cell Selection/Reselection Process

The main objective of the cell selection/reselection procedure is to allow a mobile station to choose a specific cell in the network (the one that gives the best signal, for instance) to

- Record data transmitted by the network to mobiles,
- Be ready to access the network,
- Inform the network about its movement.

This procedure requires each noncommunicating mobile to listen continuously to base stations located in its neighborhood. The data transmitted by these base stations and measurement of signals received allows the mobile to choose, at any given time, a preferred base station. This is the base station with which it would set up a link in the case of a mobile network communication request such as location updating or incoming and outgoing calls.

Selection is the procedure implemented at mobile switch-on whereas reselection occurs when the mobile is on the move after a cell has already been selected. Cell selection and reselection use the same algorithm for the choice of a cell. Note that during an active phase (i.e., in a call), it is the handover procedure that is used during the transfer from one cell to another and the selection/reselection procedure is "masked."

8.2.1.1 Selection/Reselection Cells Process Phases

After switch-on a mobile goes through the following stages:

- It searches for system carriers: The mobile will either scan all the system channels (a process that is likely to last some tenths of a second or even a few seconds), or use a list stored during its last active phase in memory in the terminal or card (the SIM card in the case of GSM).
- Select several carriers from among the most powerful.
- Collect system data from these carriers: State of cells (access barred or not), access parameters, handover parameters, synchronization (in the case of TDMA systems), location of channels in frequency and time.
- Register, if required, within the location area (see the definition of a location area in Section 8.2.2.4.) of the selected cell.

Throughout its movement within network coverage a mobile station performs the following tests and actions:

- If the cell is no longer received, a better cell will be selected.
- If it receives a paging message from the network (the case of an incoming call), it then transmits a request for re-establishment of connection with the network.
- If its location area has changed, the mobile transmits data to the network (see Section 8.2.2.4.).
- If the mobile goes out of coverage, a new network (if there is one and the subscriber has access rights) is selected.

If several networks can be selected (as is the case today where several operators can cover the same area), selection of a cell is realized automatically by the mobile or manually by the subscriber. In the former case, a list of networks arranged in priority order is used by the mobile during the selection phase; in the second case, a list of networks received by the mobile is displayed to allow manual network selection.

8.2.1.2 Parameters Used for Cell Selection/Reselection

Selection/reselection of a cell uses several parameters among which the most commonly used in the cellular systems are the following:

- The broadcast channel received signal level (the broadcast channel is a common signaling channel on which each base station transmits data for all the mobile stations located in its cell and in its neighboring cells). The mobile listens to the broadcast channels of base stations located in its neighborhood.
- The state of the cell: certain cells can be barred from access for various reasons (e.g., resource congestion, link transmission failure, cells only intended to handle handover traffic).
- The network identity: the case where several networks use the same frequency band and cover the same area.
- The geographic area: some subscriptions categorized by region are available.

Some timer parameters can also be used to avoid selection at short intervals, by a moving mobile station, of cells belonging to different location areas. This avoids successive location updating and the generation of heavy signaling traffic on the radio interface and network. These timers can also be used to direct the mobile on to a particular type of cell in the case of layered networks (macro and microcells networks). Slow mobiles are directed on to microcells and fast mobiles on to macrocells. Timer parameters are used in the GSM selection/reselection procedure.

8.2.1.3 Conclusions

Mobiles that are in the idle state carry out the cell selection/reselection process. It helps each mobile camp on a cell to collect all the parameters required for connection on that cell when a communication has to be set up with the network. This process also allows mobiles to identify their location at any time in the network and, therefore, inform the system where they have roamed. In the following section, location management procedures are dealt with more thoroughly.

8.2.2 Location Management

In fixed telecommunications networks a user is always associated with a terminal and a terminal is associated with an access point in the network. In mobile communications systems, terminals have no fixed connection. They communicate dynamically with the fixed infrastructure over radio channels. In this case, connection between the terminal and the network access point is dynamic. Yet, with most systems (GSM is an exception) a terminal is always associated with its owner.

Knowledge of the location of each mobile is one of the most important characteristics of cellular systems. This is possible thanks to the use of location management procedures. Contrary to fixed networks where a number (i.e., a telephone number) corresponds to a fixed physical address (a telephone socket, typically), the mobile terminal number becomes, from the network point of view, a logical address, but not fixed. The main consequences of terminal mobility are the need for the system to continuously identify the location of each mobile and the need for the mobile to remain "active" (in idle mode) to inform the system about its movements (by radio transmission). Mobility management entails considerable signaling traffic and/or processing over the radio interface and the network, whereas in a fixed network an inactive terminal (where no communication is under way) generates no traffic at all on the network.

Roaming is not specific to radio systems. It is a service offered by a growing number of fixed networks that are adopting intelligent network architecture. Indeed, by defining *roaming* as being "the possibility of using a telecommunications terminal at a given point on the network," it is possible to imagine that wired networks can provide a roaming service to their subscribers. The subscriber is provided with an identity (smart or magnetic card, or personal code) allowing access on any network terminal. In this way the network is able to track the location of subscribers and route their calls. The notable differences with cellular systems are twofold. It is the impossible to have a handover (a subscriber is not allowed to change from one line to another when the call is under way), and there is a discontinuity of mobility offered to the subscriber (in effect, the calls do not go through unless the subscriber is located at a network terminal). At the end of this chapter, there is a brief reference to the Universal Personal Telecommunication (UPT) concept, which is entrusted with the generalization of roaming within fixed and mobile networks.

8.2.2.1 Location and Paging

Location management schemes are essentially based on users' mobility and incoming call rate characteristics. The network mobility process has two basic procedures: *location* and *paging,* both of which fight each other for network resources and a compromise has to be reached between them. The location procedure allows the system to have knowledge of a user's location, more or less accurately, in order to route an incoming call. Location registration is also used to bring the user's service profile near its location and allows the network to provide rapid services (i.e., the VLR functions in GSM). The paging process consists of sending paging messages in all the cells in which the mobile terminal could be located. Therefore, in terms of resources, if the location cost is high (the user's location is known accurately), the paging cost is low (the paging message is only transmitted over a small area). If the location cost is low (the user's location is only known approximately), the paging cost is high (paging messages will have to be transmitted over a wider area).

In first-generation cellular systems, traffic was highly unbalanced. Less than one third of calls were incoming and the remainder outgoing. The paging process (which is only required for incoming calls) was a relatively rare event and had only a small impact on mobility management traffic. Location updating also had a low impact due to the large cells. Against the trends stated earlier, current systems experience a balance between incoming and outgoing call rates. The paging process has gained in importance and thus location management is also more critical. For instance, recent statistics reported from GSM operators show that, in the dense urban environment of Paris, location updating can be 10 times that of the call rate at peak hour traffic.

In the remainder of this section, the main location procedures used in present systems are introduced. In particular the methods used in GSM are examined. In Appendix 8A, a classification of location management methods as defined in the literature for future mobile systems is shown.

8.2.2.2 Level 0: No Location Management

In early wide area wireless systems (precellular), human operators had to process calls and user location was not managed by the system. A user was able to set up a call through any base station and paging messages addressed to called mobiles were transmitted through all base stations. The main characteristics of these systems were their very large cells and low user population and call rates. Small capacity cellular systems (with a few tens of base stations serving a few thousand users) may also not use a location management method even when the system can support it. If subscriber numbers and calling rates do not require it, location management may not be necessary. This level 0 method is as simple as it can be: no location management is implemented (i.e., the system does not track the mobiles). Search of a called user has to be made over the complete radio coverage and within a limited time. This method is usually referred to as the *flooding* algorithm [31]. It is used in paging systems because of the lack of an uplink channel, which allows a mobile to inform the network of its location. It is also used in small private mobile networks because of their smaller coverage areas and user populations.

The main advantage of not locating mobile terminals is obviously the simplicity. In particular, there is no need to implement special databases. Unfortunately, it does not suit large networks dealing with high numbers of users and high incoming call rates.

8.2.2.3 Level 1: Manual Registration

Manual registration requires the user to invoke a special procedure to indicate location if incoming calls are to be received. From the network side, this method is relatively simple to manage because it only requires the management of an indicator that stores the current location of the user. The mobile is also relatively simple, as its task is limited to scanning paging channels for detection of messages. This method has been used in telepoint cordless systems (such as the former Bi-Bop CT2 system in France). The user had to register each time a new island of CT2 beacons was encountered. To page a user the network transmitted messages first through the beacon on which registration had been made and, if the mobile did not answer, extended paging via neighboring beacons. Manual registration can be used in paging systems where an international roaming service is provided to the users (e.g., Ermes). In this case, a subscriber wishing to receive a call in the visited country will have to call a special number (over the PSTN, for instance). Paging messages are then transmitted in the relevant areas.

The main drawback of this method is the requirement for a user to register in each area when moving around. Nevertheless, this low level of ergonomics can be balanced by the low equipment and management costs that allow the operator to provide users with attractive tariffs.

8.2.2.4 Level 2: Use of Location Areas for Automatic Location Management

Currently the location method most widely implemented in first- and second-generation cellular systems (e.g., NMT, GSM, IS-95) uses *location areas* (LAs) (Figure 8.14). In these wide area radio networks location management is performed automatically.

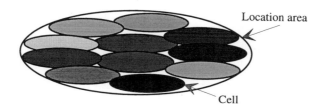

Figure 8.14 Location area structure.

LAs allow the system to track mobiles roaming in the network: approximate subscriber location is known if the system knows the LA in which the subscriber is located. When the system needs to establish a communication with the mobile (typically, to route an incoming call), paging only occurs in the current user LA. Thus consumption of resources is limited to this location area. Paging messages are only transmitted in the cells of this particular LA.

The network must have continuous knowledge of the address of the user's LA. This address is stored in a centralized database with a location pointer and the users' subscriber data.

When the mobile powers on, it registers at the network. This registration allows the network to locate the mobile and to store its LA address in the location pointer and also allows the subscriber gain access to network subscription services. When powered down the mobile may deregister but this action is not mandatory. This function is optional for the following reasons. A deregistration that is triggered when the mobile is switched off occurs for a large number of professional subscribers (a large percentage of the population on present cellular systems) after working hours. Massive deregistrations of a large number of mobile terminals, especially during peak hours, can lead to resource congestion. In addition, these professional subscribers have little chance of being called after the busy hour so the advantage of terminal deregistrations becomes very marginal compared with resource loss that may occur. To track subscribers roaming under network coverage which is split into several LAs, various location tracking methods can be used.

Periodic location updating. This method is the simplest because it only requires the mobile to transmit its identity periodically to the network. Its drawback is related to its consumption of resources, which are user independent and may be unnecessary if the user does not move for several hours. In general, this method is combined with the following one.

Location updating on LA crossing. This method requires that each base station periodically broadcasts the identity of its LA. The mobile has to listen continuously to network broadcast information (on the broadcast channel) and store the current LA identity. If the received LA number differs from the stored one, a location update (LU) procedure is automatically triggered by the mobile (Figure 8.15).

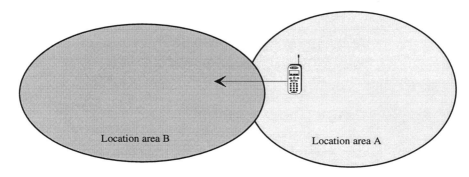

Figure 8.15 Location updating on LA crossing.

The advantage of location updating is that it only requires location updates when the mobile actually moves. A highly mobile user generates a lot of LUs, while a low mobility user only triggers a few.

A hybrid method that combines these two previous methods can also be implemented. The mobile generates its LUs each time it detects an LA crossing. Nevertheless, if no communication (related to an LU or to a call) has occurred between the mobile (in idle mode; i.e., switched on but not communicating) and the network for a fixed period (say, 3 hr), the mobile generates an LU (i.e., a periodic LU). This periodic LU typically allows the system to recover user location data if the database has failed.

8.2.2.5 Separate User and Paging Networks

Search for the user can be achieved through a network separated from the one that carries the calls. A signaling network (*paging network*) is physically separate from the users' data transport network (*user network*). This has been used for some wireless systems offering a telepoint cordless service combined with a paging service (Figure 8.16). This combining allows users to be reached even outside user network coverage, which is generally limited to high-density areas.

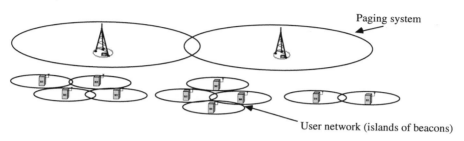

Figure 8.16 Location management with separate paging and transport networks.

8.2.2.6 Hybrid Method

The method most widely used in cellular systems is the one that combines location updating on LA crossing and periodic LUs. The mobile triggers the LU each time it crosses an LA. Nevertheless, if no communication (for an LU or for a call) occurs between the mobile and the network for a predetermined period, the mobile triggers a LU to indicate that it is still present and powered on. Using this method is justified because the network may lose the location data for the subscriber (e.g., breakdown of network equipment) or the mobile may have been out of range from the network for a period of time. In this last case the network may have triggered an implicit deregistration. This kind of hybrid method is used in the GSM system. The LU period can be adjusted according to mobility and radio environmental parameters, as well as the use of implicit and explicit deregistration mechanisms.

8.2.2.7 Location Area Size Optimization

To minimize location management costs (LU + paging traffic and processing) location areas have to be designed carefully. Two kinds of methods can be used to design optimum LAs: analytical methods and heuristic approaches.

Analytical approaches are based on restrictive hypotheses regarding cell shape, LA structure, and user movement, by usually assuming that these are regular. An interesting problem is to determine a subscriber mobility model that can approach as far as possible the movement of real subscribers. Common approaches for modeling human movements are mentioned in [32], among which is the Markovian model, also known as the random-walk model (this is a model that describes individual movement behavior) and the fluid model, which considers traffic flow as the flow of a fluid and is used to model macroscopic movement behavior. One of the first studies to address the second method has been that of Morales [33]. Mobility traces that record actual movement and behavior for certain segments of the population are more realistic than the mobility models.

In general, authors use a simple model based on fluid flow assumptions. This model is applied in [33] to determine the size of an LA. A simple method is defined, taking into account R, the cell radius, V, the mean mobile velocity, LU_{cost} the cost of LUs (in terms of the number of location updating messages required to update the location of a mobile), PAG_{cost}, the cost of paging (in terms of the number of paging messages required to find a mobile), and N (to be determined) which is the number of cells per location area. Calculations lead to the following formula that gives the optimum value of N:

$$N_{opt} = \sqrt{\frac{v \cdot PAG_{cost}}{\pi \cdot R \cdot LU_{cost}}} \qquad (8.1)$$

In real cases, cell shapes and patterns are not regular. If less restrictive assumptions are considered, the LA partitioning problem is much more complex and appears as a non-polynomial (NP) complete combinatorial problem. Thus, only empirical methods have been able to approach an optimum solution.

An approach, proposed in [34], makes use of genetic algorithms. Genetic algorithms are used to efficiently group the cells under a mobility cost function constraint. They use several processes such as elitism, linear normalization of chromosomes, and edge base crossover. Other empirical methods can be used, such as simulated annealing [35], which is currently used by some GSM operators; single-move heuristic; and steepest descent optimization.

8.2.2.8 Database Architecture for Location Management

In extended networks, location information is distributed within the network in several databases. In general there are two principal database types: the *home database* (HDB, one per network), and a *visitor database* (VDB, several in each network). These databases must have the following functionality: location tracking and registration, call delivery, authentication and verification, encryption and decryption key generation, resource and service profile management, supplementary service details, and billing.

Home database. The HDB is unique in the network. Although several pieces of physical equipment can be used to store the HDB data, only one logical entity is recognized by the network. This database is used to store information about all the subscribers on the network: name, number, access rights, security data, and the current subscriber location. Paging a mobile subscriber in a network always starts by a request to the HDB for that subscriber. In GSM the HDB is called home location register (HLR). Note that the HLR can be separate from the database where security related information is stored (see also Chapter 5). In GSM the latter is called the authentication center (AuC).

Visitor database. A network may include several VDBs. A VDB stores information about all the subscribers registered within a single LA who belong to this particular VDB. The information consists of a partial copy of the information stored in the HDB, retrieved either from the previous VDB or directly from the subscriber's HDB. The VDB contains the address of the LA in which the mobile is located. In GSM, this database is called a visitor location register (VLR).

Database organization. In general, three database types can be identified: *centralized* architecture, *distributed* architecture, and *hybrid* architecture.

Figure 8.17 presents an architecture where a single centralized database is used. This is well suited to small and medium sized networks typically based on a star topology.

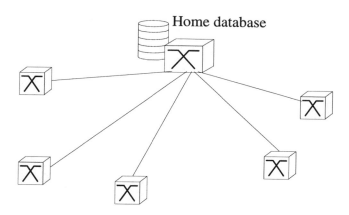

Figure 8.17 Centralized architecture.

The second type (Figure 8.18) is a distributed database architecture, which uses several independent databases according to service provider or geographic proximity.

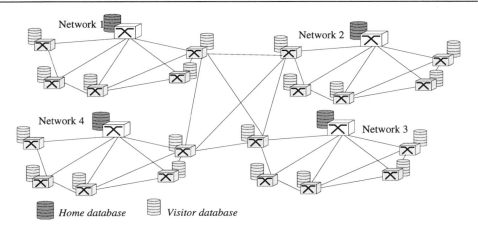

Figure 8.18 Distributed architecture.

It is best suited for large networks including subnetworks managed by different operators and service providers. The "GSM worldwide network," defined as the network of all the interconnected GSM networks in the world is such an example of a large network. This is the structure that provides the *international roaming service*. The main drawbacks of this architecture are clearly the cost of the database system acquisition, implementation, and management.

The third case (Figure 8.19) is a hybrid database architecture, which combines centralized and distributed database architectures. In this case, a central database (similar to an HLR) stores all user information. Other smaller databases (similar to VLRs) are distributed in the network. These VLR databases store portions of HLR user records. A single GSM network is an example of such an architecture.

8.2.2.9 A GSM example

The GSM standard [1] defines a database structure based on:

- HLR where all subscriber related information is stored (e.g., access rights, user location). Security parameters and algorithms are managed by the AuC, which is often considered as being part of the HLR.

- Several VLRs, which store part of the data relating to users located in its related LAs.

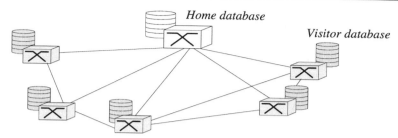

Figure 8.19 Hybrid architecture.

Location management defined in GSM combines periodic LU with LU on LA border crossing. The VLR stores the LA identifier and the HLR stores the VLR identifier.

It consists of three main types of LU procedure: the intra-VLR LU, the inter-VLR LU using the temporary mobile subscriber identity (TMSI), and an inter-VLR LU using the international mobile subscriber identity (IMSI). A fourth type, the *IMSI Attach* procedure, is triggered when the mobile is powered up in the LA where it was switched off.

In the following, the most comprehensive LU is shown which is the Inter-VLR using the IMSI. Figure 8.20 depicts signaling exchanges during this procedure. This procedure essentially consists of the following steps:

- Allocation of a signaling channel to the MS and request for an LU.

- The MS provides the network with its IMSI, which allows the new VLR (VLR$_2$) to download authentication data from the HLR/AuC consisting of the triplets (*Rand, SRES, Kc*) for the authentication and ciphering procedures.

- The VLR is then able to authenticate the MS and, if this step succeeds, updates its location at the HLR. The HLR tells the old VLR (VLR$_1$) to remove the user's data.

- Ciphering may be required if available (see also Chapter 5).

- A new TMSI is allocated to the MS and after acknowledgment of its LU request (first message sent by the MS) the channel is released.

8.2.2.10 Impact of LUs on Radio Resource Occupancy and the Number of MSC/VLR Transactions

The impact of LUs on RF resource occupancy is examined in this section for a DCS 1800 (or PCS 1900) network (urban environment, small cells, and high user densities). A number of transactions are processed by the MSC/VLR, which are caused by LUs and computed in the network. The number of transactions is defined here as the messages received or transmitted by the MSC/VLR.

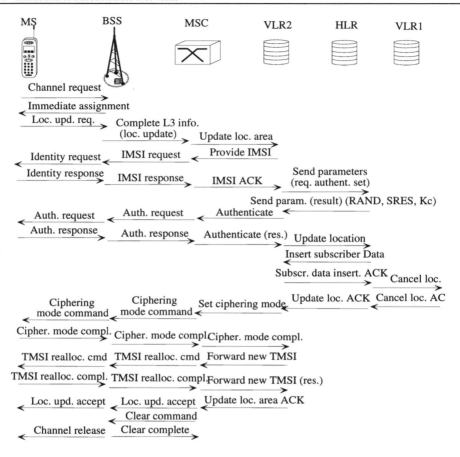

Figure 8.20 Inter-VLR LU with IMSI.

A location updating exchange uses the slow dedicated control channel (SDCCH). The SDCCH allocated to an MS consists of four timeslots (for the SDCCH blocks) every 51 TDMA multiframes (see Chapter 3). More generally a channel can be considered as a single slot in a TDMA frame which corresponds, in this instance, to one traffic channel/full rate (TCH/F). With this definition a channel can therefore accommodate 8 SDCCHs. That is, an LU consumes (during the signaling exchange) 0.125 TCH/F channel (i.e., 8 MSs can sequentially share the same timeslot in a 51-TDMA multiframe, see Figure 8.21).

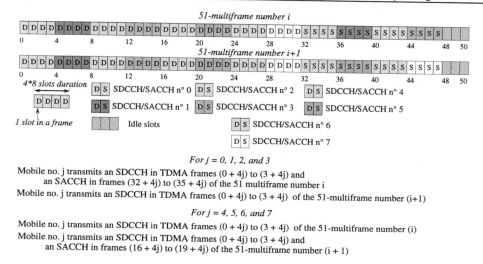

Figure 8.21 Channel multiplexing for SDDCH/SACCH transmission.

Note: Dedicated channels in GSM (voice, data, and signaling channels) are always accompanied by a slow associated control channel (SACCH). This signaling channel is used for channel measurement transmissions on the uplink and for call parameter settings on the downlink (e.g., power control).

Parameters and Hypotheses. To quantify the impact of LUs on RF usage in a cell located at the border of an LA and on transaction numbers processed at the MSC/VLR managing a set of LAs, the following parameters and hypotheses are introduced:

- ρ: density of the MSs in a cell (MSs/km^2),
- S: cell area (km^2), thus, $\rho.S$ gives the number of MSs per cell,
- R: cell radius or side of the hexagon (km),
- L: length (km) of the cell exposed perimeter (i.e., perimeter being part of the LA border),
- v: average MS velocity (km/h),
- $t_{LU}^{(i)}$: average duration of one LU in case i ($i = 1$: intra-VLR, $i = 2$: inter-VLR with TMSI, $i = 3$: inter-VLR with IMSI), equal to the time occupancy of one SDCCH/SACCH,
- $TN_{LU}^{(i)}$: number of transactions processed by the MSC/VLR for one LU in case i,
- $Pr_{LU}^{(i,j)}$: percentage of LUs in case i in the cell number j (each cell is identified by a number),

- $\lambda_{LU}^{(j)}$: number of location updates in the cell number j and per hour,
- A: average user traffic (Erl),
- N: number of cells per LA,
- N_{LA}: number of LAs managed by an MSC/VLR,
- N_p: number of cells located on the perimeter of one LA; its expression is given by [36]

$$N_p = 6 \cdot \sqrt{\frac{N}{3} - 3} \qquad (8.2)$$

- Cells are assumed to be hexagonal,
- Maximum blocking probability for the allocation of a SDCCH is 1%,
- MSs are uniformly distributed over the surface area of the cell,
- Movements of the MSs are decorrelated: the directions of their movements are uniformly distributed over $[0, 2\pi]$.

From [37], the number of location updates in an LA perimeter cell j and per hour is given by

$$\lambda_{LU}^{(j)} = \lambda_{LU} = v.L.\rho/\pi \qquad (8.3)$$

From [36], L is given by the formula:

$$L = 6.R \cdot \left(\frac{1}{3} + \frac{1}{2\sqrt{3}.N - 3} \right) \qquad (8.4)$$

Thus, the SDCCH/SACCH resource occupancy in the cell number j, from MS's LUs is given by the formula:

$$Tr_{LU}^{(j)} = \lambda_{LU} \cdot \left[\sum_{i=1}^{3} Pr_{LU}^{(i,j)} \cdot t_{LU}^{(i)} \right] \ (Erl) \qquad (8.5)$$

The number of transactions due to LUs generated in the $N_{LA}.N_p$ LA's perimeter cells (which are numbered from 1 to $N_{LA}.N_p$) and processed per hour by the MSC/VLR is given by the formula:

$$TTN_{LU} = \lambda_{LU} \cdot \left[\sum_{j=1}^{N_{LA}.N_p} \left(\sum_{i=1}^{3} Pr_{LU}^{(i,j)} \cdot TN_{LU}^{(i)} \right) \right] \qquad (8.6)$$

For numerical results, third-generation system parameters are now considered for a high-density area and during the busy hour traffic:

$\rho = 10,000$ MSs/km^2 [38], $R = 500$m, $v = 10$ km/h, $A = 0.1$ Erl, $N = 10$, $N_{LA} = 5$.

The number of accesses to the MSC/VLR is $TN_{LU}^{(1)} = 2$, $TN_{LU}^{(2)} = 14$, $TN_{LU}^{(3)} = 16$. Furthermore, practical measurements taken on a GSM network give [17] $t_{LU}^{(1)} = 600$ ms, $t_{LU}^{(2)} = 3.5$ sec, $t_{LU}^{(3)} = 4.0$ sec

Note: ρ can be the sum of ρ_1, ρ_2, and ρ_3, three user density populations of three different operators.

Radio resource occupancy. Two cases can be considered: an optimistic case (a cell where only intra-VLR LUs are generated) and a pessimistic case (a cell where only inter-VLR LUs are generated).

First case. The considered cell (number j) is located at the border of two LAs related to the same VLR. In this case, only intra-VLR LUs are processed in the cell ("IMSI Attach" procedures are ignored). (8.5) then gives $Tr_{LU}^{(j)} = 7.30$ Erl which, for a 1% blocking probability requires 14/8 = 1.75 channels (nearly a quarter of an RF channel). It consumes the same amount of resource as two user traffic calls (with a 2% blocking probability).

Second case. The considered cell (number j) is located at the border of two LAs associated with two different VLRs. In this case, only inter-VLR LUs are processed in the cell. It is assumed that there are 80% LUs using TMSI and 20% LUs with IMSI. Equation (8.5) then gives $Tr_{LU}^{(j)} = 42.46$ Erl which, for a 1% blocking probability requires seven channels (nearly one RF channel). It consumes the same amount of resource as nearly 30 user calls (with 2% blocking probability).

MSC/VLR transaction load. To evaluate the load on the MSC/VLR, the whole area managed by the same MSC/VLR has to be considered. It consists of $N_{LA} = 5$ LAs. It is assumed that one LA is located in the center of this area and that the remaining four LAs are located around the border. For the sake of simplicity, it is supposed that in all the perimeter cells that belong to the center LA and in half of the perimeter cells of the border LAs, only intra-VLR LUs are generated. For the other half of the perimeter cells of the border LAs only inter-VLR LUs are generated. The number of cells where intra-VLR LUs are generated is given by (8.2) and is equal to: $N_p + 4\dfrac{N_p}{2} = 18\sqrt{\dfrac{N}{3}} - 9$ and the number of cells where inter-VLR LUs are generated is equal to $4\dfrac{N_p}{2} = 12\sqrt{\dfrac{N}{3}} - 6$. Furthermore, it is assumed that in the generated inter-VLR LUs, 80% are LUs using TMSI and 20% are LUs using IMSI.

In this simple scenario the number of transactions to be processed by the MSC/VLR because of LUs in their LAs is computed using (8.6) and is equal to:

$TTN_{LU} \cong 12.10^6$ transactions in the busy hour.

8.2.2.11 Cellular Terminal Location

The methods presented in the previous sections and in Appendix 8A do not predict the precise location of a terminal. The location accuracy is of the order of the location area and in the most accurate case, in a cell. More precise location of a cellular terminal can be motivated for several reasons. For instance, the precise location of a cellular terminal can be linked with several types of application, such as accident reporting, navigational services, theft detection, and truck fleet management. It is necessary for these applications to use mechanisms that allow the system to know the geographic coordinates of a terminal with an accuracy of a few meters.

Currently there are several systems that determine the location of a geographic point. Three of the most well known [39] are shown below:

- *Global Positioning System (GPS)*: This system is based on a satellite network which enables location determination by using time of arrival (TOA) of several signals. A terminal can, using the temporal delays between the signals coming from several satellites, calculate its position as a function of the distances between the terminal and each satellite. Three satellites in line of sight with a GPS receiver allow the terminal to calculate its geographic coordinates.
- *Signpost navigation*: This service is available in the United States and uses a large number of radio transmitters located on the sides of highways. Each transmitter broadcasts a code indicating its latitude and longitude. The power of the signal received by a terminal allows it to determine its position relative to the transmitter.
- *Global Navigation Satellite System (GLONASS)*: This system is an initiative of the Russian government. Based on similar principles as GPS, it offers higher precision to civilian users.

Cellular terminal location services have gained importance since the U.S. Federal Communication Commission (FCC) introduced laws in June 1996 [40] imposing on cellular network operators the need to identify the number and position of the terminals in distress (E911 emergency service, *Enhanced 911*). The position of the terminal calling the E911 service must be automatically determined by the system (contrary to the usual case where the calling person has to indicate a location, which in an emergency can be a problem). The location accuracy, as specified by the FCC, must be within at least 125m in 67% of cases.

Two general approaches exist for cellular terminal location. The first one consists of implementing a location technique, which uses a system such as GPS, inside the terminal. The terminal, after having determined its position, transmits it to the network. The simplest example consists of integrating the GPS receiver in the terminal.

The second method consists of defining a network-based positioning system. The advantage of this solution is that it does not require modification of existing cellular terminals. Two position locators can be used. The first one is based on the angle of arrival (AOA), or direction of arrival (DOA), the second one is based on the time of arrival (TOA). The latter uses the arrival times of signals transmitted by the terminal to the base stations. The differences between the TOAs at different sites allow the system to determine the location of the terminal. In this method, it is necessary for the base stations to be synchronized to one another. Each base station must listen to signals transmitted by the terminal. Weighting of the received signal strength at each base station allows more accurate positioning [41]. For each pair of stations A and B, for instance (Figure 8.22), the following hyperbolic equation is determined:

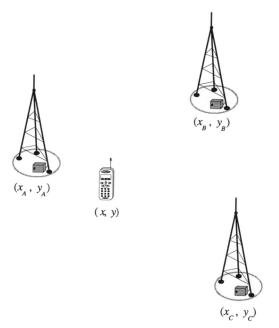

Figure 8.22 Terminal positioning in a cellular network.

$$R_{A,B} = \sqrt{(x_A-x)^2 + (y_A-y)^2} - \sqrt{(x_B-x)^2 + (y_B-y)^2}$$

Three base stations are sufficient to determine the position of a terminal with coordinates (x, y) at the intersection of two hyperbolas or three spheres (Figure 8.22).

8.2.3 Mobility Management in the Fixed and Mobile Networks: The UPT Concept

Since the advent of the call forwarding function, which allows a subscriber of a fixed network to route calls to another terminal, operators of fixed networks have been offering services aimed at making the subscriber forget that terminals are immobile. The advent of the intelligent network and the use of signaling system number 7 (SS7) have helped define a series of functions that are believed to be even closer to user needs [42, 43]. The concept of *Universal Personal Telecommunications* (UPT) allows subscribers to be totally free of the fixed link which connects their personal terminal to the network. The main UPT concepts that encompass both fixed networks and mobile networks [44, 45] are introduced.

The definition of UPT by the UIT-SG1 is as follows: "The UPT allows access to telecommunications services while allowing personal mobility. It allows each UPT user to take part in a series of personalized services, to transmit and receive calls on the basis of a personal UPT number, transparent to the network, on any kind of terminal fixed or mobile, regardless of the geographic position of the user. The limits are the terminal and network capacities and any restrictions imposed by the network operator. Calls to UPT users shall be possible from non-UPT users."

One very important aspect of the supply and use of UPT services is the concept of the UPT personal service profile within which the service facilities to which the user has subscribed are specified [46]. The UPT subscriber can personalize the UPT service with some provisos by altering the content of his or her service profile.

8.2.3.1 The UPT Concept

With UPT, the mobility that is offered to users is more global than in mobile radio systems. Association of the user with a specific terminal which is normal for fixed networks and cellular systems (excluding GSM) is eliminated and the user is identified by the network with the help of a unique personal number called a "UPT number" (*PN*). Thus, UPT helps offer personal mobility even greater than that offered by cellular networks (Figure 8.23). With this concept, a user has the possibility of connection at any network access point (or terminal) and also on any host network. In this way access to the telecommunications services to which the user has subscribed is globally available with the host terminal personally customized (e.g., with short form numbers).

The unique UPT number allocated to each subscriber allows the system to identify the user of a terminal and to provide the user with these services regardless of the terminal. It may also be used to locate the subscriber and is therefore able to route calls towards the one or many terminals where the user is registered. Just like a mobile terminal number in a cellular network, the number dialed to contact a UPT subscriber will be the same regardless of the geographic area where the call is initiated.

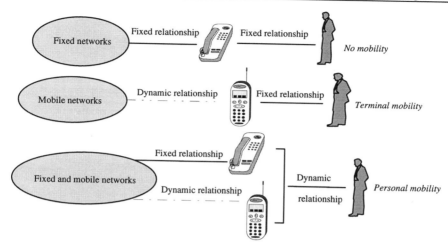

Figure 8.23 Differences between terminal and personal mobility with the UPT concept.

Thanks to the services offered by the UPT, each subscriber is described (in terms of services, location, and registrations) within the system by means of a service profile. The UPT number is considered a pointer indicating the service profile of the UPT subscriber. The profile contains static data (e.g., UPT number, supplementary services, billing options, roaming restrictions, terminal configurations) and dynamic data (e.g., location, address of the one or many terminals on which the subscriber is registered).

The subscriber identity card (SIM) can be considered a precursor for the supply of UPT-like services. It allows any GSM subscriber to use any GSM terminal after having inserted a SIM card and enter a personal code. The SIM card offers the subscriber personal mobility restricted to GSM networks (the home network of the subscriber and other networks with which the subscriber's operator has agreed roaming arrangements). The possibility of having access to the services of networks other than GSM (such as a fixed network phone card) with the SIM card then allows some personal mobility for SIM cardholders.

8.2.3.2 Main Characteristics of the UPT

The UPT is defined by its characteristics, in other words, the functions that this system will offer. The most important of them are:

- *UPT user identity authentication*: This function allows the operator (or service provider) to check the identity of the UPT user before giving access to the service required.

- *In bound call registration*: This allows the UPT user to register on any type of terminal so that all calls dialed to the user's personal number will be received.
- *Outgoing UPT call*: The UPT user has the option of calling from any terminal (fixed or mobile). It is, however, necessary to authenticate the caller for each connection attempt.
- In call delivery.

8.3 CONCLUSIONS

Mobility management (radio mobility with handover, network mobility with the cell selection/reselection algorithms, and location management procedures) is the main difference between fixed and mobile radio systems. The importance of these various procedures is even greater as one notes an increase in the impact of mobility in the telecommunications networks. This can be shown at four different levels.

First, at the *spatial level,* within a couple of decades, mobility has moved from being local (with private mobile radio, PMR) to regional mobility, then to national mobility (with the cellular networks), and finally to international mobility (with international roaming). With the advent of satellite-based global mobile personal communications systems (GMPCS), mobility will be seen at a global level.

Second, the level of *subscriber density* has been steadily increasing with a quasi-exponential growth and will exceed the most optimistic expectations. Most of the Scandinavian countries, in 1998, had penetration rates for cellular systems of more than 35% of the population. In other words, more than one in three people were equipped with a cellular handset.

Third, the *traffic per subscriber*: cellular subscribers use their terminals to send messages more often, even when located near a fixed terminal (an increase in the number of calls). High data rate transmission services will soon be made available (again increasing the flow of calls), with the help of innovations such as GSM Phase 2+ with high-speed circuit switched data (HSCSD) and the general packet data service (GPRS).

Fourth, the *level of services* offered to the subscribers. To judge by the way things are progressing with voice traffic (over 95% of the traffic in current cellular systems), mobile services are likely to progressively include facsimile, data services (e-mail), multimedia services (access to the World Wide Web), video, and maybe even interactive television.

Moreover, this trend is going to affect and influence our way of life and work habits and will become part of everyday activity. Mobile systems must then be capable of managing more generalized mobility.

APPENDIX 8A

LOCATION MANAGEMENT METHODS FOR THIRD-GENERATION SYSTEMS

The LA-based location management methods are the most developed and widely used in current cellular systems (e.g., GSM, IS-54, and IS-95), in trunk systems (e.g., TETRA), and in cordless systems (e.g., DECT, PACS, and PHS). Nevertheless, the traffic and processing generated may lead to congestion problems in high-density systems. One of the main concerns of system designers is to define methods that allow the system to reduce overhead traffic as much as possible. Indeed, traffic generated by location updates represents a significant part of signaling exchanges over the radio and network links (up to 80% of the random access channel in GSM) [37].

Future systems (or third-generation systems) are designed to serve large user populations (penetration rates of about 50% of the population in Europe, for instance). Because of scarcity of the radio resource and network processing load, it is important to design location management strategies that consume fewer resources than current systems [47, 48, 49].

Since the beginning of the 1990s, much work has been dedicated to the impact of mobility management on systems. Even though these various proposals are often very different in their principles, their common objective is to reduce the signaling traffic and the processing load generated by the location process. These proposals can be classified into two main groups (Figure 8A.1).

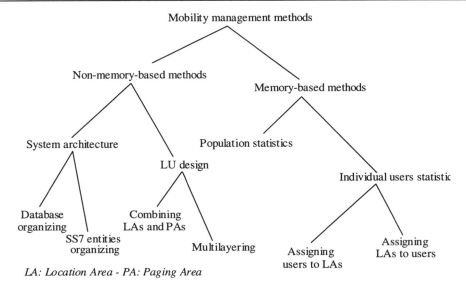

Figure 8A.1 Mobility management methods classification.

Included in the first group are all the methods based on algorithms and network architecture based on system processing capability. The second group brings together the methods based on learning processes that require collection of statistics related to user mobility behavior, for example. This second method puts the onus on the information capacity of the network.

8A.1 MEMORYLESS METHODS

8A.1.1 Database Architecture

LA partitioning, and thus mobility management cost, relies partly on the system architecture (e.g., database locations). Designing an appropriate database arrangement can reduce signaling traffic. The various database architectures described in Section 8.2.2.8 have been proposed with this aim in mind.

8A.1.2 Optimizing Fixed Network Architecture

In second-generation cellular networks and third generation systems signaling is managed by an Intelligent Network (IN) [17]. Appropriate organization of mobility functions and entities can help reduce the signaling burden in the network. The main advantage of these methods is that they reduce the network mobility costs independent of the radio interface and LA organization.

For example, in reference [50] it was proposed to use different degrees of decentralization of the control functions. Mobility costs were reduced by interconnection of modified signaling nodes. The proposition made in [51] is based on metropolitan area network (*MAN*) architecture. A higher level VLR (the VLR gateway) is used with a MAN such that most of the signaling traffic for LUs in the city area can be handled within the area network. Outside the metropolitan area the LU traffic burden is reduced as a consequence.

8A.1.3 Combining Location Areas and Paging Areas

In current systems, an LA is defined as an area both to locate or page a user. LA size optimization is therefore achieved by taking into account the two different procedures: locating and paging. Based on this observation, several proposals have defined location management procedures that use LA and paging areas (PA) of different sizes. One often considered method consists of splitting an LA into several PAs (Figure 8A.2).

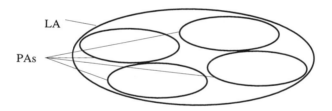

Figure 8A.2 LA and PA combining.

A mobile station registers only once, that is, when it enters the LA. It does not register when moving between the different PAs within the same LA. In the case of an incoming call, paging messages are broadcast in the PAs according to a sequence determined by different algorithms. For example, the first PA of the sequence can be the one where the mobile station was last detected by the network. The drawback of this method is the delay that can be incurred in large LAs.

8A.1.4 Multilayer LAs

In present location management methods LU traffic is concentrated in the cells at an LA border.

Based on this observation and to overcome the problem, Okasaka introduced [52] the multilayer concept (Figure 8A.3). In this method each mobile station is assigned to a given group and each group is assigned one or several layers of LAs.

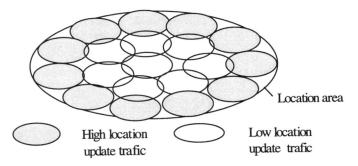

Figure 8A.3 Concentrated LU traffic in LA border cells.

According to Figure 8A.4, it is clear that group 1 and group 2 mobile stations will not generate LUs in the same cells which allow the LU traffic load to be distributed over all the cells. Although this LU method helps reduce channel congestion problems, it does not help reduce the overall signaling load generated by LUs.

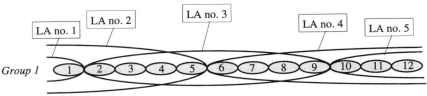

Layer 1 for group 1 = LA no. 1, LA no. 3 and LA no. 5 - layer 2 for group 1 = LA no. 2 and LA no. 4

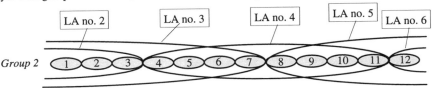

LA: Location Area

Layer 1 for group 2 = LA no. 3 and LA no. 5 - layer 2 for group 2 = LA no. 2, LA no. 4 and LA no. 6

Figure 8A.4 LA overlapping example with Okasaka's method.

8A.1.5 A Procedure for Reducing Signaling Message Exchanges

The reverse virtual call (RVC) setup described in [53] is a new scheme for delivering mobile terminal calls. It allows, under the constraint that the LA is not smaller than the VLR area, a reduction in the number of signaling messages exchanged between the called and calling databases and switches. The call setup delay is shown to be reduced by about 50% when using the RCV scheme.

8A.2 MEMORY-BASED METHODS

The design of memory-based location management methods has been motivated by the fact that systems perform many repetitive actions that can be avoided if they can be predicted. This is particularly the case for LUs. Present cellular systems achieve each day, at the same peak hours, almost the same levels of LU processing. Systems act as memoryless processes. Short term and long term memory processes can help a system avoid these repetitive actions. Some methods have thus been proposed that are based on user and system behavior observations and statistics.

8A.2.1 Short-Term Observation for Dynamic LA and PA Sizes Assigning/Adjusting

In current systems, the size of LAs is optimized by means of parameter values. In real situations, these values vary widely, dependent on the time of day and from one user to another.

Based on this observation, [54] proposes user location management based on defining multilevel LAs in a hierarchical cellular structure (Figure 8A.5). At each level the LA size is different and a cell belongs to different LAs of different sizes. According to past and present mobile station mobility behavior, the scheme dynamically changes the hierarchical level of the LA to which the mobile station registers. LU savings can thus be obtained.

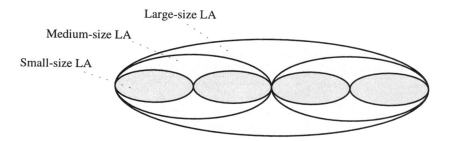

Figure 8A.5 Multilevel LAs.

A variant of this strategy evaluated in [55] consists of requiring that mobiles register in the cells on which they are camped. Registrations involve a periodic timer whose value has to be optimized. Rather than page a mobile in all the cells of an LA, it will be paged only in the cells visited during the preceding period. These are the cells on which the mobile camped during its traverse over the LA.

In Figure 8A.5, high incoming call rate and low mobility users are directed to small LAs, medium mobility users are directed to medium size LAs, and high velocity and low incoming call rate users are directed to the large size LAs.

An opposite approach considers that instead of defining a priori LA sizes, these can be adjusted dynamically for every user according to his or her incoming call rate (a) and the LU rate (u_k). In [51] and [56], a simple method for optimizing LA size (say, k cells) individually for each user is described and evaluated. In [51], a mobility cost function denoted $C(k, a, u_k)$ is minimized so that k is permanently adjusted. Each user is therefore related to a unique LA whose size k is adjusted according to his or her particular mobility and incoming call rate characteristics.

Adapting the LA size to each user's parameter values may be difficult to manage in practical situations. This has led in [57] to the definition of a method where the LA sizes are dynamically adjusted for the whole population, and not per user as in the two previous methods. Statistical information about users and mobility in the network is collected in databases and computed. Network characteristics as a function of time, place, and densities are evaluated. Results of this computation allow the network to dynamically adjust (during, for example, a day, week, month, or year) LA sizes. For instance, during the day when call rates are high it is preferable to deal with small LAs. During the night the opposite occurs, the call rate is much lower and therefore larger LAs are more suitable.

8A.2.2 Individual User Patterns

Observing that users show repetitive mobility patterns, an *alternative strategy* (AS) is defined in [58] and [49]. The main goal of AS is to reduce the overhead traffic related to mobility management by taking advantage of the users' highly predictable patterns. In AS, the system handles a *profile* recording of the most probable mobility patterns for each user. The user profile can be provided and updated manually by the subscriber or determined automatically by monitoring the subscriber's movements over a period of time. For an individual user, each period of time t_i, t_j (corresponds to a set of location areas k in the following way: (a_f, α_f) with $1 \leq f \leq k$ where: a_f is the location area the user can be located in, α_f is the probability that the user is located in a_f, and $\alpha_1 > \alpha_2 > ... > \alpha_k \left(\sum_{j=1}^{k} \alpha_i < 1 \right)$.

When the user receives a call, the system pages sequentially over the LA a_is until acknowledgment from the mobile. Note that if delay is important, parallel paging can be processed.

When the subscriber moves outside the recorded zone $\{a_1, ..., a_k\}$, the terminal processes a voluntary registration and indicates its new LA to the network. The main savings made by this method are due to the untriggered LUs when the user remains inside the profile LAs. The more predictable the user's mobility, the lower the mobility management cost. A variant of this method, called the *two-location algorithm* (TLA), was proposed and studied in [59]. In this strategy, a mobile stores the two most recently visited LAs. The same information is stored at the HLR. Obviously, the main advantage of this method relies on the reduction of location updates as the mobile goes back and forth between the two LAs.

8A.2.3 Predicting Short Term Movements of the Subscriber

The method proposed in [60] uses a process that predicts the movement of the mobile station according to its direction and velocity. The same processing and predictions are made both at the mobile station and the HLR. When movement of the mobile station does not fit with prediction, a registration is triggered by the mobile to inform the network of its real location. Otherwise, no exchange is required, which saves on LU processing and signaling.

8A.2.4 Mobility Statistics

In [61], a mobility management method similar to AS is defined. It is called *statistical paging area selection* (SPAS) and is based on location statistics collected by each mobile station, which then periodically reports them to the network. These statistics consist of a list of the average duration that the mobile station has spent in each LA. A priority rule is determined to confirm the sequence of LAs visited by the mobile. If this sequence is different from the last one reported to the network, the MS transmits it; otherwise nothing is done. The paging process is achieved in the same way as in the AS. When the mobile station moves to an area that is not on the reported list, it has to process a *temporary location registration* with the network.

In [62], the proposed method provides a means of allowing preconnection and pre-assignment of data or services to the location before the user moves into it, so that the user can immediately receive service or data. The same as the previous two methods, it is based on the user's movement history patterns. Called the *mobile motion prediction* (MMP) method, it allows the system to predict the "future" location of the user. Schematically, MMP combines two movement models: the *movement circle* (MC), based on a closed circuit-like model of user's movement behavior, and *movement track,* used to predict routine movement. MC is used to predict long term regular movement.

Finally there is the method proposed in [63], which uses a *cache* memory for reducing the search cost. The proposal is to store the location of frequently called mobiles in a local database (i.e., cache). This scheme reduces the number of queries to the HLR, thus reducing signaling traffic on the fixed network between the local database and the HLR.

APPENDIX 8B

MAIN UPT FEATURES

Authentication

- UPT user identity authentication;
- UPT service provider authentication.

Personal mobility

- In bound call registration;
- Out bound call registration;
- All call registration;
- Linked registration;
- Remote in bound call registration;
- Remote out bound call registration;
- Remote all call registration;
- Remote linked registration;
- Multiple terminal address registration;
- Variable default in bound call registration;

Service profile management

- UPT service profile interrogation;
- UPT service profile modification;
- Access to groups of UPT service profiles;
- UPT specific indications;

Call handling—calling party

- Outgoing UPT call;
- Call importance indicator;
- Calling party identification restriction.

Call handling—called party

- In bound call delivery;
- Variable routing;
- Call forwarding unconditional;
- Call forwarding on busy;
- Call forwarding on no answer;
- Call forwarding on not reachable;
- Call pickup;
- Calling party identification presentation;
- Called party specified secure answering of incoming UPT calls;
- Intended recipient identity presentation.

Follow-on

- Out bound call follow- on;
- Global follow-on.

Help desk

- Operator assisted services.

Non-UPT specified features

- Callback from recorded list;
- Variable routing;
- Display of A and B number.

REFERENCES

[1]   Mouly, M., M.-B. Pautet, "The GSM System for Mobile Communications," Published by the authors, 1992.

[2]   Proposed EIA/TIA Interim Standard – "Wideband Spread Spectrum Digital Cellular System Dual-Mode Mobile Station - Base Station Compatibility Standard," May 15, 1992.

[3]   Austin, M. D., and G. L. Stüber, "Analysis of a Window Averaging HandOff Algorithm for Microcellular Systems," *Proceedings of the Vehicular Technology Conference'92*, Denver, CO, pp. 31–35, May 1992.

[4]   Münoz-Rodriguez, D., and K. W. Cattermole, "Multiple Criteria for Handoff in Cellular Mobile Radio," *IEE Proceeding-F/Communications, Radar and Signal Processing*, pp. 85–88, Feb. 1987.

[5]   Gudmundson, M., "Analysis of Handover Algorithms," *Proceedings of the IEEE Vehicular Technology Conference'91*, St. Louis, USA, pp. 537–542, May 19-22, 1991.

[6]   Holtzman, J., "Adaptive Measurement Intervals for Handoffs," *Proceedings of the International Conference on Communications '92*, pp. 1,032–1,036, 1992.

[7]   Vijayan, R., and J. Holtzman, "A Model for Analyzing Handoff Algorithms," *IEEE Transactions on Vehicular Technology*, Vol. 42, No. 3, pp. 351–356, Aug. 1993.

[8]   Senadji, B., and S. Tabbane, "A Handover Decision Procedure Based on the Minimization of Bayes Criterion," *Proceedings of the IEEE Vehicular Technology Conference'94*, Stockholm, Sweden, June 8–10, 1994, pp. 77–81.

[9]   GSM 05.08.

[10]   Twingler, J., N. Andersen, and M. Hoenicke, "GSM - A Standard, A Platform, A Future," *Mobile Communications International*, pp. 43–46, October 1995.

[11]   Paradello, J., C. Rossignani, and C. Carpintero, "Access and Handover Methods in DECT-Like System," *Proceedings of the Vehicular Technology Conference '91*, St. Louis, MO, pp. 710–714, May 1991.

[12]   "CDMA for Cellular and Personal Communication Networks," QUALCOMM, Geneva, Switzerland, Oct. 1991.

[13]   Chia, S.T.S., "The Control of Handover Initiation in Microcells," *Proceedings of the Vehicular Technology Conference'91*, St. Louis, MO, pp. 531–536, May 1991.

[14]   Barba, A., and J.L. Melius, "Traffic Evaluation in the Decision Phase of a Handover," *Proceedings of the International Conference on Universal and Personal Communications'95*, ICUPC'95, Tokyo, Japan, pp. 354–358, Nov. 6-10, 1995.

[15]   GSM 05.10.

[16]   Mouly, M., and J. L. Dornstetter, "The Pseudosynchronization, a Costless Feature to Obtain the Gains of a Synchronized Cellular Network," *Proceedings of the IEEE Vehicular Technology Conference '91*, St. Louis, MO, pp. 51–55, May 19-22, 1991.

[17]   Lagrange, X., P. Godlewski, and S. Tabbane *Réseaux GSM-DCS*, éditions Hermès, Sept. 1996.

[18]   Pollini G., "Trends in Handover Design," *IEEE Communications Magazine*, pp. 82–90, Mar. 1996.

[19]   David K., P. Blanc, T. Kassing, M. I. Lopez-Carillo, R. Maddalena, W. Mohr, and P. Ranta, "The First Results from ADTMA's Test Campaigns," *Mobile Communications International*, Apr. 1996, pp. 56–58.

[20] Murase A., I. C. Symington, and E. Green, "Handover Criterion for Macro and Microcellular Systems," *Proceedings of the IEEE Vehicular Technology Conference '91*, St. Louis, MO, May 19-22, 1991, pp. 524–530.

[21] Hong, D., and S. S. Rappaport, "Traffic Model and Performance Analysis for Cellular Mobile Radio Telephone Systems with Prioritized and Nonprioritized Handoff Procedures," *IEEE Transactions on Vehicular Technology*, Vol. VT-35, No. 3, pp. 77–92, 1986.

[22] McMillan, D., "Traffic Modeling and Analysis for Cellular Mobile Networks," *Proceedings of the International Teletraffic Congress*, ITC-13, Copenhagen, Denmark, 1991.

[23] Chelbus, E., "Analytical Grade of Service Evaluation in Cellular Mobile Systems With Respect to Subscribers' Velocity Distribution," *Proceedings of the 8th Australian Teletraffic Research Seminar*, Melbourne, Australia, pp. 90–101, Dec. 6–8, 1993.

[24] Nanda, S., "Teletraffic Models for Urban and Suburban Microcells: Cell Sizes and Handoff Rates," *IEEE Transactions on Vehicular Technology*, Vol. 42, No. 4, pp. 673–682, Nov. 1993.

[25] Eriksson, H., and R. Bownds, "Performance of Dynamic Channel Allocation in the DECT System," *Proceeding of the IEEE Vehicular Technology Conference '91*, St. Louis, MO, pp. 693–698, May 1991.

[26] Chia, S.T.S., "Cellular Structure and Handover Algorithms for a Third-Generation Mobile System," *BT Technol J*, Vol. 11, No. 1, pp. 111–113, Jan. 1993.

[27] "The Evolution of Second-Generation Mobile Networks to Third-generation PCN - A Report for the Commission of the European Communities, DG XIII," *Multiple Access Communications Limited*, Apr. 1993.

[28] Tabbane, S., "Handover Procedures in ATDMA Project," *RACE Mobile Telecommunications Workshop*, Metz, May 1993.

[29] Amitay, N., "Distributed Switching and Control With Fast Resource Assignment/Handoff for Personal Communications Systems," *IEEE Journal on Selected Areas in Communications*, Vol. 11, No. 6, pp. 842–849, Aug. 1993.

[30] Foschini, G. J., "Channel Cost of Mobility," *IEEE Transactions on Vehicular Technology*, Vol. 42, No. 4, Nov. 1993, pp. 414–424.

[31] Lee, Y.-K., and W. S. Wong, "Location Update by Binary Cutting of IDs," *Proceedings of the International Conference on Universal and Personal Communications*, ICUPC'95, Tokyo, Japan, pp. 963–967, Nov. 6–10, 1995.

[32] Lam, D., D. C. Cox, and J. Widom, "Teletraffic Modeling for Personal Communications Services," *IEEE Communications Magazine*, pp. 79–87, Feb. 1997.

[33] Morales-Andres, G., and M. Villen-Altamirano, "An Approach to Modelling Subscriber Mobility in Cellular Radio Networks," *Telecom Forum '87*, Geneva, Switzerland, Nov. 1987.

[34] Carle, P., G. Colombo, "Sub-Optimal Solutions for Location and Paging Areas Dimensioning in Cellular Networks," *Proceedings of the IEEE International Conference on Personal Communications*, Tokyo, Japan, Nov. 6-10, 1995.

[34] Gondim, P. R. L., "Genetic Algorithms and the Location Area Partitioning Problem in Cellular Networks," *Proceedings of the Vehicular Technology Conference '96*, Atlanta, GA, pp. 1,835–1,838, Apr. 29-30, May 1, 1996.

[35] Behrokh, S., and W. S. Wong, "Optimization Techniques for Location Area Partitioning," *International Teletraffic Conference Seminar on UPT*, Italy, 1992.

[36] Alonso, E., K. Meier Hellstern, and G. Pollini, "Influence of Cell Geometry on Handover and Registration Rates in Cellular and Universal Personal Telecommunications Networks," *WINLAB Technical Report*, TR-31, May 1992.

[37]   Thomas, R., H. Gilbert, and G. Mazziotto, "Influence of the Moving of the Mobile Stations on the Performance of a Radio Mobile Cellular Network," *Proc. Third Nordic Seminar on Digital Land Mobile Radio Communications*, Copenhagen, Denmark, Sept.12-15, 1988.

[38]   McFarlane, D., and S. Griffin, "The MPLA Vision of UMTS," *CODIT/BT/PM-005/1.0*, Feb. 1994.

[39]   Rappaport, T. S., J. H. Reed, and B. D. Woerner, "Position Location Using Wireless Communications on Highways of the Future," *IEEE Communications Magazine*, pp. 33–41, Oct. 1996.

[40]   FCC NEWSReport No. DC 96-52, June 12, 1996.

[41]   Krizman, K. J., T. E. Biedka, and T. S. Rappaport, "Wireless Position: Fundamentals, Implementation Strategies, and Sources of Error," *Proceedings of the IEEE Vehicular Technology Conference 1997*, Phoenix, AZ, pp. 919–922, May 4–7, 1997.

[42]   Modaressi, A., and R. Skoog, "An Overview of Signaling System No. 7," *Proceedings of IEEE*, Vol. 80, No. 4, pp. 590–606, Apr. 1992.

[43]   Folkestad, A., S. Kleir, and C. Görg, "Impact of UPT Services on Network Performance and Call Setup Times: Distribution of UPT Service Logic in SS7," *Proceedings of ISS'95*, Berlin, Germany, pp. 97–101, Apr. 1995.

[44]   Kikuta, H., M. Fujioka, and Y. Wakahana, "Global UPT Architecture With Internationa Mobility Management," *Proceedings of the International Conference on Universal and Personal Communications'92*, ICUPC'92, Dallas, TX, pp. 229–235, Sept. 29–October 2, 1992.

[45]   Wilbur-Ham, M., and P. Gerrand, "Universal Personal Telecommunications (UPT) Services and Architecture Overview," *Proceedings of the International Conference on Universal and Personal Communications'92*, ICUPC'92, Dallas, TX, pp. 64–68, Sept. 29-October 2, 1992.

[46]   Sundborg, J., "Universal Personal Telecommunications (UPT) - Concept and Standardisation," *Ericsson Review N° 4*, pp. 140–155, 1993.

[47]   Rose, C., "Minimization of Paging and Registration Costs Through Registration Deadlines," *Proceedings of the Vehicular Technology Conference '95*, Chicago, IL, pp. 735–739, June 1995.

[48]   Tabbane, S., and R. Nevoux, "An Intelligent Location Tracking Method for Personal and Terminal FPLMTS/UMTS Communications," *Proceedings of the International Switching Symposium '95*, Berlin, pp. 114–118, Apr. 1995.

[49]   Tabbane, S., "An Alternative Strategy for Location Tracking," *IEEE Journal on Selected Areas in Communications*, Vol. 13, No. 5, June 1995.

[50]   Listanti, M., and S. Salsano, "Impact of Signaling Traffic for Mobility Management in an IN-based PCS Environment," *Proceedings of the IEEE International Conference on Personal Communications*, Tokyo, Japan, Nov. 6-10, 1995.

[51]   Xie, H., and D. J. Goodman, "Signaling System Architecture Based on Metropolitan Area Networks," *Winlab Technical Report*, TR-39, Aug. 1992.

[52]   Okasaka, S., S. Onoe, S. Yasuda, and A. Maebara, "A New Location Updating Method for Digital Cellular Systems," *Proceedings of the Vehicular Technology Conference '91*, St. Louis, MO, pp. 345–350, May 1991.

[53]   I, C.-L., Pollini, G. P., and R. D. Gitlin, "PCS Mobility Management Using the Reverse Virtual Call Setup Algorithm," *IEEE/ACM Transactions on Networking*, Vol. 5, No. 1, pp. 13–24, Feb. 1997.

[54]   Hu, L.-R., and S. S. Rappaport, "An Adaptive Location Management Scheme for Global Personal Communications," *Proceedings of the International Conference on Universal and Personal Communications'95*, ICUPC'95, Tokyo, Japan, pp. 950–954, Nov. 6–10, 1995.

[55]   Pashtan, A., and I. Arich Cimed, "The CLU Mobility Management Scheme for Digital Cellular Systems," *Proceedings of the IEEE Vehicular Technology Conference*, Atlanta, GA, pp. 1,873–1,877, Apr. 28–May 1, 1996.

[56] Kim, S. J., and C. Y. Lee, "Modeling and Analysis of the Dynamic Location Registration and Paging in Microcellular Systems," *IEEE Transactions on Vehicular Technology*, Vol. 45, No. 1, pp. 82–89, Feb. 1996.

[57] "Location areas, paging areas and the Network Architecture," *RACE II Deliverable*, R2066/PTTNL/MF1/DS/P/001/b1, Apr. 1992.

[58] Tabbane, S., "Comparison between the Alternative Location Strategy (AS) and the Classical Location Strategy (CS)," *WINLAB Technical Report*, No. 37, Aug. 1992.

[59] Lin, Y.-B., "Reducing Location Update Cost in a PCS Network," *IEEE/ACM Transactions on Networking*, Vol. 5, No. 1, Feb. 1997, pp. 25–33.

[60] Rokitansky, C. H., "Knowledge-Based Routing Strategies for Large Mobile Networks with Rapidly Changing Topology," *Proceedings ICCC'90*, New Delhi, India, pp. 541–550, Nov. 1990.

[61] Shirota, M., Y. Yoshida, and F. Kubota, "Statistical Paging Area Selection Scheme (SPAS) for Cellular Mobile Communication Systems," *Proceedings of the IEEE Vehicular Technology Conference '94*, Stockholm, Sweden, June 8–10, 1994.

[62] Liu, G.Y., and G.Q. Maguire, "A Predictive Mobility Management Algorithm for Wireless Mobile Computing and Communications," *Proceedings of the IEEE International Conference on Personal Communications*, Tokyo, Japan, Nov. 6–10, 1995.

[63] Lin Y.-B., and S.-Y. Hwang, "Comparing the PCS Location Tracking Strategies," *IEEE Transactions on Vehicular Technology*, Vol. 45, No. 1, Feb. 1996, pp. 114–121.

SELECTED BIBLIOGRAPHY

[1] Chelbus, E., and W. Ludwin, "Is Handoff Really Poissonian?," *Proceedings of the International Conference on Universal and Personal Communications'95*, ICUPC'95, Tokyo, Japan, pp. 348–353, Nov. 6-10, 1995.

[2] Chia, S.T.S., and R.J. Warburton, "Handover Criteria for City Microcellular-Radio Systems," *Proceedings of the Vehicular Technology Conference'90*, Orlando, FL, pp. 276–281, May 1990.

[3] Gudmunson, B., and O. Grimlund, "Handoff in Microcellular-Based Personal Telephone Systems," in *Third-Generation Information Networks*, edited by S. Nanda and D. J. Goodman, Kluwer Academic Publishers, 1991.

[4] Mouly, M., J. L. Dornstetter, "The Pseudosynchronization, a Costless Feature to Obtain the Gains of a Synchronized Cellular Network," *Proceedings of the IEEE Vehicular Technology Conference '91*, St. Louis, MO, pp. 51–55, May 19–22, 1991.

[5] Park, S., H.S. Cho, and D.K. Sung, "Modeling and Analysis of CDMA Soft Handoff," *Proceedings of the Vehicular Technology Conference '96*, Atlanta, GA, pp. 1,525–1,529, Apr. 29-30, May 1, 1996.

[6] Rose, C., and R. Yates, "Location Uncertainty in Mobile Networks: A Theoretical Framework," *IEEE Communications Magazine*, pp. 94–101, Feb. 1997.

[7] Senarath, G.N., and D. Everitt, "Controlling handover performance using signal strength prediction schemes and hysteresis algorithms for different shadowing environments," *Proceeding of the Vehicular Technology Conference '96*, Atlanta, GA, pp. 1510–1514, Apr. 29-30 - May 1, 1996.

[8] Seskar, I., S.V. Marié, J. Holtzman, and J. Wasseman, "Rate of Location Area Updates in Cellular Systems," *Proceeding of the Vehicular Technology Conference '92*, Denver, CO, pp. 694–697, May 1992.

[9] Spaniol, O., et al., "Impacts of Mobility on Telecommunications and Data Communication Networks," *IEEE Personal Communications*, pp. 20–33, October 1995.

[10]   Wang, J. Z., "A Fully Distributed Location Registration Strategy for Universal Personal Communication Systems," *IEEE Journal on Selected Areas in Communications*, Vol. 11, No. 6, Aug. 1993, pp. 850–860.

[11]   Xie, H., and D. J. Goodman, "Signaling System Architecture Based on Metropolitan Area Networks," *Winlab Technical Report*, TR-39, Aug. 1992.

[12]   Zoican, R., "Comparative Analysis of the Handoff Strategies for Microcellular Systems," *Proceedings of the International Symposium on Spread Spectrum*, Hanover, Germany, pp. 1,178–1,181, Sept. 1996.

CHAPTER 9

PROFESSIONAL MOBILE RADIO

Less popular with the public at large than cellular or cordless systems, the *professional/private mobile radiocommunications* (PMR) systems—*private business radio* (PBR), or *land mobile radio* (LMR), also called *specialized mobile radio* (SMR)—represent the oldest forms of mobile radiocommunication systems. PMR networks are generally very simple, in terms of equipment and management, compared with cellular systems. They provide many advantages compared with cellular networks as management and services are independent of one another. The primary advantage is that they allow direct and full control of the network, as PMR systems are specifically designed for professional applications.

The very first infrastructures to use these kinds of networks were the public safety agencies (e.g., police, fire, and ambulance), transport companies (as early as the beginning of the century on some ships), manufacturers, and utilities (those entrusted with the management of energy, water, and road networks, for instance). In 1996, the penetration rates of these systems were about 8% in the United States, 4% in the Nordic countries, and less than 2% in France, Great Britain, and Germany [1]. Annual growth-rate estimates have ranged between 3% and 5%. The number of PMR users in Europe was about 5 million in 1996. The trunked radio market reached 9 million worldwide at the end of the 1990s. Trunked systems are discussed in Section 9.4.

The PMR market has a range of applications as varied as the enterprises involved. Private mobile radios have always been considered a permanent source of innovation despite the large number of constraining regulations prevailing in most countries. These

369

regulations, along with the lack of spectrum allocated to PMR systems are at the heart of the slow progress of these systems compared with that of cellular. PMR is quite varied because of the wide range of mobile users targeted. For example, staff members on the move (e.g., maintenance, servicing) and the others present at a site or inside a building (service area, refineries, sports events, power stations) do not have the same requirements of mobility and services. Note that some cellular networks offer PMR-like services such as Radiocom 2000.

PMR systems quite often use proprietary technologies and specific protocols, which has caused frequent incompatibility problems between various systems. The size and complexity of a PMR system ranges from a single site infrastructure with a single frequency serving a few mobiles to a giant installation such as that of the national police force in France (*Gendarmerie*) which covers most of the country.

The communication needs of PMR users are quite different from simple telephony, their task being mainly to facilitate operational exchanges (e.g., information, command, acknowledgment). For these reasons, they face different constraints from the mobile systems used by the general public (e.g., public cellular or cordless systems). The task consists, first and foremost, of allowing data exchange between one or several users' while meeting the professional users' operational needs (in particular, very short access times and half-duplex communications). PMR networks may be connected to fixed networks (e.g., PSTN or X25) if regulatory constraints allow it. Regulations with regard to interconnection vary from country to country.

The detailed architecture and services provided by PMR are quite varied, an aspect that allows them to meet very specific needs such as in networks for police forces, oil companies, management of the water-supply network, and so forth. Some general characteristics are usually common and will be introduced in more detail once the history of these systems has been discussed.

9.1 HISTORIC AND GENERAL BACKGROUND

Used since the beginning of the twentieth century by specific services and organizations, especially the maritime companies and armed forces, PMRs have spread into many economic and administrative spheres. For example, in France, the first civil applications of PMR, as early as the 1950s, involved taxi companies and the ambulance service.

PMR did not witness any outstanding technical progress until the 1970s. The very simple principle upon which systems were based consisted of allocating a radio channel per fleet (a fleet consisting of one or several groups of users). This channel was used exclusively by this system and resulted in a low spectral efficiency for the system. This kind of PMR network is known as a *conventional system*. Increased pressure on the radio spectrum has led to the introduction, as early as the beginning of the 1970s, of the *trunking technique*, which allows, on one hand, an increase of system spectral efficiency by allowing a large number of users to share a pool of channels; and on the other, the ability to host several fleets of independent users on the same system. PMR systems have witnessed, with the advent of this technique [2], an important step in progress. The systems based on this particular technique are called *trunked systems*.

9.1.1 Definition and General Background

The first function of a PMR system is to establish communication between mobile terminals and dispatchers, users of private branch exchanges (PBX), or public-switched telephone network (PSTN) subscribers. PMR allows fast setup of communication, generally short and frequent, between two users (individual communications) or more, usually between groups (conference call).

Generically, PMR also includes private paging systems. In this case mobiles are only able to receive data. Paging systems are mainly used by people located inside private sites (e.g., hospitals, plants) or for operating remote control or remote alarm networks (measurements or indicators rise). Public paging networks have shown a recent upsurge in popularity with the introduction of *calling party pays* tariff plans in some countries. In this chapter only PMR systems using bidirectional links are discussed. Paging systems are dealt with in Chapter 11.

The main frequency bands commonly used for PMR two-way or one-way links are as follows:

- Decimetric waves: *UHF* (*ultra high frequency*) ranging between 300 and 3,000 MHz (example of frequencies used in practice by PMR: 403–470 and 540–520 MHz). These frequencies allow users to communicate at ranges limited to a few tens of kilometers for power transmissions of about 20W. Their reflection and diffraction properties make them useful within an urban environment.
- Metric waves: *VHF* (*very high frequency*) ranging between 30 and 300 MHz (an example is the 136–174-MHz band). At the lower end of the band they cover distances up to about a hundred kilometers, reducing to tens of kilometers at the high end, and are the most widely used in current systems.

9.1.2 PMR Categories According to User Needs

Unlike cellular networks, PMR coverage is implemented in accordance with the needs of its specific user, whereas a public cellular network is supposed to cover areas of highest population density. Very often, good coverage is difficult to obtain from one site only, usually because of shadowing. Therefore, a PMR network spreads over several sites whenever it is implemented to cover a region or a whole country. In this case, the frequency reuse principle is used together with cellular-system planning processes but occurs in a simpler way than for a public cellular network.

PMR network types, in terms of the quality of service and area covered, vary in accordance with the use for which they have been planned. To simplify things, PMR can be categorized into four groups based on coverage area: PMR without a fixed infrastructure, single site PMR, multisite PMR, and wide area-coverage PMR.

9.1.2.1 Networks Without a Fixed Infrastructure

Networks without a fixed infrastructure represent the simplest system. Mobiles communicate either directly between themselves, or via another mobile that takes the role of a base station (Figure 9.1). These systems are used for short-range communications (a few hundred meters to a few kilometers) and where the implementation of an infrastructure—a fixed base station—is not wanted. They are often used for groups of roaming users (e.g., building sites, on-the-spot intervention, follow-up of events).

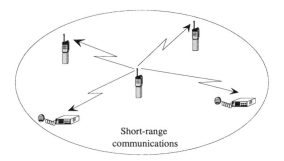

Figure 9.1 Network without a fixed infrastructure.

9.1.2.2 Single Site Networks

The coverage area of single site systems could reach tens of kilometers around a base station (Figure 9.2). The mobiles communicate over a greater distance with the help of a repeater (sometimes called a base station) that retransmits the signals. The range between two mobiles, in this case, is at least twice and often a lot more than that obtained in the previous case. This kind of configuration suits enterprises operating within limited areas: airports, industrial sites, oil rigs, power stations, for example.

9.1.2.3 Multisite Network

In the case where users operate over large areas (e.g., highways, countries, pipelines) the implementation of several base stations is necessary. These base stations are interconnected to allow communication between the different areas (Figure 9.3).

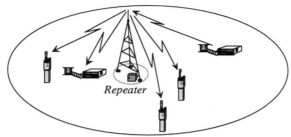

Medium-range communications via the repeater

Figure 9.2 Single site network.

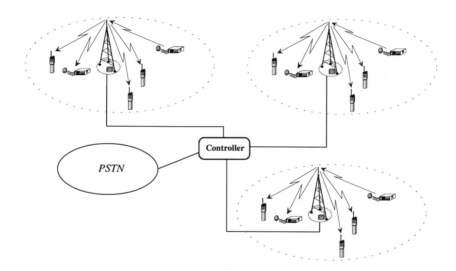

Figure 9.3 Multisite coverage network.

The method used to cover a wide area consists of installing several base stations on elevated sites. These stations are connected to a central controller by land lines (usually leased lines) or by secondary radio links (e.g., microwaves).

9.1.2.4 Wide Area Network

Wide area network configuration allows the network coverage area to be extended to a whole country, as with some military systems. The base stations covering the various areas are connected to a central controller (by cables or microwave links) to allow handover and dynamic activation of the sites (i.e., only the sites where called users are registered are made active during the call). Because these systems are often installed within an organization having one or more PMR networks, the integration of several kinds of PMR (conventional and trunk systems) is sometimes necessary. A general categorization of the different groups of systems according to the coverage areas and the user densities is presented in Figure 9.4.

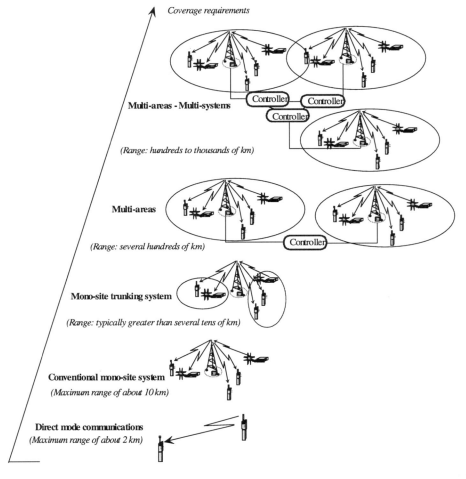

Figure 9.4 PMRS general categorization according to their coverage.

9.1.3 Categorization of Users and Organizations

As opposed to cordless or cellular systems where each user is treated as an individual by the system, PMR defines exchanges between users according to a hierarchical structure. This is often based on the structure of the organization where the groups and services reflect the shape of that organization (Figure 9.5).

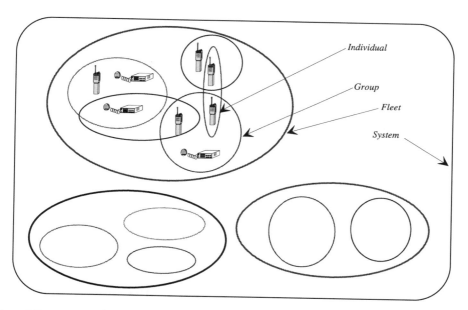

Figure 9.5 Exchange structure among PMR users.

User calls can be categorized into four groups: *private* calls (typically between two users of the same hierarchical level), *group* calls (between users belonging to the same group), *fleet* calls (between all the users of the same groups of the same fleet), and *general* calls (between all the users of the system).

9.1.4 PMR General Categorization

PMR systems are mainly designed for voice communications, but the demand to transmit data is increasing and users can transmit data across them using modems. For systems exclusively dedicated and optimized for data transfer, see Chapter 13.

Three broad groups of PMR systems can be defined:

- Systems based on the traditional techniques,
- Trunk systems,
- Radio data communication systems.

9.1.5 PMR Categorization According to Their Operation

PMR networks can be operated in several ways, according to their size, characteristics, and constraints imposed on the organizations that use them. Their management is determined by their operations and according to fleet size. Essentially, the PMR networks can be either:

- *Private*, in which case the organization owns its own infrastructure and leases the frequencies, or
- *Operated*, where, like cellular systems, the users lease a service from an operator who owns the infrastructure.

9.1.5.1 Operator Controlled Networks

Operator controlled networks go under the name of public access mobile radio (PAMR). They are operated by an enterprise to which a set of frequencies is allocated and who installs, manages, and operates the network in the same way as a cellular network operator. The services offered are then sold on to users. Users are divided into private groups allowing confidentiality of their communications. The users of these systems are mainly small or medium-size companies characterized by average size fleets and which do not need to have their own infrastructure (e.g., delivery or transport companies, health sector, after-sales-service companies).

For these users it is usually more economic to subscribe to a public trunked system. There is also a specific category of system which is operated by smaller enterprises (e.g., equipment sellers) who own and operate a single base station/repeater which is shared by several groups of users. This is known as common base station (CBS) working.

9.1.5.2 Private Networks

Private networks are used exclusively by one organization. These networks are collectively known as PMR systems as opposed to PAMR. They are carefully designed to be tailored to the needs of companies whose fleets have strategic importance. These networks are optimized and adjusted to the exact needs of their users. Among these the most significant are the police, ambulance, fire and military forces, utility companies, local authorities, or large urban or national transport companies.

The companies and organizations operating their own networks can be identified according to network coverage, helping in the separation of large infrastructures from small and medium ones. Among the large infrastructures, police, firefighters, armed forces, railroads, transport networks, and supply companies (water and power) are the most notable. In the 1990s, one of the largest PMR networks worldwide was that of the French Gendarmerie, using the RUBIS network, with 40,000 users.

Note: In the rest of this chapter the term *PMR* is used for both PMR and PAMR.

9.2 PMR SERVICES

Mainly because they are designed for operational communications, PMR services are highlighted by specific characteristics. We will deal with these characteristics before providing a survey of the main services offered by PMR.

9.2.1 PMR Characteristics

Historically speaking, the security services (police and army) were the first to be interested in using PMRs. This kind of system effectively helps their field agents to remain in permanent contact with an operational center. With technological progress and the decrease in PMR costs, users have changed their motivation for PMR use. While security was once the primary requirement, such things as economic benefits (time savings, fuel economy, better assistance for the teams) and competitiveness (faster reaction of staff, better knowledge of customer needs) have become more important.

Some of the requirements of organizations using PMR have led to the definition of specific functionality that would be hard to transpose onto the large public systems. PMR networks have a series of common features, the most outstanding of which are the following.

9.2.1.1 Half-Duplex Communications

In PMR systems, communication is usually made in a half-duplex mode. That is to say, at any one moment only one user of the group is able to talk on the channel allocated to his or her group. This discipline, which is specific to PMR systems, requires the users to adopt stringent communication discipline. The duration of communication is limited solely to the exchange of operational information. In this way, call times can be reduced by a factor of four compared with full-duplex communication. This characteristic helps to save on channel occupancy. In this way, compared with cellular systems, PMR represents a highly efficient system in terms of radio spectrum usage. Unlike cellular systems which requires each member of a conference call (i.e., a group) to be allocated their own duplex channel, a single site group only uses one radio channel and is therefore much more spectrally efficient.

Channel access is made by use of a push-to-talk (PTT) button. As the channel becomes free, the user can, by use of an appropriate selective calling mechanism, make a private call to another terminal. The allocation of a channel for the call is initiated when the user presses the PTT button of the terminal. The connection time required for the call ranges from a few tens to a few hundreds of milliseconds and happens automatically within the time from when the user presses the PTT button and the moment the communication starts. An acknowledgment tone can be used to indicate to the caller when to begin speaking. Throughout the time of the call (ranging from a few seconds to some tens of seconds), the channel is reserved for that user, and is freed as soon as the user releases the PTT button.

9.2.1.2 Call Setup Times

Call setup times generally range between a fraction of a second to one second (access time for some trunked systems can be lower than 300 ms), and short dial numbers (2 to 4 digits) reduce the length of the dialing procedure. These high-speed procedures are necessary for short (a few tens of seconds with an average of 30 sec) and frequent (several tens of calls per day per user) communications to reduce the signaling overhead.

9.2.1.3 Costs

The use of a PMR (public or private) system has for a long time been more economical than subscribing to a public cellular phone service. The service provided, the cost of the equipment, and other running and maintenance charges had characteristics that were closer to the operational needs of user organizations. The cost of a PMR network usually includes the purchase and installation of the equipment (base station and mobiles), license fees for the use of the radio frequencies, maintenance, line and site rentals, and electricity charges. Due to competition during the late 1990s, the tariffs offered by the operators of cellular systems, such as GSM, and the cost of terminals, are such that PMR is becoming a less attractive cost option than cellular services.

9.2.1.4 Important Reliability and Availability

Since many PMR systems have been designed for security- and emergency-type applications, they include features of increased robustness, reliability, and availability. Moreover, being autonomous, in other words, independent of fixed networks, they are less sensitive to external events such as natural catastrophes. Owners and users have been able to tailor their PMR systems in accordance with their own special needs (e.g., equipment redundancy and secure transmission links).

9.2.1.5 Terminal Security

Radio terminals used in PMR can be designed to meet intrinsically safe standards. As they are likely to be used in specific environments (gas distribution industries) and within specific environmental conditions (temperature and pressure), they have to meet very specific characteristics. Two standards are used: the European CENELEC and the American FM (Factory Mutual) standards. Each terminal has to meet three kinds of constraint:

- Ruggedness (according to the area, continuous, intermittent, or occasional damage),
- Protection against explosion in inflammable gas atmospheres (e.g., hydrogen, acetylene),
- Operation over a wide temperature range (e.g., up to 135°C in the case of the CENELEC standard).

When used in a military environment the equipment has to meet bump, vibration, temperature and humidity, and other environmental specifications.

9.2.2 Services Offered

Among the range of services offered by PMR, voice communication tops the list but data transmission has been steadily gaining more importance and is being enhanced by new PMR equipment based on digital techniques. Some systems are now able to offer voice and data services simultaneously.

9.2.2.1 Voice Services

These are the most widely used services today. Most of them remain specific to PMR and can be divided into two categories: private and group communications.

Private communications. These are point-to-point communications between two PMR users, or between a PMR user and the PSTN or PBX.

Group communications. These predominate the range of services offered to users and can be subdivided into two categories: those available in all PMR types (conventional and trunk networks) and those available in trunk networks only.

- *Grouped organization.* An organization (identified by a fleet or a group) distributes its agents in subfleets (or subgroups). These groups generally are based on the operational architecture of the organization. For example, the commercial people will be combined in a subfleet *Gr 1*, the delivery staff in a subfleet *Gr 2*, and the after-sales service technicians in subgroup *Gr 3*.
- *Intraorganization communications.* As a general rule, PMR groupings allow communications among users of the same fleet only. Access to a private switch (PBX) may allow them to communicate with other "fixed" users (within their

organization) and if the system and local legislation allows it, with subscribers on public networks.

Services Available in All PMR Systems:

- *Fleet or subfleet calls.* The user can contact, simultaneously, all the members of the same fleet or subfleet located in a predefined area.
- *Broadcast calls.* A user can communicate to all users belonging to his or her group regardless of location. Upon receipt of the call, mobiles of the same group automatically receive the messages. This method of operation is usually unidirectional.

Services Available on Trunked Systems:

- *Priority calls.* The allocation of priorities to certain users and/or to call types helps in handling urgent cases. A person with the highest priority can interject at any moment and take precedence over a call that is already under way (preemption). In a similar way, users (without a high priority level) can be routinely provided with priority access whenever they dial an emergency number.
- *Open conference calls.* This service allows all members of the same group to communicate on a particular channel. A signaling message transmitted on the group control channel tells them to go to a specific traffic channel. During the conference the "Go to Channel" message is continuously broadcast on the control channel and any newcomer belonging to this group is able to join it on the traffic channel.

9.2.2.2 Data Services

Services using data instead of voice are some 5 to 20 times quicker. Selective calls and dual-tone multifrequency (DTMF) are methods that have been used for data transmission. The transmission flow is relatively slow although the long duration of the tones makes it possible to decode them efficiently even under poor signal conditions. In practice transmission rates through a 12.5-kHz separated channel quite often range between 2,400 and 4,800 bps using various modulation techniques.

In the early 1990s data transmission was much more widespread in PMR than in cellular systems. Data transmission allows, for instance, the exchange of short messages, database queries, reservations, alerts, status, measurement data, and paging. Data transmission usually needs the use of an external modem. The user is able to connect a personal computer over the modem or radio terminal to a central computer.

The steadily increasing demand for the data transmission services over PMR is further enhanced by four major advantages:

- *Spectral efficiency.* The transmission of a message, regardless of the fact that it may be coded, requires a duration significantly less than that for voice communication.
- *Accuracy.* The message received, especially if it is coded, is more accurate. It could be read over again (without having to be retransmitted), printed out, stored, and then distributed to third parties.
- *Coding features.* The coding of information helps avoid any eavesdropping, an aspect that is of great interest to those enterprises that would like to keep their communications confidential.
- *The possibility of sending a message even if the recipient is absent.* Contrary to voice calls where the addressee must be available to be informed of the message content, the transmission of data messages does not automatically require the physical presence of the addressee. The terminal stores the message and the user can read it later. In this way time is saved both for the caller who does not have to monitor the call and for the addressees who has no missed messages should he or she be away from his terminal.

Three levels of application using data transmission can be identified within PMR systems in accordance with the volume of transmitted information:

- *Precoded messages.* Some systems allow the users to predefine (preformat and code) messages that are most commonly used and, by so doing, simplify the use of the mobile equipment. Each message is identified by a code and is not transmitted explicitly. This results in high spectral efficiency and confidentiality. With this kind of use the amount of data exchanged is quite low.
- *Dispatching of fields from preformatted forms.* The user is required to complete the fields shown on a screen. Only the data in these fields are sent. This limits the quantity of information transmitted over the radio channel,
- *Transparent data transmission between two computer terminals.* With this mode of operation, no error control or transmission protocols are implemented within the PMR system, which will simply transmit the information. This can lead to multiple errors due to flow constraints (a few kilobits per second) from the terminals as well as propagation problems of the radio channel. This kind of application can only be used as an adjunct in a PMR system which, in the first place, was designed for voice applications.

9.2.2.3 Other Services

- *Include call.* A user can join a conversation under way between two users of his or her group.
- *Coding.* Communications could be coded to guarantee the confidentiality of

exchanges, an aspect that is particularly interesting for the security forces and some companies.

- *"Out of coverage" indication.* The mobile informs the user that he or she is located outside the reception area of the base station.
- *Remote kill.* The mobiles can be disabled remotely and are unable to transmit until commanded to act otherwise. This kind of service is used in the environments where explosion risks are high (e.g., oil-drilling cases) or if a mobile unit has been stolen.
- *Dynamic regrouping.* In some cases, a redefinition of the groups could be achieved in a dynamic way, in other words when the operation is underway.

9.2.3 Conclusion

The characteristics, services, and users of PMRs are heavily impeded by constraints in architecture and implementation techniques. In the following sections, the two main kinds of PMR ("conventional" and trunked systems) are discussed further.

9.3 CONVENTIONAL PMR SYSTEMS

PMR systems commonly described as "conventional" are the oldest ones and also the most widespread at this time. These systems are not as efficient in terms of spectrum usage compared with the next generation of systems (based on trunking techniques). In this section, classic PMR systems, their weaknesses, and the causes leading to the implementation of trunked systems are discussed.

9.3.1 Architecture of Conventional Radio Networks

Conventional (or classical) PMR systems are based on quite simple, even elementary infrastructure and organization. Generally, conventional PMR networks are based on a star structure, where the base station (fixed or mobile) represents the center. In some systems mobiles communicate directly, one to another, and such a network could be considered as a "bus" structure. Nevertheless, the use of a base station at the center of the network remains the most widespread. Long-distance transmission is only possible between mobiles if a base station/repeater is used. It is able to transmit at high power levels (from a few tens to several hundreds of watts) and with antennas located on high sites and thus has wider coverage. The powers of the mobiles range from a few watts to some tens of watts.

Two kinds of conventional PMR system can be identified dependent on the radio resources that are offered:

- One channel (one or two frequencies) is allocated to a single fleet.
- Several fleets share one channel. An infraband signal helps select the wanted fleet and only mobiles able to decode this signal are activated.

Note: When only one channel is shared between fleets (which should be small in size to avoid congestion), one fleet only can communicate at a time. In trunked systems several fleets are able to communicate at the same time, using the pool of channels. There are conventional systems where the base station acts as a repeater and retransmits the signal without modifying it and without accessing a network. The repeater acts like a mirror, receiving the signals on one frequency and retransmitting it on another.

9.3.1.1 Simplex Networks

In a *simplex* network, the single channel is shared by all entities—terminals and base stations (if any). All the entities have the same access rights to the radio channel. The major strength of this type of network lies in its easy installation. Its second strength is that it does not require large radio resources. One disadvantage is that it can only cover small areas. In practice, the radius of coverage is usually only a few kilometers. In this kind of network, each station transmits in its turn while the others are listening. The called party replies at the end of the message with the transfer from receive to transmit being controlled by means of a pedal or a push-button connected to the microphone or radio unit. It is also possible to operate in this mode with a voice-activated switch (VOX). This kind of simplex mode is very useful when there is a requirement to operate "hands free." Simplex networks operate with a single frequency. To increase network coverage, the use of other types of base stations is necessary (Section 9.3.1.2).

9.3.1.2 Half-Duplex Networks

When the number of mobiles is very large and the operating area covers several tens of square kilometers, the use of a half-duplex system with pairs of frequencies per channel is required together with the ability for base stations to act as repeaters (Figure 9.6). This technique helps extend the range of network coverage significantly compared with the case of a simplex network. This type of network is the most common to be found in conventional PMR. With a half-duplex network, the base stations transmit on one frequency (f_1) and receive on another frequency (f_2). In their turn, mobiles transmit on the receive frequency of the base station (f_2) and receive on its transmit frequency (f_1).

The base station can be either fixed or mobile. In the latter case, although relatively rare (e.g., in public safety at the scene of an incident), the base station could be a mobile working with reversed frequencies.

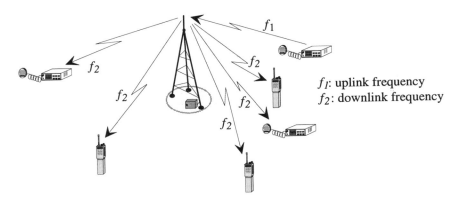

Figure 9.6 Operating principle of a duplex network.

With a half-duplex network, as with simplex networks, communication occurs alternately: one mobile transmits (on the uplink from the mobile toward the base station) and the others listen (retransmission by the base station on the downlink). Wide coverage areas are possible: from a few tens, to a few hundreds, to even thousands (airborne or maritime use) of square kilometers (Figure 9.7).

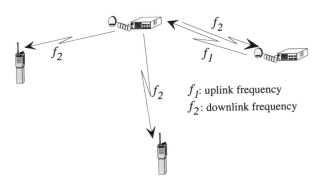

Figure 9.7 Half-duplex network where the base station is a mobile.

The base station transmitter/receiver is usually installed at a high site, connected to a fixed terminal installed in a building or may be a vehicle located on a high spot (Figure 9.8).

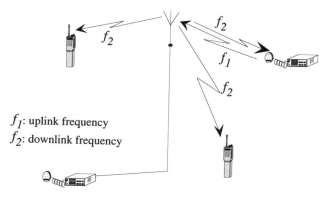

f_1: uplink frequency
f_2: downlink frequency

Antenna installed on a high spot

Figure 9.8 Half-duplex network with transmitter/receiver transfer.

9.3.1.3 Signaling

Several signaling standards have been defined such as five-tone coding (*Select Five*), continuous signaling tone (CTS), or ZVEI signaling. These signaling systems are based on audio-tones. There are three basic types: the *continuous tone* systems, the *sequential tone* or burst systems, and the *combined tones* systems [3].

Continuous tone signaling. The *continuous tone controlled signaling system* (CTCSS), also called *private line* (PL) is the most commonly used method for groups of users on a single channel. Thirty-two tones are used between 32 and 250.3 Hz. This method of signaling is used to solve the problem of intrusion on to other peoples' communications. Only receivers equipped with a tone decoder tuned to the same CTCSS frequency will be activated on receipt of a radio frequency carrier. CTCSS signaling thus helps avoid interference from the transmissions of remote network *B*, which is using the same radio channel as another network *A*, from disturbing the users of network *A*. Normally, should network *A* have activated its receivers with its own CTCSS tone, then signals from *B* will not be heard as they will be relatively too weak and the "capture effect" will attenuate them.

With *digitally controlled squelch* (DCS) signaling, also called digital PL, the transmissions consist of a low bit-rate digital signal continuously transmitted. This transmission contains a repetitive codeword representing a three-digit number (Figure 9.9). The advantage over the previous system is that there are many more possible codes (i.e., 104 codes) and the system response is fast. Signaling systems with two, five, or more tones have been specified. Among them are:

- CCIR signaling (Comité Consultatif International des Radiocommunications), originally defined by the international maritime mobile service and widely used in France and Scandinavia,
- ZVEI signaling (Zentralverband der Elektrotechnischen Industrie), used mainly in Germany, Austria, and Switzerland,
- EIA signaling (Electronic Industries Association), used in the United States,
- EEA signaling (Electronic Engineering Association), used in Great Britain.

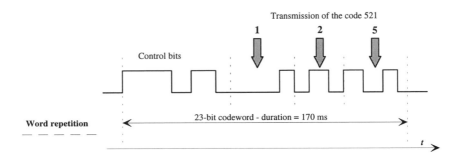

Figure 9.9 DCS codeword format.

In these systems, the signaling burst consists of a series of audio frequency tones (e.g., five) transmitted sequentially. The audio frequencies are contained in the 600- to 2,250-Hz band. The codes are characterized by the tone frequency, the tone duration, and their mode of transmission (simultaneous or sequential, with or without an interval between successive tones).

Dual tone multifrequency (DTMF) signaling is also used. It is based on a frequency matrix by row and column , which represents a standard telephone keypad with 12 buttons. In this case, a number is transmitted by the simultaneous transmission of two appropriate frequencies, one low and one high, ranging between 697 and 1,477 Hz.

In conventional PMR, signaling and voice are multiplexed on the same radio channel, a method that tends to generate low-density traffic. Besides which, as with the first systems, signaling, like voice, used to be analog (three or five tones, typically). With modern systems signaling is digital and the transmission of voice could be either analog or digital. Digital signaling messages include the following type of information: user identification, group calls, channel identification, and the like. In general, they are made up with about fifty data bits.

In some systems signaling is used to define several groups sharing the same channel. By assigning to each group a private line (PL), intragroup communication can be made confidential (other groups will not be able to transmit or receive while a call is in progress).

Each group using a channel f_i could identify itself with a subaudible frequency Δf_i (Δf_i ranges from 50 to 160 Hz). When a channel is in use with the Δf_i identifying tone, another subfleet sharing channel frequency f_i cannot use it.

Signaling can be effected in two ways: either with *inband* tones that interrupt conversations and consequently degrade the quality of service, or *infraband*, using inaudible tones.

A note on infraband signaling: Infraband signaling consists of multiplexing signaling with the voice transmission, without it being detected by the end user.

Infraband signaling can reduce the effects of interference. With a full duplex communications system, interference caused by the transmission from a remote mobile is related to the prevailing C/I ratio, whereas in half duplex the local mobile is in a receive state while the remote mobile is transmitting. The wanted carrier C is therefore at zero and the base station receives the remote signal I if it is strong enough. Infraband signaling ensures that the base station ignores this remote signal if it is not the wanted one.

Infraband signaling also helps to supervise control of the connection and avoid cases where release of the traffic channel was not correctly actioned. There are two common procedures to do this:

- *First method*: This is triggered when a mobile has remained in receive state for a predetermined period of time (a few tens of seconds, typically). The mobile automatically switches to the transmit and sends an infraband message to inform the network that it is still active. If the base station does not receive an infraband message within a given time period, the channel is released.
- *Second method*: When the mobile has been transmitting for a certain duration, it automatically reverts to receive and looks for the infraband signaling before switching back to the transmit state. If the infraband signaling is not received within a certain time (e.g., about ten seconds), the channel is released. This mechanism is called the "anti-talk" procedure. The maximum time (T) of the communication can be adjusted between two values (T_1, the minimum value and T_2, the maximum value) as a function of traffic density. The calculation of period T can be achieved dynamically by using a formula such as the following:

$$T = T_1 + (1 - C)(T_2 - T_1)$$

where C is the network load expressed in terms of percentage occupancy of the traffic channels.

The ETSI Binary Interchange of Information and Signaling (BIIS) 1200 Standard. The MPT 1327 standard (see Section 9.4) was designed as a signaling method for trunked radio networks. It was deemed necessary to define a new standard for digital signaling and for the transfer of data that would meet the needs of conventional networks. In the early 1990s the lack of compatibility between the various signaling systems impeded the upgrading of the PMR networks. Thus ETSI, prompted by ECTEL (the European Trade Association), undertook to develop the I-ETS 300 230 standard which is largely inspired from the MPT 1327 standard. The feature that makes them distinct from one another is the lack of a separate control channel. The BIIS 1200 standard was approved in August 1993 (Figure 9.10).

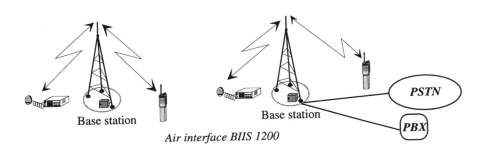

Air interface BIIS 1200

Figure 9.10 BIIS 1200 networks configuration.

This standard uses fast frequency shift keying (FFSK) modulation and exchanges information at a 1,200 bps rate. It defines the radio interface and some of the fixed equipment (base station, base station, operator station, or fixed telephone subscriber via PBX or public switch). The signaling protocol enables the linking of two or several terminals to communicate by voice or to exchange data (Figure 9.11).

The main services offered by this standard are:

- *Voice*: individual calls, group calls, broadcast calls, and connection to a PBX,
- *Data*: status messages, short messages, and long messages,
- *System*: priority calls, emergency calls, caller identification, direct mode (mobile to mobile), callback, call forwarding, for example.

Each codeword contains 48 information bits and 16 cyclic redundancy check bits. The first word is called the *address word*; it contains the address of the caller and the address of the person called as well as an operating code (Figure 9.12).

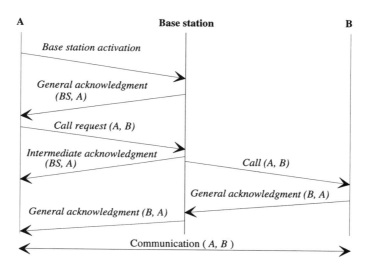

Figure 9.11 Message exchange for call set-up in BIIS 1200.

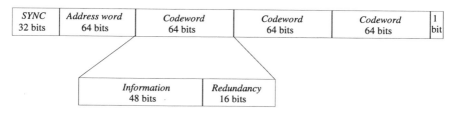

Figure 9.12 BIIS 1200 message structure.

9.3.2 Weaknesses of Conventional Radio Systems

Conventional PMR is particularly applicable for organizations that like to operate their own network so as to ensure management control. Their main drawback is their spectral inefficiency and, for the owners, the additional burden of running and maintaining the system. With increasing pressure on the radio spectrum, charges stemming from its use are increasing (increased license fees, spectrum auctions, and spectrum trading). The organization that wishes to operate classic PMR must be prepared to purchase (or lease) expensive infrastructure and also to pay recurring license fees for the use of the spectrum. This often requires a considerable budget. Advances in technology, which have allowed sharing of infrastructure, have led to the adoption of trunked networks.

9.4 TRUNK RADIO NETWORKS

Trunked networks can serve several groups of users while reducing the number of necessary frequencies (e.g., improving the spectral efficiency) and also the cost of the system infrastructure, which is jointly shared between all subscribers. This technique globally helps increase the spectral efficiency of systems and enhances service quality (notably the waiting time for access). Trunk systems can serve over a hundred users per channel against an average of a few tens in the case of many conventional systems. (This must be regarded as a mean value that depends on the traffic profile. Some conventional taxi dispatcher systems can support up to 200 mobiles per channel because dispatch messages are typically of only 5 seconds duration.)

9.4.1 History of Trunked Systems

Trunked systems have been used in the United States since the mid-1970s. The trunking method upon which they are based owes its name to the technique used in telephone systems ever since their inception. It consists of multiplexing the traffic coming from several users. Once it reaches the first switch the traffic is carried by a shared resource called a "trunk" in order to be transmitted to the second switch. The size of the trunk is determined by the probability that no two users will require making a call at exactly the same time. At the beginning of the 20th century, Erlang derived the probability functions commonly used. Since the early 1980s, trunked radio systems have been over installed all over the world because of their improved efficiency.

9.4.2 Efficiency of the Trunking Method

The theory of trunking is based on:

- The mean occupation rate of a single resource per user is generally low (especially for voice communications).
- The probability that a large number of subscribers will try to access a channel at the same time is unlikely to occur (the peaks of traffic are rare and of a short duration).

In conventional systems each user has a specific channel for all communications. This is the one allocated to the group to which he or she belongs and which is shared with other users. Any member of the group wishing to make a call has to wait until it is released before accessing it. Conversely, in trunk systems, mobile terminals have access, not to a single channel but to a group of channels and the resource is allocated to the mobile terminal only on request. Once the communication is complete the channel is released and can be accessed by any other terminal. The average use of a channel is increased without impact on the channel access delay because the traffic load is evenly distributed between various channels.

Trunked mobile terminals require extended functionality over that needed by conventional PMR terminals. Each terminal has to be capable of tuning to any system frequency. In addition the mobiles have to be able to communicate with a control center when accessing the system (e.g., to set up a call). Finally, digital techniques (notably in the case of second-generation trunking systems) added to frequency sharing and, in particular, the use of signal processing, have helped to obtain an excellent spectral efficiency as well as an increased variety of voice services and more efficient data transmission services [4, 5, 6].

9.4.3 Trunking Architecture

The main components of trunk systems are subdivided into two groups (Figure 9.13):

- A *hardware part*: control centers, radio sites, management and maintenance databases, and mobile terminals,
- A *software part*: programs for managing and operating the system and the terminals.

9.4.3.1 Fixed Infrastructure

The intelligent central unit, usually an embedded microcomputer, called the *controller*, typically manages between 5 to 20 duplex channels. A single channel (called the *broadcast channel or control channel*) is dedicated to signaling and the rest to traffic. System performance is highly dependent on the control channel, which may be dedicated, time division multiplexed with other sites, or able to carry traffic in the event of all traffic channels being loaded. The controller supervises and manages the system. It is connected to the base stations, regulates the flow of communications, processes communications requests, performs statistics, and logs the state of equipment and signal alarms.

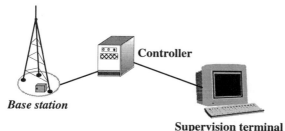

Controller

Base station

Supervision terminal
(allows the interface between the system and the human operators)

Figure 9.13 General structure of a trunk system.

The controller undertakes the following tasks:

- Location of the mobiles,
- Call processing and switching functions (including interconnect with the PSTN or a PBX),
- Traffic-channel allocation,
- Management of the subscriber database: informs the system of fleet characteristics, numbers of subfleets, identification of individual mobiles using the network, authorized coverage area, and so forth,
- Generation of billing-information: system call reports, helps to issue billing records, either from the center *per se*, or from an *ad hoc* and separate management center,
- Data-statistics: helps the system manager track system load (number of calls, duration, location) and to adjust the system parameters appropriately,
- Supervision: has algorithms and procedures to manage the various components of the system.

Base stations contain one or several duplex-radio units, each of which supports a single radio channel pair. A fixed radio unit consists of a transmitter, a receiver, a power supply, and control together with interface circuits which are necessary to connect the base station to the system. Finally, the supervision center of the system allows the user to run the system.

9.4.3.2 Signaling

Several forms of signaling are used in trunking systems. The most widespread is the MPT 1317 signaling originally defined in Great Britain. One of the forms used in France is the DGPT 2424 standard (an extension of the PAA 1382 standard), which is quite close to the MPT 1317 standard.

The EEA (Electrical Engineering Association which has since become the FEI, Federation of the Electronics Industry), a British organization, in conjunction with the British Ministry of Post and Telecommunications (MPT, but now known as the Radiocommunications Agency), jointly developed the standard at the end of the 1970s. A similar standard has been issued in France with slight differences (notably a change to the word synchronization field and a tightening of the modulation specification), known as PAA 1382.

The MPT 1317 standard is very commonly used over the rest of Europe. Its derivatives are MPT 1327, 43, and 52. MPT 1327 defines the specification of the network-signaling protocols and terminal functionality, and particularly, it defines the mapping of the codeword fields. The MPT 1343 standard defines the protocol for the use of codewords when in trunked systems. DGPT 2424 is the French equivalent of both these specifications. MPT 1352 is a type approval document for trunked mobiles.

This signaling is performed digitally using subcarrier modulation based on two audio frequencies (1,200 and 1,800 Hz). The modulation used is FFSK at 1,200 bps (Figure 9.14). The format of an MPT 1317 message is composed of a 16-bit synchronization sequence of "1s" and "0s", a word-synchronization sequence of 16 bits, a 48-bit data field, and a 16-bit cyclic redundancy check (*CRC*).

| Bit synchronization 16 bits | Word synchronization 16 bits | Data 48 bits | CRC 16 bits |

Figure 9.14 Format of an MPT 1317 message.

The combined 48 data and 16 CRC bits (64 bits) is called a *codeword*. Channel access is performed using the S-ALOHA (slotted ALOHA) method with dynamic frame lengths (Figure 9.15).

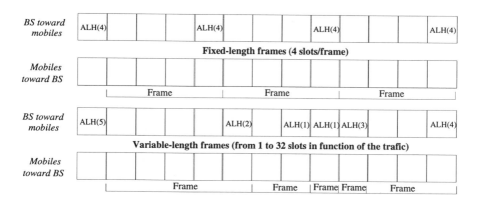

Figure 9.15 Structure of a fixed or dynamic frame structure.

9.4.3.3 Call Setup Protocol in MPT 1327

With the MPT 1327 standard, signaling is transmitted digitally (Figure 9.16). The downlink signaling messages are 128 bits in length and their duration is 107 ms. The terminals are allowed to access the channel in accordance with the slotted ALOHA protocol. The base station regularly transmits invitation messages (marked ALH for *ALOHA*) on the broadcast channel. The access procedure starts with the calling mobile transmitting an access message (marked RQS for *request*). The control center answers with an AHY (*ahoy*) message. This message is used by the calling unit as an acknowledgement, by the called unit as a paging request, and indicates to all other mobiles that the next slot is reserved. If the called party receives this message and is powered up, the unit answers by sending an ACK (*acknowledgment*) message in the reserved next slot. Upon receipt of this message and if there is a channel available, the base station transmits to the caller and the called terminals a GTC message (*go to channel*) indicating to the terminals the traffic channel that has been allocated for the communication.

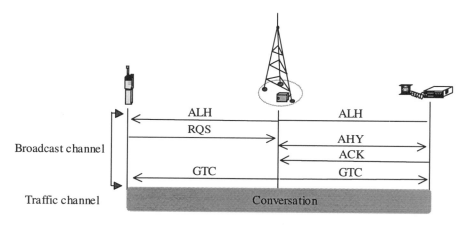

Figure 9.16 Signaling exchanges with MPT 1327.

The basic services available with an MPT 1327 system include the following:

- *Type of calls*: individual call, group call, include call, data call, status, and short data messaging.
- *Supplementary services*: emergency calls, in bound call restriction, call divert, and access to the PSTN.

9.4.3.4 Signaling Channels

Fixed-signaling channel systems. The signaling/control/broadcast channels are located at fixed frequencies known by all the mobiles. In an idle state (i.e., in the absence of communication), the mobiles listen to one of these channels and receive the paging messages that are addressed to them.

One the major advantages of this configuration is that the system is able to manage a large number of traffic channels. Mobiles have fast access to the system because they are normally synchronized to the control channel slot structure. Finally, the mobiles need not know all the channels used by the system. They simply need to know the broadcast frequencies without having to scan large portions of the spectrum.

Some systems allow the control channel frequency to be allocated to a mobile if all traffic channels are busy. In this case mobiles in idle state have to wait for its reappearance (in other words, its release). It is preferable not to do this in systems using a large number of channels because of the importance of the signaling time. Indeed, the time between the release of a channel and its reallocation to another call should be minimized; otherwise a queue of mobiles waiting for free channels may form when the system is under heavy load. For example, in a 10-channel system where the current allocation time is about 2 sec, significant capacity can be wasted if the signaling is too slow. Systems with more than 10 channels should use a dedicated control channel. A major drawback with such a method is that one channel is monopolized for signaling and the level of activity does not generally require a whole channel. Also, in case of transmission problems on the broadcast channel transmitter, there can be a single point of failure. To overcome this problem, some systems offer several different solutions. Either—as is the case with conventional systems—when a control channel fails, mobiles enter a fallback mode where they are distributed on to traffic channels by fleets (i.e., talk through mode), or an alternative control or emergency channel is predefined and its frequency is known by all mobiles.

Time shared control channel systems. The most commonplace method under this name consists of a multitude of sites sharing the same broadcast channel frequency for spectrum efficiency reasons (Figure 9.17). In this case, the broadcast channel always uses the same frequency but changes its site at each time interval.

Successive allocation of the beacon to the different sites

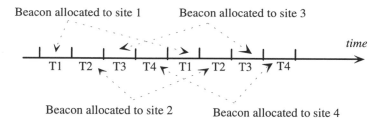

Beacon allocated to site 1   Beacon allocated to site 3

time

T1   T2   T3   T4   T1   T2   T3   T4

Beacon allocated to site 2   Beacon allocated to site 4

Figure 9.17 Time-shared control channel system (fixed frequency shared by several sites).

Rotating control channel. This is the second most common method. The control channel is always at the same site but changes its frequency according to the call arrival rate. The control channel no longer uses a fixed frequency, and the mobile that wishes to make access to the system has to scan all the channels until it detects this channel. In this case access time increases in proportion to the number of channels.

Principle of a rotating broadcast channel system on a single site. When a mobile wishes to set up a call it makes access on a broadcast channel which, when the call is set up, becomes its traffic channel. For each new call the system must select a new broadcast channel by choosing a free channel and identifying it as such. The broadcast channel carries specific signaling that allows other mobiles to detect it. As calls are set up, mobiles in the idle state are required to search for the newly identified broadcast channel.

The major advantage with this technique is that all the channels can be allocated for user traffic. The disadvantage is that at peak traffic hours, the average duration between two successive channel allocations is never less than the time required for a mobile to scan half the channels. The scanning time is generally about 90 ms per channel. The constraint is that the search time should not exceed the mean time required for call waiting. This technique is mainly valid for systems with fewer than 10 channels. The optimum is around a five-channel system.

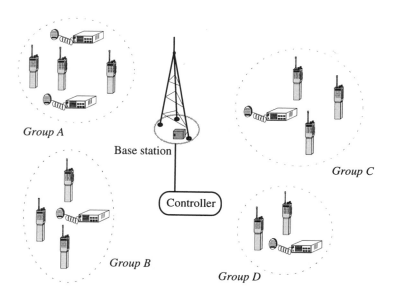

Figure 9.18 A rotating broadcast channel system with three channels for four groups of users.

Example of the use of a rotating broadcast channel at a fixed site. An example of the use of a rotating broadcast channel system is shown in Figure 9.18. This technique can be implemented using tone calling or with a subset of MPT 1327 signaling.

Suppose a three-channel system serves four groups of users as presented in Figure 9.18.

The base 1 (channel 1) transmits a tone *F* continuously to show that the channel is free. The other channels do not transmit a tone (Figure 9.19).

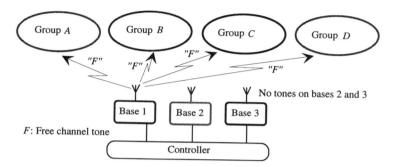

Figure 9.19 Terminals not in use: all users "locked" on the broadcast channel (channel 1).

The *F(ree)* tone is totally different from the tones of the user groups (*A, B, C,* or *D*). In their idle state, the mobiles search for this tone on the three channels, and thus, in our example, are tuned to channel 1. A mobile of group *A*, wishing to communicate with other members of its own group transmits its tone (*A*), which is retransmitted by the base station on channel 1 and sent to all mobiles. Members of group *A* decode this tone.

Base 2 now starts transmitting the *F* tone and becomes the new free channel (Figure 9.20). The mobiles of group *A* move to a receiving state and listen to channel *A*. The mobiles of the other groups (*B, C,* and *D*) search for the *F* tone and tune to this new channel.

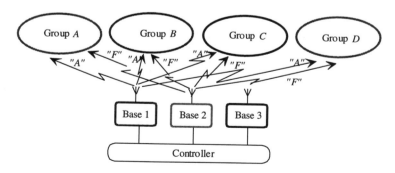

Figure 9.20 Activation of group-A mobiles.

When a mobile belonging to group *B*, *C*, or *D* wishes to establish a communication, it, in its turn, sends the corresponding tone for its group on channel 2, which then becomes busy (Figure 9.21).

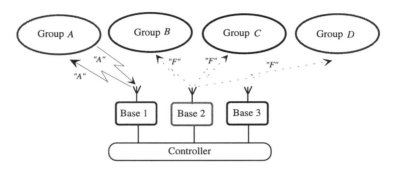

Figure 9.21 Group *A* on channel 1. Other mobiles on channel 2.

9.4.4 Engineering

PMR systems engineering is relatively simple compared with that of cellular systems. Nevertheless, with wide-coverage systems, some precautionary measures have to be observed. This is particularly the case for systems that cover overlapping shadow areas, and where synchronous or quasi-synchronous transmission methods [3] are used. Transmissions from different sites and at similar field strength are received simultaneously by any mobile that can be potentially located on any point of the radio coverage. These simultaneous transmissions can give rise to destructive interference problems entailing deep fades at several points of the coverage. In practice, the simplest solution consists of transmitting the signals in a quasi-synchronous mode. With this method, the transmitters at each site use frequencies shifted from 10 to 40 Hz, which helps to avoid points of permanent cancellation. Although co-channel interference is experienced, it manifests itself as distortion and not complete loss of signal. The major constraint of these systems consists of providing equalization mechanisms to minimize the distortion of the received signals.

An alternative used to avoid the complexities and the problems caused by these systems is that of using several channels. Each channel covers a specific area and the mobiles must therefore select the channel for the area in which they are located. The drawback with such a solution is that it is a costly one in terms of spectrum. Another alternative consists of a signal quality voting method that helps to obtain wide area coverage using only one channel (Figure 9.22). With such a method the base stations are installed on the area to be covered, and at any given time only one base station is active. Thus, the mobiles located in the coverage area do not encounter the interference problems related to synchronous systems.

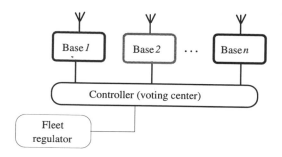

Figure 9.22 The majority voting-method principle.

The sites are connected to a control center equipped with a voting system. Its role is to evaluate transmissions coming from the sites according to the SNR, for instance, measured by each station or at the control center. The station for which the quality of signal (i.e., the SNR value) is best is chosen. This choice is carried out for each transmission and the most suitable base station is connected through to the dispatcher.

Example: Dimensioning of a single control channel

The task consists of determining the number of calls per hour. By taking the case of the MPT 1327-like system, using a duration slot of 106.7 ms, there are $(1,000/106.7) \times 60 \times 60$ slots per hour, or 33,750 slots per hour. Access on this channel is performed in S-ALOHA (in other words, with a maximum throughput of $1/e = 37\%$). However, only about 40% of these slots are available for random access as others are reserved for acknowledgements, short data messages, and system overhead control. The maximum number of accesses on this channel is $0.4 \times 0.37 \times 33,750$ or 4,966 slots per hour. By taking into account propagation problems (e.g., fading, interference), suppose that only 85% of the attempts do really have access at the first attempt. With these values, it can be deduced that the control channel can carry an average of $4,966 \times 0.85$ or 4,220 calls per hour. For a mean call duration of 30-sec, the number of traffic channels which can be supported by a single control channel is approximately 35.

9.4.5 TETRA

As with cellular networks (GSM), cordless systems (DECT), and paging systems (ERMES), ETSI has drawn up a standard for PMR systems, namely, the Terrestrial Trunked Radio system (TETRA) [7].

The TETRA standard combines two major groups of specification: the first one defines the radio and network interfaces for trunked voice and data (V+D) services; the second defines an optimized radio interface for packet data (PDO) transmission services interworking with the classic network protocols. Finally, as a subset of V+D, there is a third group called "direct mode" that specifies terminal to terminal communication on a single radio channel.

The first systems to be in compliance with the TETRA standard went live in 1997. Among systems competitive to TETRA are the enhanced digital access communications system (EDACS) designed by Ericsson, TETRAPOL, from MATRA-COM (which will be discussed in Section 9.4.7), and APCO25, an American standard based on FDMA techniques.

9.4.5.1 Architecture

The TETRA standard defines several entities (Figure 9.23):

- A fixed-network access point (FNAP) for interfacing with fixed-user equipment.
- A mobile-network access point (MNAP) for interfacing by radio with mobile terminals. When the mobile is integrated equipment without an external data port, the MNAP function is realized in software.
- The mobile terminating units (MTU). There are three classes of mobile specified with power outputs ranging from 1 to 10W. The highest power output enables the use of cell radii of about 25 kms in rural areas.

Four types of reference network architecture are defined:

- Single site without a network-switching center and with limited coverage and a low number of subscribers,

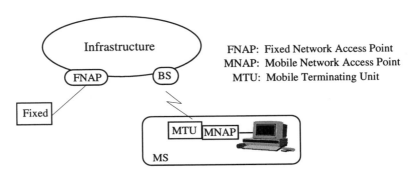

Figure 9.23 TETRA interfaces and structure.

- Single site with a switching center and gateways with other networks; for use in heavy-traffic areas (airports, industrial areas),
- Multisite with a switching center for urban and regional applications,
- Multisite with several switching centers for multiregional or national applications. In this case, the switches are interconnected through leased lines.

The main functions of a TETRA network are:

- Roaming, allows interworking with roaming terminals (i.e., coming from other TETRA networks),
- Interfacing with other networks (PBX, PSTN, PSDN, and ISDN).

Six interfaces are defined for the infrastructure:

- Radio air interface for the trunked mode of operation,
- Line-station interface,
- Intersystem interface,
- Terminal equipment interface,
- Network-management interface,
- Radio interface for direct mode communication.

The main procedures available are *handover, roaming, authentication, and ciphering*.

9.4.5.2 Radio Interface

Public safety TETRA systems operate within the 380- to 400-MHz band with a duplex spacing of 10 MHz. This band has been released by NATO in Europe on the understanding that it will be shared with the military and only used for public safety systems. Specific applications are also envisaged in the 900-MHz frequency band with a frequency duplex spacing of 45 MHz. The channel separation is 25 kHz. The frequency bands identified by CEPT for civil (private or public networks) TETRA are 410-420/420-430 MHz, 450-460/460-470 MHz, and 870-888/915-993 MHz.

Various versions of the standard use the same radio interface based on a $\pi/4$ DQPSK modulation at a 36-Kbps bit rate. The time required to set up a half duplex call is less than 0.5 sec. The TETRA radio interface is based on the FDMA/TDMA method with a frame structure of four slots per radio carrier. The traffic and signaling channel multiplexing is realized using a four-level structure (slot, frame, multiframe, and hyperframe).

Burst structure. The radio bursts are based on the structure reproduced in Figure 9.24.

Several types of structure for the transmitted data are defined, four on the uplink and four on the downlink:

On the uplink the following bursts are:

- *Normal uplink burst* (NUB), composed of 512 bits and carrying voice or data traffic as well as signaling.
- *Control burst* (CB), composed of 162 bits and used for the transmission of signaling. Two bursts of this kind can be transmitted in one slot with the proviso to observe suitable guard times.
- *Linearization burst* (LB), which allows terminals to linearize their transmissions the first time they access a new radio channel. This type of burst does not carry data.
- *Slot flag* (SF), which corresponds to one of the two training sequences (located in the middle of the burst). The SF is used to indicate the presence, or lack of one, or two logical channels on blocks 1 and 2 of the burst (Figure 9.24).

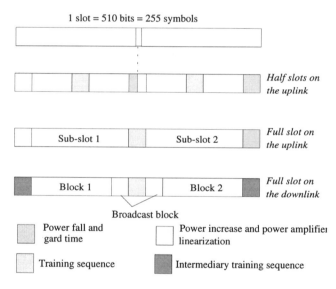

Figure 9.24 Bursts structure in TETRA.

On the downlink channel, the following bursts are defined:

- *Normal downlink burst* (NDB), composed of 512 bits carrying voice, data, broadcast messages, and signaling,
- *Synchronization burst* (SB), composed of 216 bits, characterized by a long synchronization sequence, preceded by a pure sinusoidal wave of 2.1 ms allowing the synchronization of the mobile to the base station,

- *Broadcast block* (BBK), composed of 30 bits (14 useful bits) and used for AACH (see Section 9.4.5.2.2),

- *Slot flag* (SF).

The slots are organized in four-slot frames with each slot having a fixed duration of 14.167 ms. A system or user communication requires at least one slot (Figure 9.25).

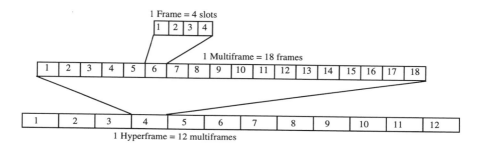

Figure 9.25 Slots, frames, multiframes, and hyperframes in TETRA.

At the higher level, an 18 TDMA frame multiframe is defined. The first 17 frames carry user data and the 18th carries signaling. Finally, above the multiframes, a hyperframe structure is defined to allow the mobiles to listen to adjacent cells in a controlled manner.

The transmit and receive operations are shifted by a two-slot period (see example reproduced in Figure 9.26).

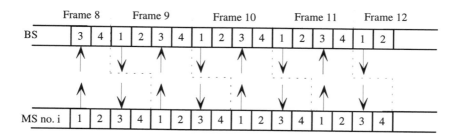

Figure 9.26 Slots combination on the uplink/downlink channels.

Logical channel structure. TETRA defines logical channels (see Table 9.1) at the lower medium access control (MAC) layer level and at the upper MAC layer level (Figure 9.27). To obtain higher bit rates (28.8 Kbps, 19.2 Kbps, or 14.4 Kbps), up to four concatenated slots located on the same frequency carrier can be allocated for the same communication.

The BSCH is used by mobiles for cell acquisition (the SYNC message is transmitted on that channel). BNCH (Table 9.2) is then received by the mobile for frequency synchronization which allows it to receive and decode system information (e.g., messages called SYSINFO and indicating the frequency of the central frequency carrier, the number of secondary control channels, the cell access parameters). The mobile can then be synchronized on MCCH (located on slot 1 of the main carrier) or on a secondary control channel (which may be on the same or a different carrier).

Table 9.1

MAC-layer logical channels lower level

Logical channels	Abbreviation	Uplink/Downlink
Signaling channel	SCH	↑↓
Access assignment channel	AACH	↓
Broadcast synchronization channel	BSCH	↓
Stealing channel	STCH	↑↓
Common linearization channel	CLCH	↑
Traffic channel:	TCH	↑↓
7.2 Kbps	TCH/7.2	
4.8 Kbps	TCH/4.8	
2.4 Kbps	TCH/2.4	

FACCH is used during the allocation of a traffic channel (TCH) for one or several mobiles. They listen to the logical FACCH to identify that the slot (channel) has been allocated to them. The allocated channel then becomes a TCH. At the end of the call the logical FACCH is obtained by slot stealing on the TCH in order to transmit close down signaling. Thus the FACCH and TCH channels alternately use the same slot and are mutually exclusive from each other. A similar mechanism is used in GSM (see Chapter 12).

Table 9.2
MAC-layer logical channels upper level

Logical channels	Abbreviation	Uplink/Downlink	Use
• Signaling			Exchange of information with idle mobile stations
* Common: *common control channels*	CCCH	↑↓	
- *Main control channel*	MCCH		
- *Extended control channel*	ECCH		
* Associated: *associated control channels*	ACCH	↑↓	Exchange of information with mobile stations in the dedicated mode
- *Fast associated control Channel*	FACCH		
- *Stealing channel*	STCH		
- *Slow associated control channel*	SACCH		
• Broadcast		↓	System information broadcast
Broadcast control channels	BCCH		
- *Broadcast synchronization channel*	BSCH		
- *Broadcast network channel*	BNCH		
• User plane		↑↓	Speech or data information exchanged in circuit mode
Traffic channels	TCH		
- *Speech traffic channel*	TCH/S		
- *Speech/data traffic channel*	TCH/7.2 ;		
- *End-to-end specific data*	TCH/4.8 ;		
	TCH/2.4		

STCH is also obtained by means of slot stealing on an established TCH, both on the uplink or downlink channels. It is used for the transmission of fast data (notification of an urgent call, for instance).

SACCH is jointly used with a TCH or an FACCH. It uses the slots located on frame 18.

9.4.5.3 Bearer Services and Teleservices

TETRA defines the bearer services and teleservices which rely on three specification groups, each of which define a particular transmission mode and are as follows (Figure 9.28):

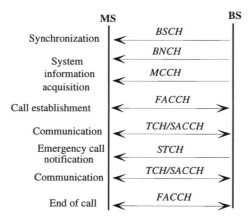

STCH: stealing channel; SACCH: slow associated control channel

FACCH: fast associated control channel; MCCH: main control channel

Figure 9.27 Use of the logical channels during communications.

	Speech		**Data**				
			Circuit			**Packet**	
Bearer Services	Circuit mode unprotected 7.2 kbps		Circuit mode protected up to 19.2 kbps	Circuit mode unprotected up to 28.8 kbps	Connection-oriented X25 ISO 8208	Standard connection-less ISO 8473	Special connection-less
Teleservices		Clear speech 4.8 kbps					
		Encrypted speech					

Figure 9.28 Bearer services and teleservices in TETRA.

- Voice and data transmission (*V+D*),
- Data transmission in packet mode; packet data only (*PDO*),
- Direct terminal to terminal transmission; direct mode (*DM*) (a mode for direct communication between terminals with or without control from the network).

Note: With DM, and contrary to previous trunking modes and dependent on implementation, the user may select the channels manually. The specifications relating to these three modes are based on the same radio interface for transmission (e.g., modulation, frequency, and transmission/reception). The interworking of the various modes is guaranteed from layer 3. Layers 1 and 2 are specific for each mode.

Direct-mode transmission is used in two cases: either when the mobile station is outside network coverage or to communicate in a secure manner. In the latter case, simultaneous interworking with the trunked infrastructure is possible with the help of a gateway terminal, TETRA trunk/DM.

The bearer services provided by the specifications (*V+D*) and PDO are the following:

- Voice transmission in circuit mode,
- Data transmission in circuit mode,
- Packet transmission in circuit oriented mode,
- Packet transmission in connectionless mode,
- *Ad hoc* connectionless packet mode transmission for the transmission of status and the short data messages (*SDMs*).

The (*V+D*) services have been arranged to offer a mobile multimedia platform capable of simultaneously handling voice, data, and imaging.

The speech services are:

- Individual calls (point-to-point),
- Group calls (point-to-multipoint),
- Acknowledged group calls,
- Broadcast calls.

Among the supplementary services available are: list search call, conference call, call forwarding, call transfer, call barring, call report, call holding, calling/connected line identity presentation, calling/connected line identity restriction, short number addressing, and access priority.

9.4.5.4 Call Management Procedures

In the idle state, mobiles are permanently listening on the MCCH to obtain system data and to detect incoming calls. There are two ways of setting up calls, either with or without presence checking.

In the case of call setup with presence checking, a mobile (M_1 in Figure 9.29) that wishes to communicate with a mobile M_2, transmits an access request message *u-setup*. On receipt, the base station responds by issuing an acknowledgment, *d-call proceeding,* and sends, simultaneously, a *d-setup* message to mobile M_2 to check its presence. Mobile M_2 answers, if available, by a *u-connect* message in the slot indicated in the *d-setup* message (slot 2 of the uplink channel in the case of Figure 9.29). On receiving this acknowledgment the base station allocates a traffic channel to mobiles M_1 and M_2 by simultaneously transmitting the *d-connect* and *d-connect-Ack* messages.

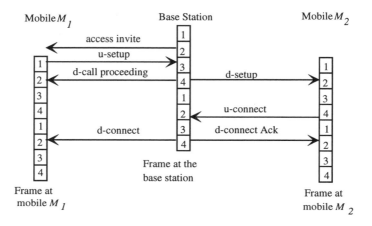

Figure 9.29 Example of an individual call setup in TETRA.

Call setup without presence checking is used for very fast call setup. In this case, radio resources are allocated without checking for the presence of the required parties. This mode can be used for group calls or individual calls. With this mode the traffic channel is immediately allocated to the call by the base station. The base station sends back the telegrams including the number of the channel allocated: *d-connect* for the calling and *d-setup* for the called parties.

9.4.5.5 Trunking Methods Used

According to traffic conditions, various trunking methods can be used to optimize spectral efficiency and minimize access time delays to the system. TETRA thus makes the distinction between *message trunking*, *transmission trunking*, and *quasitransmission trunking* modes. The network operator or manufacturer can choose the mode. From the mobile station point of view, there is no difference between the different trunking strategies as each mobile station just follows the network call setup and call maintenance instructions.

Message trunking. This strategy relates to the allocation of a traffic channel. The same traffic channel is allocated for the duration of a call, which may consist of several and separate transactions deriving from various terminals (Figure 9.30). The traffic channel is returned to the pool when the call is explicitly cleared.

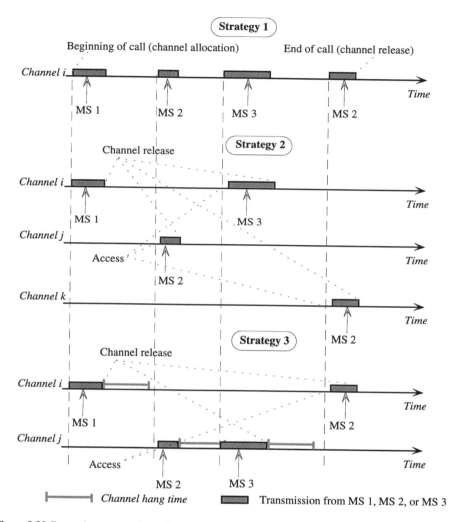

Figure 9.30 Comparison among the various trunking modes.

Once the traffic channel has been allocated, the users benefit from minimum access delay because the channel is always allocated to them. This method helps reduce the signaling and processing at the infrastructure level. Its drawback is that the channel remains allocated to a call even during silences, which leads to a decrease in the channel spectral efficiency.

Transmission trunking. With this strategy, the traffic channel is only allocated during the transaction time (i.e., for each activation of the pressed button, see Figure 9.30). The advantage of this technique is that the channel is used very efficiently. It is useful in the query-response type of communications. Its drawback is that when the system is heavily loaded, channel access delays can be considerable.

Quasitransmission trunking. With this mode the traffic channel is released at the end of each transaction only after a channel hang time (Figure 9.30). During this time the traffic channel could be allocated to a new transaction that is part of the same call. If the time expires before a new transaction within the same call occurs, the channel is placed back in a pool of channels that are available for general use. This strategy offers a compromise between the two previous methods. The channel hang time has to be carefully chosen to obtain the maximum benefit from this method.

9.4.5.6 Location Management

As is the case with cellular systems TETRA network coverage is divided into location areas, each of which contains one or several cells. A terminal will be paged only in the location area where it is registered. TETRA defines an implicit registration mechanism that helps register a terminal without having to issue explicit registration messages. This kind of registration can be triggered by any message that conveys the mobile identity (e.g., call request, response to paging, cell change request).

9.4.5.7 Security Functions

Access control. Each TETRA terminal has a unique terminal (TETRA) equipment identifier (TEI). This allows the infrastructure to deactivate certain terminals (for example, those that have been reported stolen). The authentication mechanism is similar to that used in DECT. It entails two algorithms: TA11 and TA12 (Figure 9.31).

Confidentiality. Traffic confidentiality is realized by the use of an alias subscriber identity (ASSI), substituting for the individual TETRA subscriber identity (ITSI). This temporary identity can be modified at each new registration. It should be noted that the identity of a group (GTSI—group TETRA subscriber identity) cannot benefit from the same protection procedures. Indeed, this would require a more complex management strategy to inform all the subscribers to a group of GTSI changes.

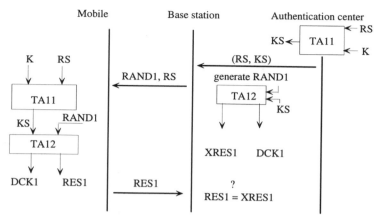

DCK1: derived ciphering key; KS:tTemporary session key; RS: random sequence

K: authentication key; XRES1: authentication algorithm result

Figure 9.31 Authentication mechanism.

TETRA defines two levels of ciphering (Figure 9.32):

- A basic level using the radio interface encryption,
- A higher level using end-to-end encryption with end to end key-management.

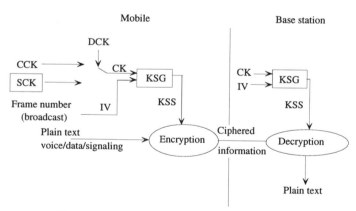

SCK: static ciphering key; DCK: derived ciphering key
CCK: common ciphering key; KSG: key stream generation
KSS: key stream segment

Figure 9.32 Ciphering principle in TETRA.

The higher level is specified for use by groups of users such as public safety organizations. Mechanisms for end-to-end authentication and key management are not defined by the standard. Organizations are free to use their own algorithms.

The encryption on the radio interface is realized by XOR functions (modulo 2 addition) between a ciphering sequence and the plaintext. The synchronization of the ciphering sequence is achieved through numbered frames, which are repeated over a 60-sec period. The ciphering key should be changed after that period, otherwise the key-sequence is repeated and could allow third parties to retransmit the intercepted messages obtained from a previous period without being detected by the addressee.

To ensure continuity of the security functions when terminals roam between the various location areas, two methods are available. The first allows the mobile to recover the common ciphering key (CCK) of the new area before leaving the previous one. The second method requires the registration of the mobile in the new area to allow it recover the new CCK. The mobile's derived ciphering key (DCK) is transmitted to the new area through the fixed network. If key transfer is not possible, the mobile is required to authenticate itself in the new area in order to generate a new DCK.

9.4.5.8 Modes of System Operation

According to traffic conditions (e.g., low, heavy, packet data) different operating modes are defined: *normal mode*, *extended mode*, *minimum mode*, and *time-sharing mode*.

Normal mode. In the case of normal traffic load, each site is allocated four or five radio frequency pairs (in other words, 16 to 20 channels). The main common control channel used (MCCH) is present in slot 1 on all frames between 1 and 18 on one of these frequency pairs. It is used for all forms of common control signaling.

Extended mode. It is used in large networks where more than one common control channel is required. The MCCH is then available in slot 1 of each frame and also one or more extra control channels are defined in other slots.

Minimum mode. This mode is used in low-traffic areas where a frequency pair per base station is allocated (in other words, three voice channels and a control channel or four voice channels). With this structure when a base station has allocated all slots to traffic, frame 18 is used for common control.

Time-sharing mode. Discontinuous transmission allows the splitting of the same radio carrier among several sites. This kind of implementation suits low-density traffic areas where the allocated radio spectrum is limited.

9.4.6 TETRAPOL

TETRAPOL is a voice and data digital transmission system. The TETRAPOL specifications were defined in 1995 by the MATRA-COM company based on the ACROPOL system specifications produced by the same company for the French Interior Ministry [8, 9].

The general architecture of the TETRAPOL system is quite similar to that of the TETRA system. It is shown in Figure 9.33.

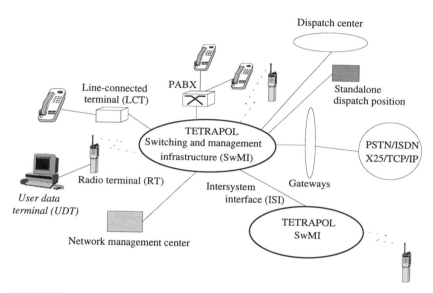

Figure 9.33 TETRAPOL general architecture.

9.4.6.1 Radio Interface Characteristics

TETRAPOL is an FDMA digital system. The modulation used is GMSK at a bit rate of 8 Kbps. The frequency bands that can be used by TETRAPOL equipment are located in VHF (68–88 MHz) and UHF (380–512 MHz and 830–930 MHz). A duplex interval of 10 MHz is envisaged in the 400-MHz band. The channel separation is either 10 or 12.5 kHz with an evolution expected towards 6.25 kHz separation.

Two groups of mobiles are defined: handhelds with 2W transmitter power and a transportable or fixed terminal with an output power of 10W. The sensitivity of the receivers vary from –113 dBm at the base station and –111 dBm for mobiles inclusive of fading. The system operates with a faded C/I threshold of 15 dB. The transmitter and receiver specifications comply with the ETS 300 113 standard.

9.4.6.2 Frames and Logical Channels

TETRAPOL uses 20 ms duration bursts of 160 bits. A superframe structure of 200 frames (duration of 4 sec) is also defined for transmission periodicity. Error detection is obtained by using convolutional coding with interleaving.

Five types of bursts are defined:

- *Voice* bursts (transmitted on the uplink and downlink channels),
- *Data* bursts (transmitted on the uplink and downlink connections),
- *Random* access bursts (transmitted on the uplink connection),
- *Training* bursts (transmitted on the uplink connection) used for the equalization procedure,
- *Interrupt* bursts (transmitted on the downlink connection) used to order a mobile to stop transmitting.

In the same way, five groups of frame are identified:

- *Voice* frame (152 coded bits, 123 information bits) transmitted in the voice burst,
- *Data* frame (152 coded bits, 71 information bits) transmitted in the data burst,
- *Data high-rate* frame (152 coded bits, 100 information bits) also transmitted in access bursts,
- *Random access* frame (75 coded bits, 16 information bits) transmitted in the random access burst,
- *Interrupt* frame (64 coded bits, 5 information bits) transmitted in the interrupt burst.

For the voice or data frames, the coding structure is based on the use of a convolutional code accompanied by a detection and correction code. Each frame contains a flag allowing the type of frame, user data, or signaling to be identified.

TETRAPOL defines several kinds of logical channels for signaling and traffic (Table 9.3). At least one signaling channel is used at each base station.

The RACH channel is used by the mobile to access the network. The dynamic random access channel helps obtain the immediate transmission of data such as a channel request, and message status.

The stealing channel, used for the transfer of fast signaling, is obtained by stealing voice frames.

Table 9.3
List of the logical channels in TETRAPOL

Channel	Uplink (↑)/Downlink (↓)
* Control	
- Random access channel	↑
- Random access answer	↓
- Dynamic Random access channel	↑
- Signaling and Data channel	↓↑
- Broadcast Control channel	↓
- Paging channel	↓
- Stealing channel	↓
* Traffic	
- Voice or data channel	↓↑

9.4.6.3 Modes of Operation and Offered Services

TETRAPOL can operate in three different modes (Figure 9.34):

- The *network mode* allows the mobiles to communicate with other mobiles or fixed users through a network infrastructure (e.g., base stations, switching centers) covering wide areas.
- The *repeater mode* uses an autonomous portable base station and allows communication between mobiles located outside the coverage of the fixed base station.
- The *direct mobile* mode allows direct communications between mobiles that are close to each other. This can be used in a tactical context or in the absence of infrastructure. It allows the establishment of ciphered communications and emergency calls.

Note that mobiles are able to handle dual-mode operation (on the network and direct) and a synchronized simulcast mode by using the global positioning system (GPS) as a reference.

In the three modes, the users can benefit from all or part of the following services:

- *Voice services*: broadcast call, emergency call, group call, private call, conference call, open channel call (on one or several preset sites), and point-to-multipoint communication within a defined area.

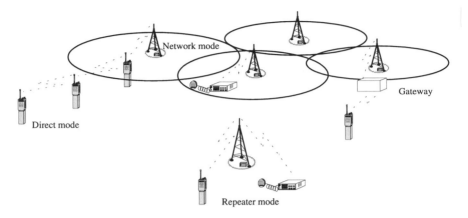

Figure 9.34 Network and direct functioning modes in TETRAPOL.

- *Data services*: bearer services (circuit and packet switched connection or connectionless modes), broadcast with acknowledgment, short messages, status, access to TCP/IP), and applications of X400 messaging.
- *Supplementary services*: priority access, automatic call, call barring, identification of the calling or connected line, call transfer, DTMF mode, and call interception.

9.4.6.4 Network Procedures

Procedures defined in the TETRAPOL networks are the following: attach/detach, status indication (terminals powering on and off), call duration limitation, call reestablishment, call protection (against preemption by network resources and the resultant risk of interruption), dynamic group number assignment (dynamic change of mobiles within a group), group merging, migration (to another network), presence control, power-saving mode, access priority (during a call), roaming (inside the same network), registration (location updating at the cell level), power control, and profile management of user services.

9.4.6.5 Protection Mechanisms

TETRAPOL is a network that was designed, from the very beginning, for security applications. For this it offers enhanced security (Figure 9.35).

The various mechanisms defined to ensure confidentiality of communication and access control include the following: intrusion detection, ciphering, use of a password at terminal level, mutual authentication (mobile and network), secure key management, use of temporary identities, terminal disabling, control of the terminal identity.

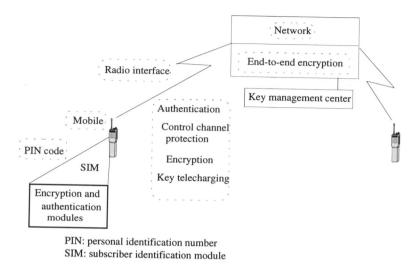

PIN: personal identification number
SIM: subscriber identification module

Figure 9.35 Security mechanisms in TETRAPOL.

9.4.6.6 Present Situation and Perspectives

In 1997, over 15 TETRAPOL networks were operational or ordered. These were for various applications as follows:

- Security (national networks in France, the Czech Republic, Asia, and America),
- Airports (e.g., Stuttgart),
- Transport: railroad (SNCF), urban (Berlin bus),
- Third party (public access) network: Mobilcom (Swiss operator).

The TETRAPOL industrial forum (composed mainly of Siemens, Bull, AEG Nortel), operating in the TETRAPOL technical working group (TWG), presented the TETRAPOL specifications to the ITU and ETSI in 1996 and 1997 to make the TETRAPOL interfaces publicly available.

In 1997, over 300,000 professionals were using, or about to use, TETRAPOL worldwide. New specifications were issued in 1997 to define the interfaces between TETRAPOL networks, as well as between TETRAPOL networks and GSM networks. TETRAPOL, like GSM with its MoU (memorandum of understanding) group, also have their own group of users called the TETRAPOL User's Club (TUC).

9.4.7 Conventional PMRs Comparison/Analog Trunk/Digital Trunk

Table 9.4

PMR characteristics comparison

	Conventional mobile radio	Analog trunks	Digital trunks
Voice	Analog	Analog	Digital
Modulation	Analog	Digital	High-speed digital
Signal quality	Several shadow areas	Shadow areas are suppressed by using cellular techniques	Optimum quality on the whole coverage area with signal digitization
Voice services	Some available services	Several services	Numerous services available
Data services	Some services	Data services with bit rates less than 1.2 Kbps.	All data services with bit rates under 8 Kbps.
Privacy	None, unless implemented by an external device	One channel per communication or privacy ensured by an external device	Integrated devices insuring the privacy and the voice and data ciphering
Coverage	Single site	Adapted to any service area by use of cellular techniques	Adapted to any service area by use of cellular techniques
Spectrum efficiency	Nonoptimum frequency reuse	Good frequency reuse	Optimum frequency reuse

9.4.8 Short-Range Business Radio

Recently, and because of the demand of PMR users to implement simpler systems and procedures, several countries have released some frequencies where their use does not require the acquisition of a license. France has opened a general license for three simplex channels (446.9500 MHz, 446.9750 MHz, and 446.9875 MHz) with 12.5-kHz separation. Terminals are allowed to transmit in these bands with a limited transmission power of 500 mW and can communicate over a distance of 2 km maximum. These systems are aimed at allowing single site local communications. The main characteristics of these short-range business radio (*SRBR*) systems are as follows:

- Communications range up to about 1 mile,
- No need for using a conventional base station and separate antenna,
- Single frequency channel required per network,
- High reuse factor,
- Simple licensing procedure,
- No individual frequency planning.

9.5 PMR EVOLUTION

Like most of the other radio mobile systems, PMR is witnessing rapid progress. It is envisaged that this progress will remain significant for some years to come. In 1997, there were about 20 million PMR users in the world, half of them in Europe. Most of these employ analog systems and the transfer towards digital systems will be progressive thanks to the advent of standards such as TETRA. The very quick introduction and high growth of cellular systems should, nevertheless, have an impact on this rate of progress. The correlation between the costs of the two systems and services available on each type of network has increasingly been the determining factor for users when making a choice.

Until recently the main reasons that have impelled an organization to resort to a PMR system rather than to cellular were related to cost (classic PMR has been relatively cheap), practical functionality (the working practices of an organization correspond to the services offered by PMR), and control (users are eager to keep close control of their communications for reasons of confidentiality but also for their budget).

Over the last few years, other motivations have come into effect:

* With an increased awareness of costs and because of the considerable decrease in cellular charges, PMR has become less attractive,
* Outsourcing, which forces enterprises to resort to subcontracting services such as to external computing bureau.

Mass migration to cellular systems is not envisaged in the near future. PMR users have expressed a preference to adopt third-party networks as large enterprises have dropped the idea of owning and running their own networks. A small market segment such as local administrations, industrial sites, ports, and airports are suitable for a mixed communications solution. This could be a cordless system operating within a site premises and PMR or cellular for coverage outside [1].

On the other hand, because of the considerable decrease in the cost of cellular terminals, the general public is readily accepting these. This trend, if it continues, will make the cellular networks the most economical tool for potential or even current users of PMR. Moreover, with the advent of services offered by phase II+ of the GSM system it will, as is the case for Radiocom 2000, be able to include services that have been unique to PMR, such as group calls, fast call setup, closed user groups and the like [10]. Compared with TETRA, GSM phase II+ will also be offering similar services at around the same time [11].

REFERENCES

[1]   Péruset, P., B. Fino, and M. Nicole, "The Professional Mobile Market in a Competitive Telecommunication Environment," *6th CEPT Radio Conference*, Paris, Nov. 13-15, 1996.

[2]   Heneine, J., H. Azemard, M. Pierrugues, R. Prévot, and A. Shatz, "Réseaux de Radiocommunication à Ressources Partagées," *Commutation et Transmission*, Numéro Spécial "Communications avec les Mobiles", pp. 65–74, 1993.

[3]   Hanson, D. A., "Conventional Private Mobile Radio" *Personal & Mobile Radio Systems*, edited by R.C.V. Macario, P. Peregrinus, 1992.

[4]   Robles, J., "TN 100 : Réseau Privé de Radiocommunications à Ressources Partagées," *Commutation et Transmission No. 2*, pp. 71–78, 1993.

[5]   "STARSITE system," *MOTOROLA specifications*, 1996.

[6]   Loussouarn, Y., "DIGICOM 7 : Réseau Radioélectrique à Ressources Partagées," *Commutation & Transmission No. 1*, pp. 19–28, 1991.

[7]   "Designers' Guide, Part 1: Overview, Technical Description, and Radio Aspects," *Transeuropean Trunked Radio System*, RES06(95à034), RES6.1(94)060, Ver. 0.0.10, Jan. 1996.

[8]   TETRAPOL Forum, "TETRAPOL Specifications; Part 1: General Network Design; Part 1: Reference Model," PAS 0001-1-1, Ver. 3.1.0, Dec. 1996.

[9]   TETRAPOL Forum, "TETRAPOL Specifications; Part 17: Guide to TETRAPOL Features; Part 1: Technical Report," PAS 0001-17-1, Ver. 0.0.5, May 1997.

[10] Brydron, A., "Breaking the Mould," *Mobile Europe*, Oct. 1995, pp. 93–96.

[11] Whitehead, J.F., "Distributed Packet Dynamic Resource Allocation (DRA) for Wireless Networks," *IEEE Vehicular Technology Conference 1996*, Altanta, GVA, pp. 111–115, Apr. 29-30, May 1, 1996.

SELECTED BIBLIOGRAPHY

[1]   Britland, D.E.A., "Trunked Mobile Radio Systems," *Personal & Mobile Radio Systems*, edited by R.C.V. Macario, P. Peregrinus, 1992.

[2]   Gourgue, F., "Air Interface of the Future European Fully Digital Trunk Radio System," *Proceedings of the IEEE Vehicular Technology Conference '93*, Secaucus, NJ, May 18-20, 1993, pp. 714–717.

[3]   Harrison, D., "TETRA - A Digital Encoded Speech PMR Trunking System," *Mobile Europe*, Apr. 1994, pp. 17–20.

[4]   I-ETS300 230 (Oct. 1993): Radio Equipment and Systems (RES). Binary Interchange of Information and Signaling (BIIS) at 1,200 bps.

[5]   Prévot, R., "Five Years of PAMR in France," *Mobile Europe*, pp. 17–20, Feb. 1995.

CHAPTER 10

CORDLESS SYSTEMS AND APPLICATIONS

Cordless phones, also called *wireless phones*, are the most widespread and commonplace radio systems in general use. In some countries, especially the United States, their use exceeds that of fixed telephones [1]. They are also characterized by a wide variety of standards, even though standardization activity has started to produce some commonality.

Historically speaking, cordless phones first appeared in the North American market at the end of the 1970s. From there, rise in use grew dramatically, most often without regard to national laws in many countries. In Great Britain, for example, some cordless systems illegally imported from the United States were operating in the 1.7-MHz and 49-MHz bands, which were already used by maritime radio communications systems and TV services. The quality of communication and security features offered by these early systems were often very poor. In the 1980s this compelled several countries to standardize both analog and digital systems.

The current cordless systems generally fall into two main categories: first generation *analog* systems (e.g., CT1) and second generation *digital* systems (e.g., CT2 and DECT). The common denominator between the two generations is the system design, allowing low-power bidirectional communication between a handheld terminal and a fixed point, a few hundred meters from one another, and with low user mobility. The first cordless systems, used only for voice communication, were generally incompatible with each other and were designed to provide communication in relatively small residential or professional environments. Yet, with the growth of the CT2 and DECT standards, capacity improvements, voice quality, encryption of user information over the radio interface, and handover between BSs have all helped encompass more applications and environments, such as professional use in offices, telepoint areas, and for distribution to fixed subscribers (e.g., *wireless local loop* applications; Section 10.1).

From the user point of view progress in wireless system design is such that the functionality available is very close to that offered by cellular systems including full duplex telephony, handheld terminals, and roaming. From the service point of view, the differences are most noticeable at the level of mobility offered. Cordless systems have short range (a few tens to a few hundreds of meters) and can only serve mobiles with a low movement speeds. One of the advantages of these systems is the long battery life of the terminals because of their low transmitter power. From an operators point of view cordless systems differ from cellular systems by their lower complexity and their "free for all" organization, that is, cordless terminals can co-exist without necessarily being coordinated.

The main objective of these systems is to supply a means of communication that is cordless, economical, primarily designed for residential use, and offering a voice quality similar to that of a fixed terminal. Cost constraints for the terminals and maximization of battery life are the most important design factors. These characteristics make these systems significantly less costly than cellular systems.

The primary use of cordless systems has been for residential telephone applications where a terminal communicates with a single base station directly connected to the public switched telephone network (PSTN). The other two classic applications are access to a telephone network for multiple terminals over a public (telepoint application) or private (PBX) exchanges. The latter application is sometimes called, not quite accurately, a *wireless PBX*. The fourth application that has been growing steadily over the last few years is the wireless local loop (WLL).

Early in this chapter the basic principles and characteristics of cordless systems, including four applications, are introduced. Discussion of the most important digital cordless system standards follows.

10.1 BASIC PRINCIPLES AND APPLICATIONS

Most of the cordless telephone systems are designed for domestic use. They are generally composed of two pieces of equipment: a base station (also called *radio fixed part* in DECT, for instance) and a cordless terminal (called *radio portable part* in DECT). The base station interfaces with the PSTN in the same way as a fixed phone terminal but is also connected to a power socket. The cordless terminal communicates with the base through a duplex radio link, in either Time Division Duplex (TDD) or Frequency Division Duplex (FDD) mode and is operated in the same way as a fixed telephone. The on hook/off hook, short form and normal dialing operations can generally be performed from the cordless terminal.

10.1.1 Characteristics

The usage and design constraints of cordless telephones are quite specific to these systems and generate very special characteristics, which are introduced below.

10.1.1.1 The Picocell Indoor Environment

Propagation characteristics related to the indoor environment were briefly introduced in Chapter 2. Some are revisited in this section. The most complex of these applications is the professional cordless telephone. In the professional environment it is necessary to meet coverage constraints higher or equal to 99% in propagation conditions which are very often complex (e.g., obstructions such as filing cabinets and moving people).

Propagation and traffic characteristics. In an indoor environment, radio waves reflect from and penetrate through walls and floor. The mobility of people (who are also as much a potential reflector) around the transmitters/receivers is an additional source of problems. The power of the signal at any given point will consequently vary to a greater extent, relative to the transmitter to receiver distance, than in the outdoor case. The propagation loss L also depends on the building structure. Its variations are particularly significant because of the furniture and moving people.

The picocellular environment is characterized by being more unpredictable both for propagation and traffic capacity than the macrocellular situation. The distribution of signal fading in the indoor environment, residential or office, is very complex.

Diversity techniques can be used to mitigate this kind of signal degradation [2]. In addition, using very small cells complicates the planning of these systems which tend to have very uneven traffic distribution. In fact, unlike cellular systems it is extremely difficult to allocate a frequency plan to cordless systems (see Chapter 7).

Traffic densities in these environments can reach very high values: ranging from 150 Erl/km^2 in residential areas to 10,000 $Erl/km^2/floor$ in an office environment. These densities can vary considerably.

In such a difficult context two techniques help provide good radio coverage. The first of these methods consists of using multiple small antennas installed around the areas to be covered with each antenna giving three-dimensional coverage. The second method consists of using leaky feeders radiating the signal and laid along corridors or in the ceiling. The first method is the most commonly implemented in office sites.

Flexibility constraints. Cordless systems designed for professional and residential applications must necessarily be simple to install and operate. Frequency planning, site measurement and engineering work which require the use of skilled human resources, normally undertaken for cellular systems, should not be necessary for implementation of cordless systems in an office or home environment. The same reasoning is valid for any change or addition to base stations, which should all be easily fitted and by people who are not specialists.

Thus, the complexity of the propagation environment coupled with the need for ease of implementation and simple system management has motivated the adoption of random self-organizing systems. Channel allocation is dynamic (DCA, see Chapter 6) and decisions are generally based on simple C/I ratio measurements. The DCA technique helps, during the implementation phase to identify the propagation and frequency reuse characteristics of the environment by supplying preliminary measurement data. Moreover, the system automatically and immediately adjusts itself by introducing of new terminals or a new system. Nonetheless, in the case of a public telepoint or WLL application, an engineering phase is necessary for dimensioning of the system and the determination of radio base station sites.

Mobility and security. Different location management methods (see Chapter 8) can be defined for cordless systems. The first generation analog systems, because the terminal is connected to its corresponding single base station, only requires the simplest of location management methods, that is, search for coverage close to the base station.

Cordless digital systems can also use this method (the wireless PBX application) or in the case of public applications, resort to manual location or location area crossing methods. Note, a paging network separate from the cordless network can be used to locate the user (see Chapter 8 [2]). The last three methods are used in telepoint applications.

The latest cordless telephone standards (DECT, PHS, and PACS) provide a handover function that allows greater mobility than in previous systems. In CT1 and CT2 systems, a communication established through a given base station can only be continued and completed through the same base station.

The speed of mobiles is fairly limited (pedestrians for DECT) compared with cellular systems, even if some systems like PACS can cater to mobiles moving at several tens of kilometers per hour. Because of the propagation characteristics that make the connection between a cordless terminal and base station variable, especially during a handover, most of the systems define handover procedures, which include a macrodiversity phase (e.g., DCT 900 and DECT). This minimizes the probability of a call interruption during the handover.

From a security point of view, the first wireless systems offer little protection to the user. No authentication algorithm (therefore control of access) and no ciphering (and, therefore, no confidentiality) were defined. The problem of terminal identification in this kind of system gives rise to illicit line capture with the corollary of telephone bills that cannot be reconciled by the authentic subscriber. The simpler systems help solve this problem by making use of codes, two or three digits, for instance, randomly preallocated during manufacture or which can be later programmed by the user. The code is stored in the base station and its corresponding cordless terminal and exchanged and verified before any communication establishment. This handshaking technique provides minimal protection, but fraudsters equipped with scanners capable of intercepting the exchanged identities (codes) can negate its efficiency. Improvement of these protection techniques to prevent illicit access to wireless connected lines has been obtained by using authentication mechanisms based on the challenge response technique where no confidential information, that can be intercepted and used by third parties, is transmitted over the radio channel (see Chapter 5).

10.1.2 Applications

Cordless system applications are traditionally oriented towards three kinds of environment: residential (household terminal), professional (wireless PBX), and public (telepoint). A fourth category, the wireless local loop, has been steadily emerging mainly because of the potential gains, in terms of time, implementation, and cost (equipment and maintenance). This is true especially in countries where the telecommunication infrastructure density is low or is in need of a fast update.

10.1.2.1 Domestic Application

The domestic application, first both in numbers of users and initial implementation, is also the simplest with regard to equipment, installation, and protocols used. At the beginning the objective was to provide mobility inside or in the immediate vicinity of a residence (for example a house with garden). The cord (or wire) connecting the receiver to the telephone terminal was replaced by a radio connection, hence, the word "cordless".

This equipment consists of a fixed terminal, and, typically, a single portable terminal. The fixed terminal is directly connected to the public telephone network, much as a fixed telephone, and operates as an interface between the wireless portable terminal and the network (Figure 10.1). The implementation of such equipment, if in compliance to the standards required by the telephone network operator, is totally transparent to the network, and more particularly to the billing, which remains unchanged from that of the fixed telephone service.

As mentioned above, the first cordless systems offered only limited confidentiality and security. Users geographically close to each other and using the same frequencies can be mutually jammed and intercept each other's communications. In terms of security one terminal could use the other's line and incur unwarranted telephone charges. To guard against interference, systems using several subchannels (where a search of the channel with the least interference can be undertaken manually or automatically) have been designed. Residential cordless systems, even the simplest ones, have to implement network functions at the cordless terminal to interwork correctly with the PSTN and offer basic services to the user. Thus, even the simplest household terminal includes the following functionality:

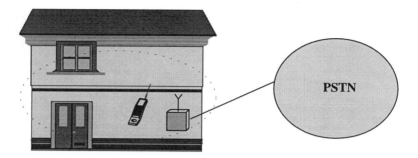

Figure 10.1 Domestic application.

- *Protocol conversion.* The subscriber's connection may use electromechanical signaling (pulse dialing) or DTMF to access the PSTN so an interface is required. In digital cordless systems the signaling data exchanged over the radio channel uses the OSI model (see CT2 and DECT systems). Message transmission for call setup is performed at layer 3 (network layer). For this purpose the terminal, even in the case of a simple domestic unit, requires the integration of a set of protocol entities and protocol conversion for interfacing between the transmission modes of the PSTN and those of the cordless system.
- *Voice conversion.* Digital wireless systems use speech coding for transmission by using techniques that are a variant of PCM (see Chapter 4). As the PSTN uses—at subscriber-line level—analog transmission, a decoder is necessary at the terminal.
- *Registration.* Systems, such as CT1, CT2 or DECT, assign unique identities to the terminals to avoid illicit use of the PSTN line. Several domestic products only allow connection of the cordless terminal after a registration mechanism using this identity.
- *Functionality.* In digital cordless systems, intelligent functions can be added. For instance, connection of several portable terminals to the same base terminal, an *intercom* function (between two portables without using the PSTN line). It is quite likely that residential cordless systems will integrate functionality similar to that of the cordless PBX.

10.1.2.2 Cordless PBX

The wireless PBX application is an extension of the domestic application. One or several fixed terminals installed on a worksite (Figure 10.2) can serve several handsets. The system is connected to a private branch exchange (PBX) to allow access to the PSTN and other PBX users. Consequently, the wireless PBX application requires a more complex infrastructure than is the case for residential use.

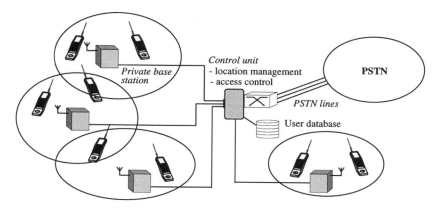

Figure 10.2 Wireless PBX application.

Three major features characterize the wireless PBX:

- It possesses several lines to an external network (e.g., PSTN or ISDN).
- It manages significant traffic between internal extensions.
- It is used by people who are very mobile.

To do this management functions for mobility, security, and authentication of users for access to external lines have to be supplied.

The PSTN is accessible through a private switch and the handsets are connected to the PBX through a controller. This controller can be either connected to a PBX via a cable, or included in a special PBX. Some mobility management functions can be included in the wireless PBX application. The system can manage user mobility with the help of a small database located at the controller and located at the interface between the PBX and the radio terminals. The controller permanently monitors the handsets' locations and pages a terminal only in the cell where it is located. Handovers can be accommodated between base stations connected to the controller, which ensures terminal mobility. This application is particularly useful for environments where the routing of cables is complex, even impossible, as is the case for some manufacturing plants or when telecommunication services need to be provided temporarily for exhibitions, political, or sports events. Similar to the residential application, a wireless PBX is transparent to the public network operator connected to the PBX.

10.1.2.3 Public Telepoint

Contrary to the two previous applications the public terminal application requires management from the public fixed telephone network operator (Figure 10.3). From the subscriber's point of view all that is required is a subscription to the cordless service. The telepoint application also requires supplementary equipment such as a management center. The base stations are implemented in public places where traffic density is estimated to be of importance to pedestrian subscribers; mainly, streets and squares, shopping centers, bus and train stations, and airports. Location of subscribers is supplied as a supplementary service. The alerting of a subscriber (who has been located manually or automatically) is implemented by the transmission of paging messages from the base station at or near the detected location.

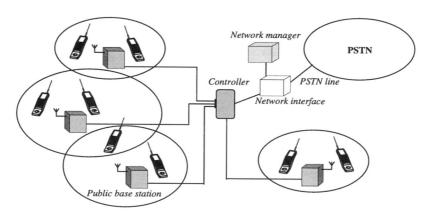

Figure 10.3 Public telepoint application.

The functionality needed for a public access base station is: authentication, collection and down-loading of billing information to the network center, a listing of fraudulent terminals, traffic, failure, and maintenance reporting.

The network management center includes the following functionality: queries to the various base stations for the collection of billing records, maintenance and traffic data, up-loading of the fraudulent terminals list to the base stations, management of billing and new subscriptions, and various security functions.

When the infrastructure covers an urban area, in-bound calls are accepted and handover is offered, then the cordless telepoint application is similar to that of a PCS network (micro-cellular). The main drivers for the use of this wireless technology compared with cellular are equipment cost, ease of management, and implementation.

10.1.2.4 Wireless in the Local Loop

Even though this application is not limited to cordless telephony standards, it has appeared, from the very beginning, as a direct application of this type of system (Figure 10.4). This is the reason why it is introduced within the framework of this chapter. There are other methods, notably cellular and proprietary standards that also provide wireless local loop connectivity.

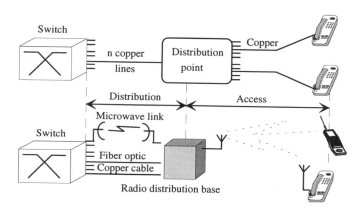

Figure 10.4 Wire or radio access for the local distribution of subscribers.

The interfaces used for WLL application can be based on cordless type standards, cellular system standards (e.g., AMPS, TACS, NMT, GSM, IS-54, PCS 1900, IS-95), or proprietary methods. There are a large variety of products, frequency bands, and multiple-access techniques (TDMA and CDMA) used for this purpose.

The WLL application is an alternative means of access for local distribution of the telephone service by replacing the copper wires. Figure 10.4 shows two cases where this transmission medium is used.

The WLL application can be used to supply services as varied as the classic telephone (POTS), fax, ISDN (64 Kbps), telepoint services, and even wideband applications (video, ATM).

Advantages. Radio distribution to fixed subscribers offers many advantages compared with cable distribution systems. The most significant advantages are the following:

- Fast deployment,
- Great flexibility: the cable infrastructure requires long term dimensioning whereas cordless infrastructure can be adjusted quickly according to demand,
- Initial investment and costs for operating and maintenance of a radio distribution service is lower than for wire line systems,
- Suitable for use in urban areas with high-traffic density as well as in rural areas for a few and scattered subscribers,
- Simplified planning: the implementation of a WLL does not require accurate

knowledge of subscriber characteristics (mobility and traffic) which allows for a flexible planning strategy and system deployment,

- Stand alone characteristics: WLLs have shown their operating capabilities as autonomous communication systems in the case of natural catastrophes where telecommunications service providers have deployed this kind of system very quickly during rescue operations [3],
- Compared with a mobile cellular system, a WLL system has some unique features including no handover, no HLR/VLR database queries, and the use of directional antennas. These characteristics all lead to simpler control logic.

Connection with the PSTN can be over various carriers including cellular networks, microwave links, satellites, copper cables, and fiber optics. This helps to implement a WLL system in any area without regard to the type of environment (e.g., urban, rural, mountainous).

Structure. An example of architecture for this kind of application using a DECT bearer is reproduced in Figure 10.5.

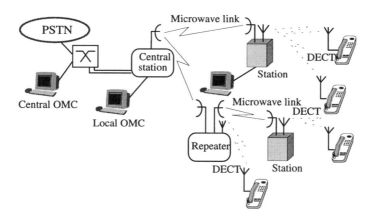

Figure 10.5 Example of architecture for a WLL based on the DECT standard.

In the example presented in Figure 10.5, the use of local base stations (or *repeaters*) allows hops of several tens of kilometers from the central station. The various elements in this architecture are based, for most of the functionality, on cellular network equipment. The local controller is an interface card for interworking with the radio interface standard (e.g., DECT, CT2). The operations and management function is achieved by a microcomputer: it configures the network, manages the alarms, and measures performance. The network access controller located at the switch allows connection to the telephone center (over an analog or digital connection), the management of radio terminals, and connection to the local controllers. Finally, the management system supervises all the system functions; in other words, the direct management of the network access controller, the remote controllers, and connection to the public network.

The use of WLLs is fully justified for a country with poor telecommunications infrastructure for the reasons described above. In countries that already have dense telecommunications networks this technique offers second-line access to subscribers who are already equipped. It also allows new private operators a means of entry to offer telecommunications services with rapid implementation.

A WLL can be implemented in two different ways [3], as shown below:

- *Direct connection to the fixed network* (Figure 10.6). This can happen with the proviso that enough capacity is available at the central office switch (denoted CO in Figure 10.6). With this application, terminals are connected to the CO, which continues to provide billing functions, and data management, as well as dialing, signaling, and progress tones.

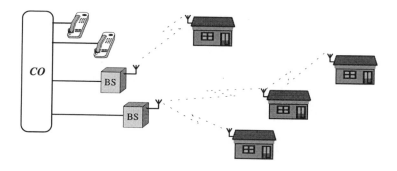

Figure 10.6 WLL terminals connected to a CO (switching center).

- *Terminals connected to CO through a PBX* (Figure 10.7). This kind of implementation is more closely related to the wireless PBX. A local private operator (e.g., a company operating a shopping center, an airport, or a cultural or sports center) can install it.

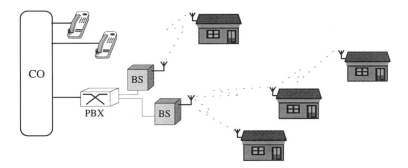

Figure 10.7 WLL terminals connected to a private branch exchange.

Planning. The complexity of the mobile radio environment implies that there are a certain number of advantages in the implementation of a fixed radio distribution network compared with a cellular mobile system. These advantages are as follows:

- *Fixed to fixed propagation connection.* The propagation loss is usually based on 20 dB/decade (or a propagation path-slope $\gamma = 2$) because connections are generally line-of-sight (LOS). As a comparison, in the mobile case, this factor is taken as 40 dB/decade (or $\gamma = 4$). Furthermore, the C/I ratio required in a WLL application is often lower than that required in cellular.
- *Antennas (base and subscriber) are located on high sites (roofs of buildings).* With a WLL system, the received signal is less exposed to fading than a fixed-to-mobile connection. The E_b/N_0 mean ratio for a 30-kHz channel in fixed-to-fixed communications conditions can be 14 dB, whereas in the mobile case it is 18 dB to allow for the fade margin.
- *Reduced reuse distance.* The fixed-to-fixed connection can use directional antennas at both ends, a factor that reduces interference. A reduction in the reuse distance results, which increase the available capacity.
- *No handovers.* The radio connection can be planned to reduce interference. As the connection does not alter after installation (assuming there is no increase in capacity and/or cell splitting), the design of a WLL system is much simpler than a mobile system.

Intracell handover procedures are used to shift from one channel to another within the same cell. They have no impact at network level and do not require any specific management. Finally, they allow more dynamic management of resources ensuring a better quality of service. The channel has stable characteristics because the received power is fixed.

In the context of a WLL, "worst case conditions" (i.e., worst propagation condition cases) can be ignored thanks to an engineering system that is relatively simple and consists of correctly positioning antennas (especially at the subscriber's side) or using diversity techniques.

10.2 EXAMPLES OF CORDLESS SYSTEMS

10.2.1 CT2

10.2.1.1 Historical Overview

In 1984 several British companies started to work together under the auspices of the Department of Trade and Industry to define a digital cordless system. This resulted, in 1987, in the allocation of 40 duplex channels each of 100 kHz bandwidth in the frequency band 864 to 868 MHz. This was followed by the publication of a technical standard for cordless telephones (see also Table 10.1). This standard is known under the name of the *Common Air Interface (CAI)* and published under the reference MPT 1375 [4]. It comprises four parts, which are detailed in the following section.

The first part defines the radio interface. Exchanges between the terminal, called the *cordless portable part* (CPP), and the fixed base station, called the *cordless fixed part* (CFP), take place on the same channel in both directions in Time Division Duplex (TDD) mode. Each communicating part (CPP or CFP) transmits at 72 Kbps for 1 ms, receives the data transmitted by the other part for 1 ms, transmits again, and so on. Taking into account switching and guard times the system is capable of two-way communication at a mean rate of 34 Kbps in each direction.

Table 10.1

CT2/CAI radio interface characteristics

Frequency band	864-868 MHz
Channel bandwidth	100 kHz
Number of carriers	40
Number of duplex channels	40
Modulation	GMSK
Duplexing	TDD
Channel bit rate	72 Kbps
Speech coding	*adaptive differential pulse code modulation* (ADPCM) at 32 Kbps
Maximum output power	10 mW
Channel allocation	DCA

The second part of the CAI standard specifies the way in which the available radio resources are allocated to voice, signaling, and synchronization channels. Layer 1 defines the initialization of the connection, whereas layer 2 specifies the handshaking procedure (for authentication) and the FEC and ARQ schemes.

The third part specifies the high-level signaling protocol, which defines a set of messages covering user dialing, terminal alerting, and authorization for access to the telepoint service.

Finally, the fourth part of the standard determines the content of the voice channel (B-channel) combined with acoustic and audio specifications. It contains the specifications for coding and decoding of voice data on to the B-channel.

10.2.1.2 Radio Interface

A data unit (called a "frame") lasts 1 ms and is transmitted by each end (CFP and CPP) alternately (Figures 10.8 and 10.9).

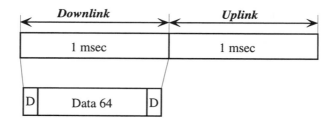

Figure 10.8 Frame structure in CT2.

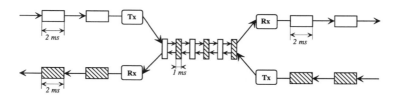

Figure 10.9 Transmission/reception process in CT2.

The portable terminal is synchronized to the base unit in master-slave mode. The base transmits its first burst at any time, and the portable unit synchronizes on these transmissions and transmits its data between these bursts.

Power control is used with two defined power levels for each terminal (a normal level between 1 to 10 mW and a lower level, which is 12 to 20 dB less).

10.2.1.3 Main Procedures

As in ISDN, the CT2 standard defines B-channels (for the transmission of user voice and data) and D-channels (for in-band signaling). In addition, CT2 specifies a synchronization channel SYN used for bit and burst synchronization between transmitters and receivers.

The coding and decoding specifications for transmission of analog information over the B-channel are subdivided into two categories:

- Specifications relating to the conversion of analog signals to a digital format and their compression,

- Specifications for the remaining analog parts.

The analog/digital conversion specifications are based on the ADPCM algorithm defined by the CCITT G.721 recommendation.

The logical channels can use the physical channel in four different ways, according to four multiplexing modes MUX1.2, MUX1.4, MUX2, or MUX3 which are successively used during a call (Figure 10.10).

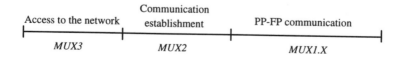

Figure 10.10 Use of different multiplex modes during a call.

MUX1.X modes for the transmission of user data. The MUX1.2 and MUX1.4 modes help determine various forms of flow on the D-channel. MUX1.2 mode is defined as the transmission of a burst of 64 bits on the B-channel (at an equivalent bit rate of 32 Kbps) and 2 bits on the D-channel (at an equivalent bit rate of 1 Kbps). It is used after the call is setup and does not require a SYN channel (Figure 10.11).

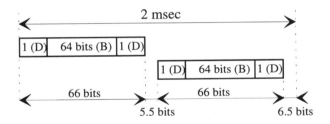

Figure 10.11 MUX1.2 mode.

Similar to MUX1.2, MUX1.4 differs in that the burst has a length of 68 bits and contains 4 bits of D-channel information (its capacity is doubled and reaches an equivalent bit rate of 2 Kbps). This mode is only used if the transmitter and the receiver are both capable of 68-bit bursts (Figure 10.12). The use of MUX1.4 puts more stringent constraints on the switching times of the radio equipment and is not necessarily implemented in all terminals.

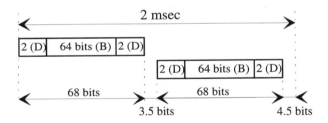

Figure 10.12 MUX1.4 mode.

MUX2 modes for call setup. The MUX2 mode consists of using 66-bit bursts that include 32 bits for the D-channel and 34 bits for the SYN channel (Figure 10.13). This mode is used at the beginning of a communication to allow the two ends to become synchronized and exchange data at a high bit rate (equivalent to 16 Kbps) before switching to B-channel operation. The synchronization channel (SYN) comprises of 10 preamble bits (101010...) followed by one of three 24-bit synchronization patterns, called CHMF, SYNCP, and SYNCF.

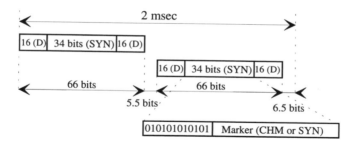

Figure 10.13 MUX2 mode.

MUX3 mode for access of the base station terminal. The MUX3 mode is used for a call setup request on the uplink channel and for a connection re-establishment request during communication (Figure 10.14). It consists of transmitting 10-ms bursts from the portable part during access attempts with the fixed part. A burst transmitted in MUX3 mode contains a SYN synchronization channel and a D-signaling channel. This 10-ms burst is split into 5 bursts, each of 2-ms duration, which are further divided into four identical subbursts sent 4 times in a frame. This complex structure of the MUX3 is used to sort out synchronization problems. The CFP has a receiving window of 1 ms, so one of the subbursts must be received complete regardless of the respective synchronization of the CFP and CPP. The CPP switches to receive for 4 ms after the 10-ms transmission period to detect an acknowledgment from the CFP.

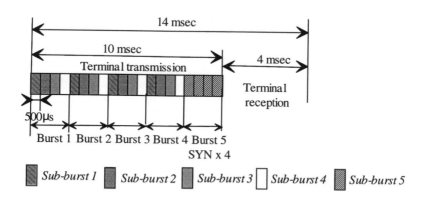

Figure 10.14 MUX3 mode.

The first four bursts contain 20 D-channel bits and 16 bits of preamble in each sub-burst. The fifth burst contains a SYN channel. This repetition of data ensures that the CFP is able to receive the data regardless of the position of its receiver window.

The CPP answers in MUX2 mode which is detected during a normal receiving slot time of 1 ms. All exchanges are performed thereafter in MUX2 which results in very temporary use of the MUX3 mode.

The SYN channel uses a binary sequence that has a good autocorrelation function, which assists synchronization at the burst level. These sequences are only carried on the *SYN* channel and are present only in the MUX2 and MUX3 modes. Two different sequences are used to differentiate the scanning search and synchronization phases: CHM, used to set up a connection, and SYN, used when a radio connection is already set up. Each sequence is divided into two sub-types to identify transmission from portable or fixed parts. Consequently, portables are unable to identify sequences from other portables, which helps avoid any kind of error. The CHMF (CHM fixed part) transmitted by the fixed terminal has all its bits inverted relative to the sequence transmitted by the portable CHMP (CHM portable part). The same thing holds for the SYNF and SYNCP markers.

10.2.1.4 Layers 2 and 3 Procedures

Layers 2 and 3. Call setup and authentication procedures are achieved at layers 2 and 3, and are based, to a large extent, on ISDN specifications.

Layer 2, which manages the data link on the D-channel, is subdivided into two sub-layers: a lower part for the establishment and measurement of radio links (fixed size packets) and an upper part for the transfer of upper layer data (variable size packets). The upper part of layer 2 is essentially based on the D-channel link access protocol (LAPD) specification from ISDN. At level 2, the acknowledgment of messages is made with the help of a bit that indicates the sequence number of the expected packet and is included in the received packet (piggy-back mechanism) which gives rise to a 1-bit sliding window protocol.

Layer 3 uses very simple signaling messages based on the Q.931 recommendations of ISDN. These messages can be classified into various categories:

- Transmission of the digits dialed at the portable and data display of these digits,
- Generation of alerting (received call, errors),
- Selection of outgoing call types for residential, telepoint, wireless PBX, emergency, and other functions such as recall and standby,
- Progress tone,
- Connection or disconnection of the audio channel with the control tone,
- Registration of the portable part,
- Authentication of telepoint calls,
- Indication of terminal and fixed part functionality,
- Unspecified messages.

Call setup. Call setup is performed in two stages. The first phase is the establishment of the radio link, described in Section 10.2.1.2, with selection of a common channel at both ends and synchronization of equipment. The second phase is call setup with authentication of the receiver and then establishment of communication (Figures 10.15 and 10.16).

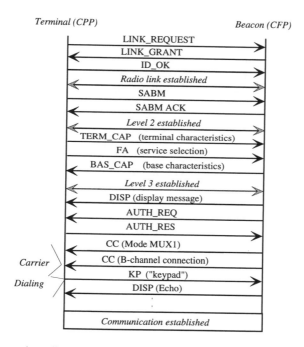

Figure 10.15 CT2 outgoing call setup.

Communication setup is subdivided into several phases:

- Search for a B-channel,
- Periodic transmission of a message (in MUX3 mode) until receipt of an acknowledgment from the other end,
- Exchange of the radio link setup messages with mutual synchronization of equipment,
- Establishment of the level 2 protocol,
- Exchange of level 3 messages for the acquisition of equipment characteristics,
- Authentication (optional), then display of a welcome message,
- Transition to MUX1 mode for establishment of the B-channel. The subscriber will then hear dial tone (in case of an outgoing call) or the caller (in case of an incoming call).

Setup of an outgoing call. The portable part performs the following actions:

- Scans the 40 channels and selects a free channel,
- Transmits a request message on the LINK-REQ channel (in MUX3 mode),
- Searches for the LINK-GRANT answer (in mode MUX2) after triggering a 5-sec counter *Tpmax*:
  - If the terminal receives an answer, it will move on to MUX2 mode and set up the link.
  - If the terminal does not receive any answer and t < *Tpmax*, it will repeat its transmissions a maximum of five times on another channel.
  - If the terminal receives no identifiable answer and *t* > *Tpmax*, it will repeat the procedure.

After connection setup, level 2 setup is achieved by transmission of a SABM message by the portable handset and waits for the receipt of an SABM-ACK message. When acknowledged, the two ends exchange level 3 messages, which include equipment and network characteristics. After authentication the RPP moves to the MUX1 mode on command from the CFP, and the user receives dial tone.

Fixed part functionality. The base station continuously and sequentially scans all the 40 channels. Messages transmitted by the handset are expected to be in MUX3 format. The detection principle is as follows: the base station locks on to a channel on detection of a carrier and starts looking for the CHMP marker. If a marker is not detected another channel is examined. If it is detected then the fixed part decodes the D-channel and the radio connection is set up. If the identification found in the field LID (link identification code of the D-channel) is recognized, the CFP replies with a LINK-GRANT message. The LID is used to identify a fixed part or a service, referenced to the CFP-CPP link or as an identifier of the CFP). The LINK-GRANT message is formatted in MUX2 and contains:

- SYNCF in the synchronization field,

- The PID (*portable identity*) of the calling handset, and a call reference number in the LID field.

Upon receipt of a SABM, the base answers with an SABM-ACK message, and the level 2 protocol is set up. After receipt of the handset characteristics the fixed part sends back its characteristics and authentication exchanges begin. The fixed part goes off-hook on the PSTN line and commands transition to MUX1. The call can then start.

Security. In CT2 two kinds of authentication are defined: authentication of the handset or authentication of the handset and network (mutual authentication). Encryption is not specified in the standard (see Chapter 5).

Incoming call. For an incoming call subscribers are paged for 5 sec. The handset answers a paging message with a link request message, which includes its identity.

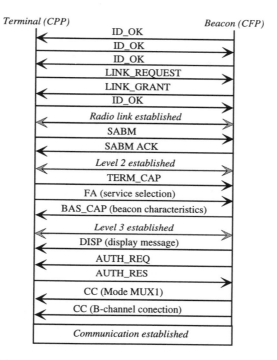

Figure 10.16 CT2 incoming call setup.

10.2.1.5 CT2 Evolution

Compared with other digital cordless telephone standards (e.g., DECT, PHS, and PACS), CT2 suffers from the major drawback of not offering handover and automatic location management. It is true that this additional functionality was expected in an enhanced version of CT2 (CT2+) but which never happened. As a telepoint application, most of the networks using the CT2 standard no longer exist. After an unfruitful attempt in Great Britain, all four licensed networks have now closed. It is also the case for *Birdie* of *Deutsche Telecom* and *Greenpoint* of *PTT Netherlands* [5].

10.2.2 DECT

10.2.2.1 Introduction

During the latter part of the 1980s, two main nonproprietary specifications existed in Europe for cordless systems, CT2 and DCT900 (developed by the Scandinavian countries).

These two systems were incompatible with little possibility of reaching an agreement between the European nations for a common standard as they operated in different segments of the 800–900-MHz band. In 1989, CEPT defined a European standard (DECT) that would be accepted by the majority of countries. DECT (Digital European Cordless Telecommunications system) [2] later had its acronym changed to Digital Enhanced Cordless Telecommunications to make it more acceptable as a world standard as GSM had done previously.

The main objective of the DECT standard was to define a cordless service within a high-density environment with heavy traffic and mainly in professional environments (typically offices). DECT standardization started in 1987 parallel with CT2. In 1992, the DECT specifications (ETS 300 175 and ETS 300 176) were adopted as the European standards for cordless telecommunication systems. The DECT standard can support several different applications: residential cordless telephones, wireless PBXs, telepoint, and even radio in the local loop.

The DECT specifications are organized in eight parts numbered 0 to 7. Part 0 defines the layer structure of the standard with the protocol architecture. Parts 1 to 4 describe the functionality of the four lower layers (physical, MAC, DLC, and network) in detail. Each part includes several subsections describing services, messages, and procedures. Part 5 introduces the management entity that controls the four lower layers and parts 6 and 7 specify the constraints related to the transmission of voice and data.

Starting from these specifications, a layered architecture is defined. It includes two common layers for user data and signaling: the *physical layer* (PHL) and *medium access control* (MAC) responsible for the allocation of radio resources and multiplexing of user data and signaling. The *data link layer* (DLC) ensures a reliable connection even during a handover. The *network layer* (NL) is responsible for call routing. The *management entity* (MGE) monitors the four lower layers and manages the handover procedure (e.g., if the quality of a connection falls below a given threshold, a handover may be triggered by the MGE entity). The layered architecture of the DECT system is represented in Figure 10.17.

The PHL manages the spectrum resources. The MAC manages the following three functions: radio channel setup and supervision, multiplexing of error control information, and the supply of reliable point to point connections. The DLC layer ensures maintenance of reliable data links. In the DECT system the radio base stations consist of RFPs (Radio Fixed Parts) and the mobile terminals, PP (Portable Parts). A DECT network can be interconnected to a larger network: PSTN, ISDN, X25, and GSM, for example (Figure 10.18).

RFPs belonging to the same system must be synchronized at the bit, slot, frame, and multiframe levels.

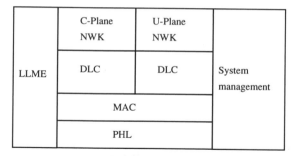

Figure 10.17 DECT system layered architecture.

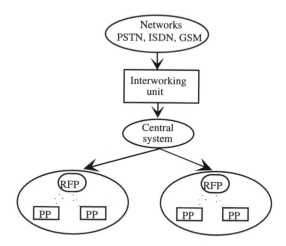

Figure 10.18 Example of DECT system architecture.

10.2.2.2 Radio Interface

Each carrier is structured into 10-ms frames, each frame containing 24 slots. These slots, numbered 0 to 11, are reserved for the downlink, and slots numbered 12 to 23 are reserved for the uplink (Figure 10.19 and Table 10.2).

Table 10.2
Characteristics of the DECT radio interface

Frequency band	1,880–1,900 MHz
Channel bandwidth	1.728 MHz
Number of carriers	10
Number of duplex channels	120
Modulation	GMSK
Duplexing	TDD
Channel bit rate	1,152 Kbps
Speech coding	ADPCM at 32 Kbps
Maximum output power	250 mW
Channel allocation	DCA

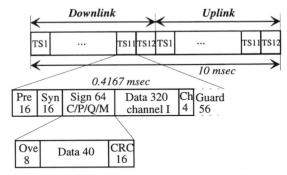

Syn: synchronization; Ove: overhead; Pre: preamble

Figure 10.19 Frame structure in DECT.

A data burst contains two fields, a 32-bit synchronization field and a 388-bit data field. GMSK is used to modulate the 420 bits on to the carrier. The modulation data rate is 1,152 Kbps. Guard intervals of 52 μs are used at the beginning and end of each burst.

The channel structure (called *bearers* in the DECT standard) of the radio interface is represented in Figure 10.20.

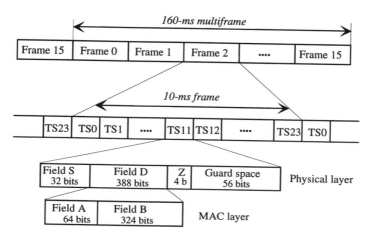

Figure 10.20 DECT channel structure.

Each 5-ms frame (uplink and downlink) supports 12 independent channels for voice and data.

Upper layer information is grouped into logical channels: I (user data), C (control channel), P (paging channel), Q (system information broadcast channel), and N (channel for the exchange of the handset and base station identities). The C, P, Q, and N channels are multiplexed on 48 bits in each burst. They are protected by a CRC code (cyclic redundancy check) for the detection of errors. In the case of the C-channel, an ARQ protocol is also used to recover bits in error.

The DECT system defines three types of slot usage:

- Use of a *whole slot*: for the transmission of voice or data,

- Use of *a half slot*: for voice transmission with a coder at a reduced bit rate,

- Use of a *short slot*: for the transmission of data in connectionless mode.

Transmissions can be performed in simplex mode (one or two slots in one direction) or in duplex mode (one slot in each direction).

The DECT standard defines bearers (equivalent to channels) that correspond to a single service at the physical level. In this case, a bearer represents a single physical channel, or a part of it, used for a call. Three kinds of bearer are defined: Short/Long, Simplex, and Duplex/Double Duplex.

10.2.2.3 Main Procedures

Handset procedures:

- *Synchronization.* A terminal must be synchronized at multiframe level with at least one RFP before attempting access. A terminal arriving in a cell first determines a list of physical channels in accordance with received signal strength and determines the most powerful RFP, which it then selects. It then synchronizes on this transmitting RFP.
- *Acquisition of system information.* The terminal listens to the system information broadcast by the RFP (RFP and network identities).
- *Determination of a channel list.* The terminal, by listening to the channels, sets up an ordered list of channels with the least interference (the best error rate).
- *Connection setup.* The terminal may attempt access on any free slot where the interference level is below a threshold (−93 dBm). Then, it can send an access request to the selected RFP.

The classification of DECT communication channels is based on CCITT recommendations (see Table 10.3).

Table 10.3

Structure of the logical channels in DECT

Channel	Information	DECT channel names
BCCH	System information broadcast	Q Channel
PCH	Subscriber paging	P Channel
SCCH	Signaling before traffic channel allocation	M Channel
UPCH	Packet data transmission	G_F Channel
FACCH	Fast signaling associated to a traffic channel	C_F et CL_F
SACCH	Slow signaling associated to a traffic channel	C_S et CL_S
TCH	User traffic information	I_P (protected) I_N (nonprotected) SI_N (short Message Connectionless Service)

DECT offers a mechanism called *fast setup scan sequence*, which allows any CPP to set up a connection rapidly. The portable terminal synchronizes to the scanning sequence of a selected RFP but scans the various carriers one frame in advance. Thus it is able to detect free slots on a carrier X and in frame N and then tries to set up a connection in one of the slots of frame $N+1$.

Outgoing call. An outgoing call request is transmitted by the handset on a physical channel which is adjudged to be free or whose interference level as perceived by the terminal is the lowest (Figure 10.21). The message includes a field indicating the number of physical channels, in multiples of half-slots, that the RPP wishes to use. If the RFP correctly receives the access messages then, half a frame later, it sends back a confirmation message. A pilot channel, which might be either a half rate channel or a full rate channel, is setup between the RPP and the RFP. For voice communications the pilot channel is sufficient to convey traffic. This channel is always set up in duplex mode, be it for voice or data.

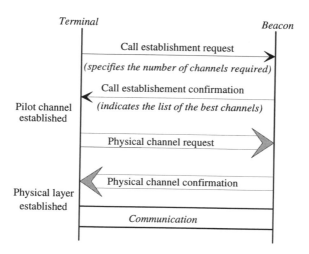

Figure 10.21 DECT call setup procedure.

The pilot channel can be used for further setup requests relating to supplementary channels. In this case the RPP sends a list of available channels to the RFP, which may be able to offer these channels by combining data received from the RFP and its own signal measurements. Then the RPP sends access packets on a number of channels corresponding to the requested capacity. These access packets define whether the channels requested are simplex (for asymmetric transmission) or duplex (for symmetric transmission). The RFP sends a channel confirmation on those channels for which it has received a demand which it can accommodate.

At channel release the RPP transmits a release message on the channel to be physically released. Acknowledgment from the RFP is not necessary because of timeouts used in the algorithm (Figure 10.22).

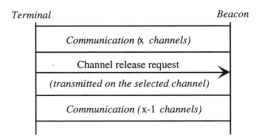

Figure 10.22 Channel release procedure in DECT.

Each RFP transmits on at least one broadcast channel. The RPP searches for this channel and synchronizes on it. Paging messages are transmitted on this channel. This technique helps the mobiles avoid having to scan all the channels. The DCS algorithm (see next section) helps avoid costly frequency planning. Even so, a minimal engineering phase is required to determine the number and location of base stations for telepoint or WLL systems.

Channel selection procedure. Each RFP can choose any of the 10 carrier frequencies available. Any slot used by a RFP(1) on a single carrier cannot be used by another RFP(2) on the same carrier frequency in the same area. Nor is it visible to RFP(2) and for this reason, it is called a *blind slot*. As far as RFP(2) is concerned it could be considered free for use by terminals but must not because it is already in use. A mobile terminal wishing to make access consults its channel list produced during the scanning sequence and communicates the busy slots to RFP2. The blind slots will be shown as busy in this list and this is temporarily stored (in the *blind slot information register*). The selection of a communication channel is realized according to a DCA-like algorithm, called here *dynamic channel selection* (DCS), also used in CT2 and DCT900.

After selecting a channel from the list, a portable terminal measures the channel during the next two frames before attempting access (Figure 10.23). If the values of RSSI on these frames have increased by more than 12 dB, the procedure is started over again.

Busy	Do not try	
b(n)		RSSI
	Possible	level
b(1)	candidates	
b(0)		- 93 dBm
Idle	Always candidates	

6 dB ↕ (to the left of b(1)/b(0))

Figure 10.23 Selected channels list.

In the case of a collision the terminal must repeat the access procedure after x frames. The value of x is a random variable with a uniform distribution on the interval $[0,\ 2k - 1]$ with k (≤ 10) being the number of successive collisions.

The handover. A handover is triggered when the quality of the radio link falls below a certain threshold. For the RPP, it consists of maintaining for a short period of time, two connections with two different RFPs. The data are transmitted over the two links until release of the original link.

In DECT the handover procedure is decentralized. Channel supervision is undertaken by the mobile and base station, which measure the RSSI and BER. In this case the RFP transmits its measurements to the mobile and the decision to trigger a handover is taken at the mobile. This method is called mobile controlled handover (MCHO).

The radio criteria for handover triggering and the selection of a target terminal are:

- The signal strength of the target RFP is greater than 12 dB above that of the current RFP.
- The C/I ratio with the target station is higher than the C/I ratio with the current station.

Thus, when handover is triggered, the establishment of a second connection, together with the current connection, is made and lasts for a few tens of milliseconds (the exact duration of this macrodiversity phase is not specified in the standard). The duration of a handover in DECT is about 100 ms. A handover can occur either intercell or intracell.

Two kinds of handovers can be identified: the fast HO *"make before connection"* reformed at the MAC and DLC layers and used in groups of close cells, and the HO between different cells of the same network or between different networks. This type is controlled by the network layer. In the idle state, each RFP transmits on at least one slot which can be a traffic slot (used for an RFP call to a RPP) or a *dummy slot*. These slots contain the identification of the system and of the RFP. During the channel-scanning phase the terminals listen to these slots. They can select one of them during handover (Figure 10.24).

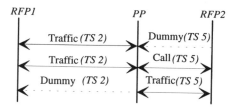

Figure 10.24 Example of handover in DECT.

Location management and security. The location management process is similar to that in GSM. It requires the use of two databases: a *home database* (HDB) and a *visitor database* (VDB). These databases are used, in addition to location management, for billing. Location management can be achieved manually, by the user, or automatically, by the terminal.

Security is guaranteed by an authentication procedure at two levels (see Chapter 5). The related data and algorithms are implemented in hardware in the receiver or in a detachable module called DAM (*DECT authentication module*), equivalent to the SIM card in GSM. The DAM contains the user identity, the authentication and ciphering keys, access rights, and authentication algorithms.

Four kinds of authentication are provided:

- The authentication of an RPP by an RFP,
- The authentication of an RFP by an RPP,
- Mutual authentication,
- Authentication of the user by introduction of a personal code.

Ciphering is optional and operates in a similar way to GSM.

10.2.2.4 Services Offered

DECT is a more advanced system than CT2 and includes an ability to offer ISDN access and supplementary functionality such as handover and more sophisticated signaling. While the system does not provide any special teleservices, it offers several bearer services to the user. The bearer services are, for example, the *transparent unprotected services* (TRUP) for voice transmission, the error corrected service (FEC), and *secondary rate adaptation services* (*SRAP*), which use the conversion procedures for primary flow as defined in the recommendation V.110 of the CCITT, marking a new generation of DECT equipment dedicated to data transmission. These products allow the exchange of data between computers (e.g., personal and peripheral computers) in office environments [6].

10.2.3 PHS

The *personal handyphone system* (PHS) is a Japanese cordless system standard. A main objective of PHS is to offer a cordless access system to the telephone network with several applications, while also striving to keep the cost below that of a cellular system.

The PHS standard specifies residential, public, cordless PBX and radio in the local loop applications. As much as possible PHS uses the available network (e.g., PSTN) to minimize the cost of the PHS fixed network. Using this system, several network configurations can be designed.

10.2.3.1 Historical Overview

In April 1993, the Japanese telecommunications technical advisory council submitted its final report for the radio interface standard to the Ministry of Posts and Tele-communications. This standard was referenced under the name RCR STD-28 [7] having been defined by the Research and Development Center for Radio Systems (RCR). The interface between the base station and the digital network is standardized under reference JT-Q921-b and JT-Q931-b by the Telecommunication Technical Committee (TTC). As is the case with DECT, the PHS interface is based on ISDN specifications. The main differences are related to location updating, authentication and handover procedures.

The PHS recommendations are the following:

- Radio interface: *RCR STD-28 Personal Handy Phone System* (RCR Standard Version 1),
- Network interface: JT-Q921-b Second Generation Cordless Telephone System Public Cell Station-Digital Network Interface Layer 2 Specification,
- JT-Q931-b Second-Generation Cordless Telephone System Public Cell Station Digital Network Interface Layer 3 Specification,
- JT-Q931-a Second Generation Cordless Telephone System Public Cell Station Digital Network Interface PHS Service Control Procedure,
- Inter-Network interfaces: JT-Q1218 Inter-Network Interface for Intelligent Network and JT-Q1218-an Inter-Network Interface for PHS Roaming.

Initially the PHS was developed for private use (residential or cordless PBX) purposes. Today, it is used in offering public PCS-like services. The professional and residential cordless telephone applications were introduced in mid-1994, the public cordless telephone application at the end of the same year.

The first public network was launched commercially in 1995. Early growth in Japan was tremendous: from zero in July 1995, it witnessed the signing up of 3.2 million subscribers in one year. The basic principles of PHS are low cost, support of multimedia services, mobility with bidirectional communication, and small sized handheld terminals with long battery life.

10.2.3.2 Radio Interface

The PHS radio interface characteristics are reproduced in Table 10.4 and the structure of the PHS frame is reproduced in Figure 10.25.

Table 10.4

PHS radio interface characteristics

Frequency band	1,895-1,906.1 MHz (residential applications)
	1,906.1-1,917.1 MHz (public applications)
Channel bandwidth	300 kHz
Number of carriers	77
Modulation	π/4 QPSK
Duplexing	TDD
Channel bit rate	384 Kbps
Speech coding	ADPCM at 32 Kbps
Maximum output power	80 mW
Channel allocation	DCA

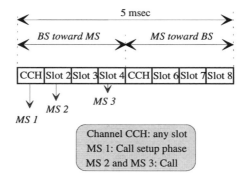

Figure 10.25 PHS frame structure.

Each carrier supports four multiplex channels (four times two time slots). The control channels used in Japan are the following: 1,898.450, 1,900.250, 1,903.850, 1,905.650 MHz for private applications and 1,916.750, 1,917.350, 1,917.950 MHz for public applications.

10.2.3.3 Handover and Call Setup Procedures

Two kinds of handover are defined in PHS: the *recalling type* handover, with a link transfer to the new base station similar to a call setup (communication disconnected from the old base station and then call setup with the target BS), and the *TCH switching type,* with a change of channel by synchronization of signals on two bases. This kind of handover is much faster than the previous one (Figure 10.26).

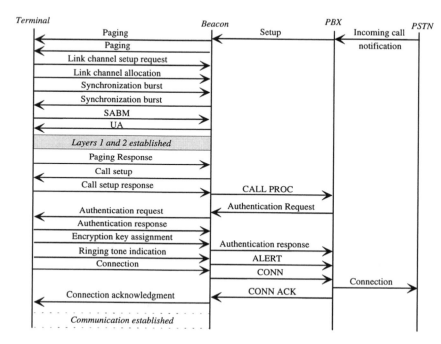

Figure 10.26 PHS call setup procedure.

10.2.3.4 System Architecture

The PHS architecture is based on microcellular techniques relying on the available fixed network (Figure 10.27). By integrating the PHS with the PSTN, some available network elements, such as local switches or subscriber local loops can be used. Each base station can cover an area with a radius from 100 to 300m. One of the unusual features of PHS is the ability of its terminals to operate in back-to-back mode allowing two users to communicate directly up to about 100m. No base stations are required. Because it does not go through, and therefore is not billed by a network operator, this function is popular at exhibition and building sites.

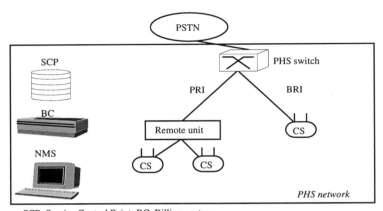

SCP: Service Control Point; BC: Billing center;
NMS: Network Management Systems; CS: Cell Station

Figure 10.27 Example of PHS network configuration.

10.2.4 PACS System

The personal access communications system (PACS) is an ANSI radio interface standard defined by the Joint Technical Committee (JTC) group of the Telecommunications Industry Association (TIA) and the Alliance for Telecommunications Industry Solutions (ATIS) [8]. This system is optimized for a low-speed pedestrian use although it has been proved that it can function at speeds up to 90 km/h.

PACS offers voice quality close to that of the fixed system with transmission delays less than 5 ms over the radio link. It is a microcellular system but with low complexity, which allows implementation of frequency reuse and, therefore, is able to deal with high traffic densities. It can also be applied to distribution in the local loop for medium range environments (in other words, for distances up to a few kilometers.) Services offered by a PACS network are individual messaging service, data services in circuit/packet mode, and a combined voice and data service.

10.2.4.1 Origin

PACS is a system that was defined in the United States, and is based on the wireless access communication system (WACS) designed by Bellcore. It includes functions of mobility, data services, interoperability between regulated and unregulated frequency bands, maintenance and management, radio port control unit (RPCU), handover, ISDN, authentication, and ciphering. While WACS was initially defined for fixed applications, the radio interface specification allows it to be used for PCS-like applications.

10.2.4.2 Radio Interface

The quasi-static autonomous frequency assignment (QSAFA) method, used for frequency allocation, relies on a DCA-like decentralized channel allocation algorithm (see Section 10.2.4.4).

PACS uses frequency duplexing for full-duplex bidirectional communication. A version that was not submitted for a license was defined and based on time division duplexing and used only one frequency per channel. The main characteristics of the PACS radio interface are reproduced in Table 10.5. The system broadcast control channel (SBC) includes three kinds of logical channel: the AC (alerting channel, equivalent to the PCH in the GSM), the SIC (system information channel, equivalent to the BCCH in the GSM), and the PRC (priority request channel) for emergency calls (Figure 10.28).

Table 10.5

PACS radio interface characteristics

Frequency band	1,850-1,910 MHz for the uplink
	1,930-1,990 MHz for the downlink
Channel bandwidth	75 kHz
Number of carriers	200
Modulation	$\pi/4$ QPSK
Duplexing	FDD
Channel bit rate	384 Kbps
Speech coding	ADPCM at 32-Kbps
Maximum output power	200 mW
Channel allocation	QSAFA

On the downlink channel the 14 bits of the SYNC channel are used for synchronization. In the uplink direction 12 guard time bits are used to avoid overlapping between packets transmitted by various mobiles.

The slow signaling channel SC (*slow channel*, equivalent to the SACCH in the GSM system) contains additional synchronization patterns, word error indications, signaling information, and user data.

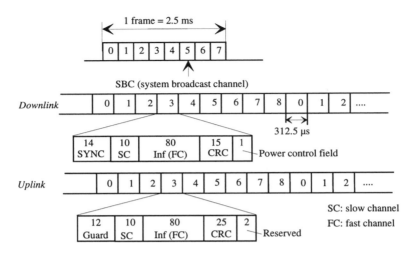

Figure 10.28 Structure of the PACS radio interface.

The traffic channel, FC (*fast channel*) contains user data (voice and data). The 15 CRC bits provide burst synchronization and error detection of the SC and FC channels. The power control channel (PCC) control bit allows adjustment of the terminal transmission power. The 80 bits of the FC are used every 2.5 ms per frame and equate to a bit rate of 32 Kbps which is enough for a good-quality voice coder. PACS offers channels equivalent to 16 Kbps and 8 Kbps with a burst every 2 or 4 frames.

10.2.4.3 Network Architecture

The network architecture of PACS is quite similar to that of the systems previously introduced (Figure 10.29).

With PACS, the terminal, called a *subscriber unit* (SU), communicates via the A interface with the base station, called the *radio port* (RP), which is connected to the RPCU through the P-interface. The interface between the RPCU and the network is called the C-interface. PACS integrates a network management module called *access manager* (AM), which takes care of network control such as access to remote databases for visitors and assistance with call setup. The AM entity can be located in either the radio system or in the network.

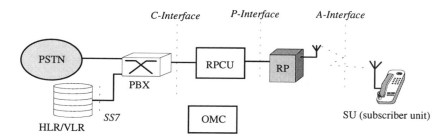

Figure 10.29 Architecture of a PACS system.

10.2.4.4 Main Procedures

Handover. The handover procedure in PACS is called *automatic link transfer* (ALT). It is controlled by the mobile using measurements obtained during the supervision phase. A mobile forwards a handover request directly to the target base station (*forward handover*). The advantage of this method is that handover signaling is performed on the new link and is, therefore, more reliable.

PACS defines five types of handover:

* The *intracell handover* (which takes a few milliseconds),
* The *inter-RP* and *intra-RPCU handover*,
* The *inter-RPCU* and *intraswitch handover*,
* The *inter-switch handover*,
* The *inter-RPCU handover* controlled by various access management centers.

These different kinds of handover are transparent to the mobile, which conveys the same request in all cases.

Mobility and security features. PACS uses a location management process that is similar to that of GSM using location areas. It is similar to the one defined in the IS-41 standard. Deregistration is explicit; in other words the network deregisters mobiles that have left a location area (Figure 10.30).

Security is ensured by means of an authentication protocol, which includes the use of public or secret keys (Figure 10.31). PACS also includes a ciphering procedure. Authentication uses a call counter which enables detection of clones (see Chapter 5).

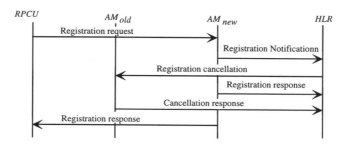

Figure 10.30 SU registration and deregistration call flow.

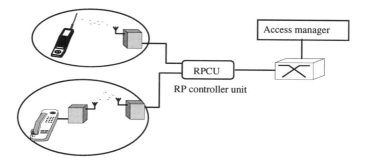

Figure 10.31 Example of a PACS network.

Services. PACS allows exchange of text messages, electronic paging, group III fax, voice, video, and data in nontransparent circuit mode with data ciphering and integrity assured by a flow control protocol. The data bit rate is 32 Kbps and can fall to 28 Kbps in bad transmission conditions. Packet data transmission is managed by a contention protocol using the DSMA mechanism.

Frequency planning. The radio frequencies are not planned manually, as is the case for current cellular systems. The automatic frequency allocation mechanism is the QSAFA [9], which selects frequency pairs without centralized coordination between base stations. Frequency allocation is controlled by the RPCU on behalf of its associated base station transceivers while frequency planning in a PACS network is realized automatically by the base stations. The QSAFA allows each RPCU to choose the frequencies used by its associated RPs without centralized control from the network. This procedure includes two main steps:

- *First phase.* RPCU sequentially transmits a command to each RP to move to receive mode. Then, the RP listens to all downlink channels in turn.
- *Second phase.* The frequency on which the interference level is lowest is chosen by the RPCU, which transmits an order to the RP to tune to this frequency and start its transmissions.

The frequency allocation procedure is repeated by all RPs one by one until no RP requires any change in frequency for two successive cycles. As the RP transmitter on the downlink channel has to be switched off throughout the period of frequency search measurements are conducted during low traffic periods. Simulations have shown that for 256 RPs using 16 frequency pairs the allocation process stabilizes in about five iterations.

Call setup procedure. In Figures 10.32 and 10.33 the message flow for call origination and call termination cases are shown.

10.3 CONCLUSIONS

All second-generation cordless systems, mainly DECT, PHS, and PACS, integrate functionality that makes them very close to cellular systems but can be more cost effective than microcellular systems such as DCS 1,800/PCS 1,900 when catering for areas with high-density traffic. The main differences between the two types of system, as far as the user is concerned, are likely to be blurred by the roaming and handover abilities of cordless systems. From the system point of view, the architecture of cordless systems lies essentially on an available network (PSTN, ISDN, X25, and even GSM). It is one of the major differences from cellular systems, which are designed as autonomous networks to bridge the gap with fixed networks.

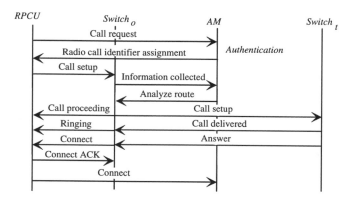

Figure 10.32 Successful call origination call flow.

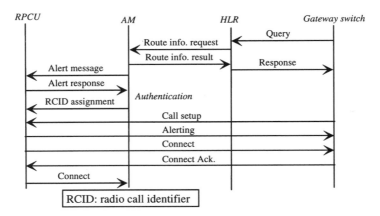

Figure 10.33 Successful call termination call flow.

The characteristics of cordless systems radio interface management, with their use of dynamic channel allocation techniques and time division duplexing, are related to the pico-cellular situation which future cellular systems might themselves adopt (notably the DCA methods). Moreover, cordless systems use rather simple signal processing algorithms: no FEC error control, and no complex mechanisms to fight against multipath propagation such as spread spectrum or equalization. These are the main reasons that make these systems less costly compared to cellular.

Experiments conducted with cellular and cordless systems together with the current trends prove that cordless techniques better suit indoor applications and high density traffic areas (from the point of view of capacity offered and economy of use). As for cellular systems, these are more convenient for wide area coverage with lower traffic densities. In fact, a combination of both systems offers better roaming between various environments and is considered a solution to the coverage problems of cordless systems and the capacity problems of cellular systems. Dual mode terminals (DECT-GSM or PHS-GSM) can offer the subscriber a quasi-comprehensive transparency between the two kinds of environment.

Among the most significant trends related to application of cordless systems there is the case of fixed subscriber radio distribution (WLL or RLL). It is coupled with numerous uses in developed countries in the current deregulated context with the advent of new operators setting up telecommunication infrastructures rapidly. In developing countries there is a pent-up demand to be equipped with telecommunications infrastructure as soon as possible, which can be satisfied by both cellular and cordless technologies. The cordless PBX application has also witnessed considerable growth and the telepoint application has been suggested as one of the alternatives likely to offer PCS-like services.

REFERENCES

[1] Cox, D., "Wireless Personal Communications: What Is It," *IEEE Personal Communications*, pp. 20–35, Apr. 1995.

[2] Tuttlebee, W. H., *Cordless Telecommunications in Europe*, Springer-Verlag, 1990.

[3] Garg, V. K., and E. L. Sneed, "Digital Wireless Local Loop System," *IEEE Communications Magazine*, pp. 112–115, Oct. 1996.

[4] Department of Trade and Industry Radio Communication Agency, "Common Air Interface Specification To Be Used for the Interworking between Cordless Telephone Apparatus Including Public Access Service," MPT 1375, May 1989.

[5] Clancy, D., "At the End of the Line," *Mobile Europe*, pp. 19–20, Dec. 1996.

[6] Clancy, D., "DECT plays the data card," *Mobile Europe*, pp. 74, Mar. 1997.

[7] Research & Development Center for Radio Systems (RCR), *Standard-28*, Japan, 1992.

[8] Noerpel, A. R., Y.-B. Lin, and Sherry H., "PACS: Personal Access Communications System - A Tutorial," *IEEE Personal Communications*, pp. 32–43, June 1996.

[9] Noerpel, A. R., "PACS: Personal Access Communications System: An Alternative Technology for PCS," *IEEE Personal Communications*, pp. 138–150, Oct. 1996.

SELECTED BIBLIOGRAPHY

[1] British Standards Institution, "Apparatus using cordless attachments (excluding cellular radio apparatus) for connection to analog interfaces of public switched telephone networks: part I and II," BS 6833, London, 1987.

[2] Department of Trade and Industry Radio Communication Agency, "Performance specification - radio equipment for use at fixed and portable stations in the cordless telephone service operating in the 864 to 868 MHz," MPT1334, Apr. 1987.

[3] ETSI RES-3, DECT Reference Document, RES 3 (89) 42, 1989.

[4] Hamano, T., "PHS: The technology and its prospects outside Japan," *Mobile Communications Internationa*, pp. 54–56 *l*, Nov. 1995.

[5] Negrat, A. M., and R. C. V. Macario, "Cordless Telephones in Transition", *Proceedings of the IEEE Vehicular Technology Conference 1996*, Atlanta, GA, pp. 456–461, April 28-May 1, 1996.

CHAPTER 11

PAGING SYSTEMS

Definitions
Paging systems provide the simplest mobile radio communications service. A paging system offers a limited communications service because messages can only be delivered in one direction. The paging service was defined by the CCIR in 1982 in its recommendation 584 as "a person selective call system, nonvoice, unilateral, with a tone and without message or with a numeric or alphanumeric predefined message." This definition, as will be seen in this chapter, covers most of the paging services now available.

Originally a paging system's main application was for paging or alerting people, hence its name. The message received by the mobile terminal consisted of a beep meaning, "someone is trying to get in touch with you" [1]. Today, messages transmitted by paging systems can be either a beep signal, a numeric message (e.g., a telephone number to be dialed), an alphanumeric message (an explicit message), or a voice (a message that has been recorded directly by the caller).

Preliminary systems and progress
Paging services were originally used for local and private applications (e.g., hospitals, worksites). One of the first paging systems was implemented in a London hospital in 1956, alerting the hospital staff silently without disturbing to the patients. The receiving services (*pagers*) were simple, consisting of receivers that were tuned to a single RF frequency. The first paging systems, introduced as early as the 1950s, used the 30 to 50-kHz band. The release of frequencies at 27 to 42 MHz and 470 MHz helped give a boost to public paging

systems in the 1960s and 1970s. These systems were of a low capacity with an average of 20 receivers per system and with a maximum of a few hundred receivers.

The first wide area system was implemented by network operators in the United States and Canada in the early 1960s. An example of this was the *SWAP* (*System Wide Area Paging*). In Europe, the first extended public systems were installed in Holland and in Belgium (1964) then in Switzerland (1965). The receiving equipment was generally installed in vehicles.

Rapid growth dictated the transition to systems able to manage a few hundred addresses (using two tones chosen from seventy) to systems allowing the management of several thousands of messages (using five-tone coding). These preliminary and fully analog tone systems experienced some errors caused mainly by the low-redundancy signaling format.

The evolution of paging systems could be subdivided into three phases. To illustrate, in France in 1975 a public paging system was introduced with the advent of the Eurosignal service (a light signal accompanied by a beep). The people interested in this kind of service included professionals entrusted with demanding tasks in the health, security, transport, and after-sales sectors.

The next phase was highlighted by an enrichment in the number of functions offered with the Alphapage and Operator networks. These new systems offered, besides the beep (or tone-only) service, the ability of transmitting numeric and alphanumeric messages.

The third phase comes as a natural progression with the adoption of the Ermes and FLEX standards, which appeared in mid-1990. The Ermes networks, besides the additional text and longer message services, also offer the option of international roaming.

11.1 CONCEPTS AND BASIC PRINCIPLES

Paging systems can be subdivided into two groups:

- The wide area public systems (which can cover an area or a country),

- The on-site systems (which often cover only one site, which extends from some tens to some hundreds of square kilometers) [2].

Public paging systems use several medium to high power transmitters and are able to cover a city, an area, or even a country. In the main, calls out to the receivers are transmitted through the public telephone network or over a data transmission network. They can be stored for a few seconds or a few minutes before being transmitted as part of a block (*batch* transmission).

Private or on-site systems (e.g., those covering a hospital, an airport, a building site) have a limited range and are capable of serving a few receivers. Calls are often transmitted by a manual operator or through a private switch with immediate message transmission.

11.1.1 Architecture

The simplest paging systems include interface equipment for access and routing, a distribution and transmission system, an operations and management center, and, finally, a set of receivers—usually called *pagers* (Figure 11.1).

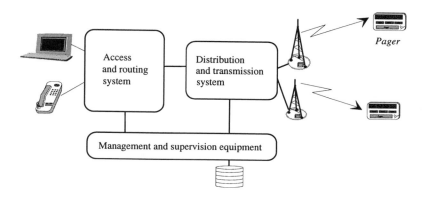

Figure 11.1 General architecture of a paging system.

11.1.1.1 Interface Equipment

Interface equipment enables the interfacing of the paging system with other networks (e.g., PSTN, X25) and is part of the access and routing system. The access and routing system function consists of collecting messages from either a telephone set or a computer terminal. Most telephones use dual tone multifrequency (DTMF) signaling that allows the telephone keyboard to be used to input the message directly. The user calls the service number (a unique number used by all callers) then "over dials" the address of the receiver and the numeric message when requested. The transmission of alphanumeric messages is more complicated and may require the use of different equipment (such as a computer terminal connected to a data transmission network) or additional procedures (predefined coded messages).

11.1.1.2 Distribution and Transmission System

This distribution and transmission system ensures the transmission of messages through to the pager via the base stations. It includes the distribution subnetwork located between the heart of the network (management and supervision center) and the transmitter sites. It includes the radio subnetwork, consisting of base stations and the base station controller (BSC). The latter receives messages from the center, codes them, and sends them to the transmitter. It also performs the base station supervisory functions.

11.1.1.3 Management and Supervision Center

This center, considered as the heart of the network, constitutes the network controller (*paging network controller* in Ermes terminology). Its main task consists of looking after the following functions:

- Management of the subscriber database and associated services,
- Interfacing with other networks, call management (call input, sending of messages to the appropriate areas),
- Management of an area controller (*paging area controller* in Ermes terminology),
- Supervision of the radio network (receipt of alarms, remote control),
- Management of the operating and maintenance center (OMC).

The management and supervision part includes, among others [2]:

- A database of authorized addresses with the permitted services and areas for each pager,
- A lookup table cross-referencing the dial in number with the pager addresses where this is necessary,
- Control procedures for transmitter powering on and off,
- An operating and maintenance system (e.g., traffic analysis software, activation or deactivation of links, error diagnosis),
- A billing system.

If the system is capable of transmitting voice messages it should integrate, in addition to the structure reproduced in Figure 11.1, an audio connection to allow the system to transmit voice calls [3] (Figure 11.2).

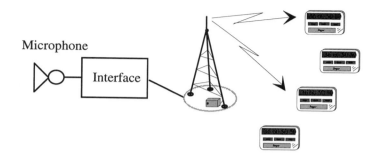

Microphone

Figure 11.2 Equipment for the transmission of voice messages.

11.1.1.4 Receivers

The paging receivers, or pagers, are small and light enough to be carried in a handbag, the belt, or in a pocket. One main characteristic of the terminal that represents its main advantage compared to other mobile radio systems is that each pager is identified by at least one address which allows it to intercept and decode its messages. Upon receiving the message that includes its address, the pager can notify the user by means of a tone, a light signal, or a vibration. A receiver can be provided with one or several addresses; or conversely, a single address can be shared by a group of pagers. Because it functions only as a receiver the paging terminal power consumption is quite low. Thus, a simple battery can provide autonomy for several days, a few weeks, or even months.

The main disadvantage with paging is that receipt of a message is not acknowledged. The system and the caller have no way of knowing whether his or her message has been received unless the called party responds.

11.1.2 Signaling Methods

The transmission of messages between transmitters and receivers in early paging systems was based on sequential tone signaling. Recent systems, on the other hand, use digital signaling. It should be noted that some paging systems use the same signaling as in bidirectional PMR systems (see Chapter 9), such as the ZVEI, CCIR, and EEA signaling standards.

11.1.2.1 Sequential Tone Signaling

Sequential tone coding consists of transmitting audio tones in a sequential and continuous manner [2]. The first tone of each call is the longest (it lasts 100 to 130 ms) and is used as a preamble. Each following tone typically lasts 30 to 40 ms and must differ from the previous one (these differences are detectable by audio filters). Thus, the need for accurate synchronization is avoided.

For example, a five-tone decimal system uses eleven different tones, one for each of the ten digits, and another to indicate the repetition of a digit. The address 22133, for example, is represented by the following sequence 2, R, 1, 3, R.

Among the tone systems a distinction should be drawn between those that use two sequential tones and those that use five or more. All the frequencies that modulate the carrier are located in the audio band between 100 Hz to 3 kHz. A paging system with only two sequential tones has 40 or 70 frequencies defined in the band 100–3,000 Hz. Two-tone systems reach a capacity ranging from 1,000 to 3,500 addresses and are able to manage call rates ranging from 0.3 to 1 call per second. To have higher capacity and bit rates it is necessary to resort to more tones (five or greater). These tones are selected from among a set of 11 tones that provide a base 10 signaling system. A five-tone system offers an addressing capacity of 99,999 ($= 10^5$ 1) users.

The first digital signaling systems were introduced in 1970 in Canada. Their major strength stems from the extra functions they are able to provide (e.g., multiple pager addresses, text messages). The major digital signaling systems are POCSAG, GSC (Golay sequential code), MBS (Sweden), NTT code (Japan), and that defined in the Ermes standard. The POCSAG and Ermes standards are more fully discussed in Section 11.2.

11.1.3 Transmission Channels

The transmission of messages (base station to the paging receivers) can be carried on either a dedicated channel or a subcarrier.

11.1.3.1 Transmission on a Dedicated Channel

With this kind of configuration, the paging system uses one or several channels that are exclusively allocated to it. The Alphapage and Ermes systems, for example, use dedicated channels.

11.1.3.2 Transmission on a Subcarrier

Where no frequencies can be dedicated to the service or the capacity of the system does not warrant special frequencies, it is preferable to use the radio resources of another system. A subcarrier of the host system may be modulated by the data to be transmitted.

The ITU has standardized the subcarrier transmission in its recommendation 643, known under the generic name of RDS (*Radio Data System*), and paging is defined as one of its applications (see also the case of the bidirectional data transmission network CDPD in Chapter 13). For instance, in France the operator paging system uses this technique by transmitting on a subcarrier of the Radio-France broadcasting network.

11.1.4 Services

Paging systems offer several kinds of basic service, including audible alarm services (beep or tone), alert services with a voice message, and systems that display numeric or alphanumeric message.

For example, the share of each of these services in 1995 in Europe was about 60% for numeric only pagers, 30% for alphanumeric, and 10% for tone only.

11.1.4.1 Alert or Beep Services

These are undoubtedly considered as the most simple. The pagers emit a tone (beep) or a light signal on receipt of a message. Having received such a signal, the user can dial a predetermined number (e.g., that of his or her company) or to go to a specific place (e.g., headquarters, emergency service) to find out why he was called.

Although limited, a beep service allows a given person to know that someone wishes to be contacted. The disadvantage is that the user is compelled to find a telephone or to move (e.g., to the monitoring center, the emergency service) to find out the origin and the reason for the alert.

11.1.4.2 Voice Message Services

Voice message services are considered an improvement over tone-only messaging as the called party receives a voice message. Yet, this service too has disadvantages. First, the transmission time is limited to a few seconds per call (hence the need for a precise and concise message), and message is not always sufficiently intelligible, and finally, this service is not confidential as the message may be heard by a third party.

11.1.4.3 Numeric Messages

Numeric messages consist of sending via a DTMF telephone handset, or through an operator, a message made up of digits, a telephone number to be called, or a preset code.

11.1.4.4 Alphanumeric Messages

Alphanumeric messages are the most complete service and the one that offers the highest level of comfort to the end user. Using a computer it allows the transmission of a message composed of alphanumeric characters which are able to clearly indicate the purpose of the call.

The display of a numeric or alphanumeric message (made of figures 0 to 9 and the letters A to Z) allows the user to save and read it in full confidentiality.

One of the most useful applications is the use of pre-coded messages. The pre-coding of messages helps reduce the transmission time of text by some 60%. In addition, coded messages offer subscribers better confidentiality.

11.1.4.5 Applications

The paging systems' service applications are as many, varied, and specific as the users and groups of people using them [4]. Yet, a classification into three groups can be attempted:

- *Traditional applications*: emergency, request to call back a specific subscriber (tone service) or multiplicity of subscribers (numeric service),

- *Applications to increase the productivity of a company*: management of agenda and travel, intra-enterprise information,

- *New applications*: broadcast of intra- or interenterprise information, combining with other products or mobile services (e.g., cordless telepoint, cellular), or electronic mail (voice or written), private use, telemetry applications (automated delivery of constantly updated information such as stock market reports).

11.1.5 Engineering

Due to their simplicity, paging systems are much easier to install and operate than cellular networks. Even in the case where a frequency reuse scheme is implemented it is done in a more rudimentary way. One of the advantages in having a one-way link in paging systems is the ability to optimize it and take advantage of this asymmetry. The use of high power transmitters (several hundreds of watts or kilowatts) at high sites assist in having low-complexity, low-consumption small-sized receivers.

For wide area coverage paging systems (WAPS) several fixed radio stations [5] transmit quasi-synchronously at powers of several hundred watts (see Section 11.1.5.2.) and cover cells with radii of several tens of kilometers. For example, at the POCSAG bit rate of 512 bps, time delay differences of 500 μs between the various propagation paths can be tolerated from each transmitter, and this allows the transmission of the same message from different sites. The signals are sent to the different transmitters in such a way as to make their emissions synchronous. In overlapping areas, the level difference between the signals coming from different transmitters is insufficient to create capture effect and the signals can interfere with each other. Perfect synchronization is not necessary (it would be highly complex to obtain because of propagation time differences), and a gap of one-quarter of a bit can be tolerated.

11.1.5.1 Area Splitting

To reuse frequencies, the WAPS are often divided into areas, each covered by one or several base stations. A user may elect to have service only in certain areas. Thus, messages destined for any given subscriber are only transmitted in the areas selected and paid for by that subscriber. In 1985, for example [6], the Eurosignal system was divided into three areas in Germany, using two frequencies and 45 sites. Time-frequency resources can be shared between areas and may be implemented by frequency-division or time-division methods. These methods can also be used in combination.

Frequency resource splitting (frequency-divided network). A frequency reuse scheme is implemented in paging systems in much the same way as in the cellular systems, but is much less complex (i.e., with relatively unstructured frequency reuse patterns). In this case, various areas use different frequencies (Figure 11.3). One disadvantage is that the terminals will become more complex in order to tune to the various frequencies.

Time resource splitting (time-divided network). Time division can be applied when the call rates are low enough not to require continuous transmission from the base stations. Different areas can use the same frequency, such as when two areas that transmit simultaneously are separated by areas transmitting at different times (Figure 11.4). The duration of the time periods allocated to one area is made according to the call rate in that area. An area "A" transmitting twice as many calls as another area "B" is allocated a transmitting time period twice as long as the one allocated to "B".

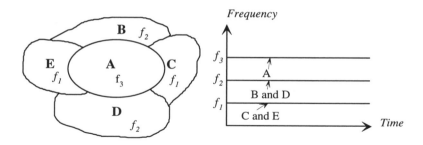

Figure 11.3 Example of frequency-divided network.

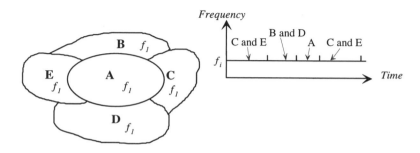

Figure 11.4 Example of time-divided network.

11.1.5.2 Quasi-Synchronous Transmission

In the WAPS, quasi-synchronous transmission is achieved with radio frequencies slightly shifted to avoid fading caused by destructive interference. The tolerance for maintenance of signal synchronization and successful decoding at the receiver varies from 1/10 to 1/4 of a bit duration. This quasi-synchronous constraint implies continuous control of the network to adjust time jitter in the equipment or over the transmission paths (the leased lines, for example, where the route can be modified by the fixed network operator [7]).

Quasi-synchronous transmission requires that several (or even all) the stations transmit the same signal at the same time. This synchronization is implemented by programming the transmission delays either in the control center or at the base stations.

Noting the parameters shown in Figure 11.5:

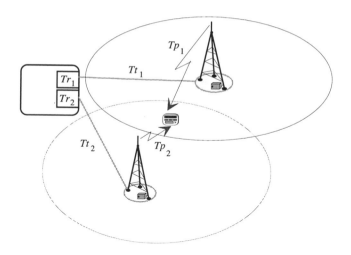

Figure 11.5 Reception of a message from two adjacent sites.

- Tp_i: propagation time of signal between the base station i and the pager,
- Tt_i: controller to base station i propagation delay,
- Tr_i: delay to be introduced at the control station for the base station i to transmit synchronously with other stations.

Transmission will be synchronous in the overlapping areas, if

$Tr_i + Tt_i + Tp_i = \text{Constant} \pm \Delta t_{max}$
With $\Delta t_{max} = 1/4$ of bit duration, typically.

If the pager is located in an area where it can receive signals at the same level from different sites, the message has to be decoded with the same accuracy as when it derives it from a single signal. This only happens if the time difference between the signals is less than the margin Δt_{max}.

Example 1: Synchronization of a paging network
Suppose a paging system where the distance between two transmitter stations A and B is equal to 200 km and the transmission delay is

100 µs between the controller and A,

80 μs between the controller and B.

If the overlapping area represents 20% of each base station coverage, transmissions from the two stations at the limits of this area should be capable of being decoded, in other words, distances $d_1 = 200 - 0.2 \times 200 = 160$ km and $d_2 = 200 + 0.2 \times 200 = 240$ km from each station. Transmissions coming from the other station must arrive within the quasi-synchronous window.

In this case, the parameters determined above have the following values:

$Tp_A(d_1) = Tp_B(d_1) = 16 \times 10^7/3$ sec,

$Tp_A(d_2) = Tp_B(d_2) = 8 \times 10^7$ sec,

$Tt_A = 10^2$ sec,

$Tt_B = 8.10^3$ sec.

The transmissions are within the limit of the overlapping areas (in the worst case) if the following relation is satisfied:

$Tr_A + Tt_A + Tp_A(d_1) = Tr_B + Tt_B + Tp_B(d_2) \pm \Delta t_{max}$

In the case of the POCSAG system operating at a 512 bps bit rate, the value of Δt_{max} is about 0.5 ms. The time constraints to be observed are then expressed as follows:

$Tr_A \cong Tr_B$ $0.197 \times 10^4 \pm 0.5 \times 10^3$ sec

The synchronization of a paging network is achieved over two stages:

The first stage consists of estimating the propagation delay of a signal between the controller and the relevant base stations. The time delays to each of the base stations are subsequently adjusted so that each transmits the signal simultaneously. In other words, the communication path to the base station which exhibits the longest delay is taken as the reference and additional delay is added to all other paths to equal this reference. Generally, this fixed network delay is slightly greater than the "over air" propagation delay that is identified in the second stage.

The second stage (which is really "fine tuning") identifies, by means of a radio propagation prediction software tool, the areas where synchronization problems can occur. Two techniques that help reduce these problems further are:

- Add time delay or advance (if possible) to the transmitters at the base stations,
- Add a mechanical or electrical tilt (see down-tilting, Chapter 7) to the antennas to minimize multipath problems.

11.1.5.3 Dimensioning

The dimensioning of a time-divided paging system aims at determining the size of the various areas and duration of the time periods allocated to each area. For a frequency-divided network, the task consists of determining the number of radio channels allocated to each area. As is the case with the other systems (e.g., see Chapter 7), this dimensioning takes into account peak time traffic parameters, the geographic distribution of subscribers and the addressing capacity of the system.

A paging system can be installed, operated, and extended at a relatively low cost compared with cellular systems. Apart from voice paging systems, which are greedy in terms of channels, paging systems use the radio spectrum very efficiently.

Example 2: Dimensioning of a paging system

Assume a population of 15 million people and a paging service penetration rate expected to reach 2%. Suppose a call rate of 0.3 messages per subscriber per hour at peak time and messages of 80 bits each. The bit rate of a system such as POCSAG is 512 bps. Assume an average transmission duration per message of $80/512 \cong 153$ ms.

The average traffic per subscriber at peak time is therefore, $0.3 \times 0.156/3{,}600 \cong 1.3 \times 10^2$ mErl.

The overall traffic volume for the population being considered is ($15 \times 10^6 \times 2/100 = 3 \times 10^5$) and is therefore about $1.3 \times 10^5 \times 3 \times 10^5 \cong 3.9$ Erl.

If the required service quality is 5% for message blocking, the minimum number of channels required is 8 (see Erlang table in Appendix 7C).

11.1.5.4 Coverage Constraints

As against bidirectional systems where the presence of some holes in coverage can be accepted by subscribers, a paging service where the caller has no means of finding out whether a message was received must be a service in which subscribers expect to receive a call with a probability of about 99%. A paging system should, therefore, provide a high signal level at any point of its coverage and, more specifically, inside buildings. For example, a paging receiver should be able to receive messages inside a movie theater downtown, at the bottom of a cave, in the countryside, or on a train. These high-density coverage constraints are rarely satisfied with cellular systems.

11.2 EXAMPLES OF ONE-WAY PAGING SYSTEMS

In this part, three paging systems are introduced: Eurosignal, POCSAG, and Ermes.

11.2.1 Eurosignal

Standardization of the Eurosignal network for Western European coverage was achieved by the CEPT. In 1968, subcommittee work led to the definition of the Eurosignal system, the specification of which is contained in Recommendation TR6 approved in 1976. Germany opened its service in April 1974 and France in December 1975.

11.2.1.1 Radio Interface

Eurosignal uses four frequencies located around 87 MHz. Signals are amplitude modulated. To distribute traffic on the four frequencies, each country has been subdivided into service areas (seven, for instance, in the case of France).

Receiver addresses contain six digits, and each digit corresponds to the transmission of a specific frequency (tone) for 100 ms. Tones sequences are separated by a time interval of 200 ms (Figure 11.6). Thus, it is possible to address one million different receivers. The use of additional frequencies allows the use of more receivers.

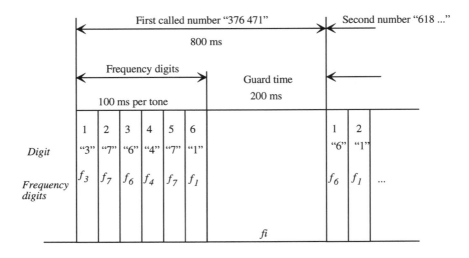

Figure 11.6 Eurosignal message format.

Tone frequencies are between 313.3 Hz and 1,153.1 Hz. The call rate is 1.25 calls per second and the paging capacity is 10^6 with a base 10 system. It offers the possibility for an extension of 7.5×10^6 with a base 14 system. When two successive digits of the address are identical, a tone called the *repeat frequency* (f_1 = 1,062.9 Hz) is used to separate the adjacent tones. Each paging signal is followed by a pause of at least 200 ms during which the transmitter is modulated continuously at a frequency of f_1 = 1,153 Hz. Each signal is transmitted simultaneously from all the transmitters in the network. At the paging receiver, a 3-sec tone signal, repeated after 30 sec, is triggered. A light signal, lit upon receipt of the signal, remains until the user intervenes.

The spectral efficiency of the Eurosignal system is very low. Compared with that of POCSAG, the transmission of an individual page requires 0.8 sec of radio time in Eurosignal versus 0.003 sec with the POCSAG system at 1,200 bps.

11.2.1.2 Architecture

The architecture of a Eurosignal network includes the three basic elements of a paging system; their designation is, nonetheless, specific to this system (Figure 11.7).

It includes the following elements:

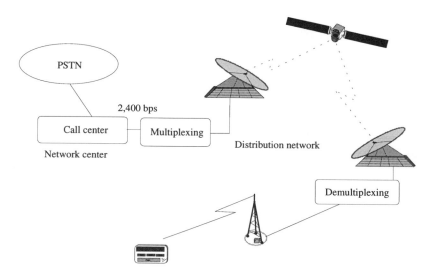

Figure 11.7 Architecture of a Eurosignal network.

- The *Network Call Center* is connected to the PSTN at subscriber private line level. It includes a call logging recorder, a minicomputer, an audio tone generator, and output interface modules, connected to each of the network transmitters, which adjust the level and phase of the frequencies to be transmitted,
- The *Distribution Network* includes satellite links,
- The *Radio Stations* operate in each of the service areas and transmit simultaneously.

11.2.2 POCSAG

The Post Office Standardization Advisory Group (POCSAG) system was defined in the mid-1970s in Great Britain by the British Post Office. In February 1982, the POCSAG code was adopted by the CCIR and recommended for international paging (CCIR recommendation 584) under the name of CCIR Radiopaging code No. 1 (RPC 1). It supports the transmission of digital alphanumeric messages.

11.2.2.1 Basic Principles

The major characteristics of the POCSAG standard are as follows:

- 18-bit addresses,
- Paging capacity of 250,000 receivers per channel,
- Radio frequency independent,
- Digital messages rate of 5 calls per second (10 digits per message),
- Alphanumeric messages rate of 1.07 calls per second (40 characters per message),
- Bit rate of 512 bps,
- Separation between channels of 12.5 kHz or 25 kHz,
- NRZ modulation,
- Message loss rate of 2%,
- BCH (32:21) coding.

Each POCSAG receiver is identified by at least a 21-bit address known as its *radio identity code* (RIC). Each codeword $m(x)$ is composed of 21 information bits corresponding to the coefficients of a polynomial whose terms go from X^{30} to X^{10}. This polynomial is modulo-2 divided by the generator polynomial $g(x)=X^{10}+X^{9}+X^{8}+X^{6}+X^{5}+X^{3}+X+1$ to generate the check sum. The complete block made of the information bits followed by the control bits corresponds to the coefficients of a polynomial that is integrally divisible modulo-2 by the generator polynomial $t(x) = m(x).g(x)$.

The message transmitted on a POCSAG channel is composed of a preamble followed by several batches of codewords. Each batch includes a synchronization word followed by eight frames (of D duration in Figure 11.8), each containing two codewords. The frames are numbered from 0 to 7 and the receivers are divided into eight batches. Once synchronized, each receiver listens (according to its address) to only the frame that it is referenced to and this allows it to save on battery power by remaining off for the rest of the time (i.e., 7/8 of the time on average). The format of a message is represented in Figure 11.9.

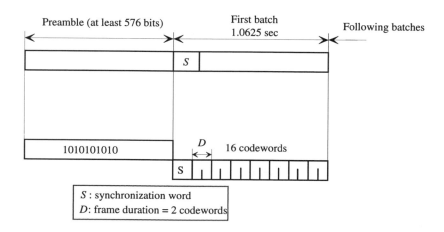

Figure 11.8 Format of a POCSAG frame.

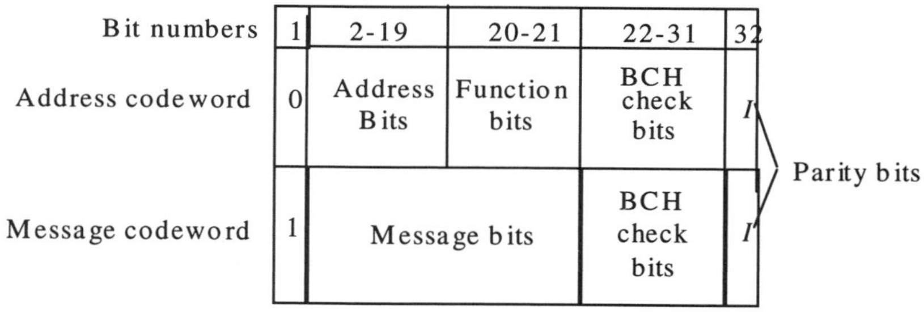

Bit numbers	1	2-19	20-21	22-31	32	
Address codeword	0	Address Bits	Function bits	BCH check bits	I	
						Parity bits
Message codeword	1	Message bits		BCH check bits	I'	

Figure 11.9 Format of address and message codewords.

A receiver listens to a specific frame determined by the three least significant bits from its 21 identity bits. The transmission of a simple codeword triggers a tone signal (beep) at the receiver.

Numeric or alphanumeric messages require the transmission of several successive codewords. These messages follow the pager address codeword and can be of unlimited length with the proviso that a synchronization sequence must be transmitted after each batch of 17 consecutive words. A numeric or alphanumeric message is terminated by an address codeword that may be that of the following message.

The numeric-only format transmits its messages as four-bit binary characters, which include decimal digits, spaces, dashes, parentheses, and the emergency symbol "U." A codeword contains 20 message bits. Five decimal characters can, therefore, be transmitted in each codeword. The transmission of a 15-digit message requires the transmission of three message codewords in addition to the address codeword. If the transmission bit rate is 1,200 bps (as is the case for the Alphapage network) and the transmission of a codeword has a duration of $\tau = 32/1,200 = 26$ ms, then the transmission time of a complete message is $T = \tau.4 \cong 0.1$ sec.

In the alphanumeric format a character is coded as seven bits. A message is divided into batches of 20 adjacent bits. A message of 80 alphanumeric characters requires, therefore, the transmission of about 28 message codewords (a total of 560 bits divided by 20).

11.2.2.2 Example: French Alphapage Network

At the end of 1987, France Telecom installed the Alphapage system. In accordance with the POCSAG standard, Alphapage permits the transmission of alphanumeric messages up to a maximum of 80 characters per message, numeric messages up to 15 digits, or the sending of a simple alert beep [8].

Alphapage uses dedicated frequencies (as is the case in most countries), the frequency band chosen was at 450 MHz for the quality of transmission in urban areas and its propagation characteristics (in-building penetration).

The basic architecture of the network includes the following entities (Figure 11.10):

- A *network controller*. It supervises the following items and monitors alarms in case of equipment failure,
- A *management center*. Its role consists of managing the subscription database and controlling the access subsystem (see below). It also manages the subscriber billing records,
- An *access subsystem*. It collates messages to be sent to pagers from three different sources:
  Rotary or line break dial telephone terminal for transmission to a tone only pager from direct dial in,
  DTMF telephone terminal for the over-dialing of numeric messages after being connected to the server,

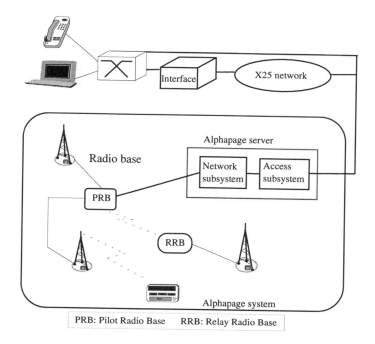

Figure 11.10 General architecture of the Alphapage network.

Computer terminal (Minitel or similar) for the formatting of alphanumeric messages.

The access subsystem receives the updated customer's file which has been downloaded from the management server and verifies the queued access requests before transmission of the messages to the radio subsystem (see below). The access subsystem is connected to the PSTN and an X25 network and is made up of four types of equipment for network access, user dialog, message management, and system interfacing.

The *radio subsystem* (RSS) formats the messages coming from the access subsystem into POCSAG frames and relays them on to the transmitters as frames containing different messages. This means that the RSS contains coding equipment for queuing and formatting the access subsystem messages for the radio interface. After formatting the messages are transmitted to the distribution network, which has two levels:

- The first level are the links between the RSS and the PRBs (Pilot Radio Bases). These connections can be cable, microwave links, or satellite leased lines. The PRBs play the role of a prime concentrator and relay out to the RB (Radio Base).
- The second level is made of microwave links PRB-RB. Each RB receives from its PRB messages to be transmitted to the pagers. Each RB covers one cell and transmits the messages to the mobiles using one of six frequencies located on the 466-MHz band.

11.2.3 Ermes

The *Enhanced Radio MEssage System* (*Ermes*) is a digital standard that has been adopted internationally by many countries in the world. The services offered by Ermes networks, such as international roaming, are much richer than those offered by preceding systems.

The basic services available on an Ermes network are as follows:

- Tone only (at least eight alert addresses),
- Numeric message (20 to 16,000 digits),
- Alphanumeric message (400 to 9,000 characters),
- Transparent data (up to 64 Kbits).

Several types of message addressing available with Ermes include individual calls to each pager, group calls to several pagers using a common RIC (*Radio Identity Code*), group calls to pagers with different RICs, group calls to several pagers using multiple RICs.

Additional available services are message numbering (allows the user to check if all the messages have been received), message storage and retrieval (messages are stored for a certain period), call diversion (messages can be temporarily diverted to another pager), deferred delivery (messages can be transmitted at a preset time), international roaming (user's messages can be routed to one or more visited Ermes networks), closed user groups, and three levels of priority.

11.2.3.1 History of the Ermes Standard

Ermes specifications started their definition in 1987 when ETSI decided there was a need for a pan-European paging system that allowed European roaming, multiple operators in the same country, a higher bit rate than POCSAG, a harmonized set of services, and common pagers in all Ermes networks [9]. To reach these objectives, Ermes has not been developed just as a signaling system, as the POCSAG standard had been, but as a communications system with all necessary equipment and interfaces.

From the end of the 1980s several European operators worked on defining the Ermes standard. From 1988 to 1990 the first work on the technical standard was started within the CEPT committee (CEPT RES4). In 1989 the recommendation CEE Com (89) 166/3 for the introduction of the radiopaging in the EEC was published. On January 24, 1990 a memorandum of understanding was signed by 23 signatories. From 1990 to 1993 ETSI took care of the technical standard development within a technical committee called *Paging Systems* (ETSI TC PS) with three subcommittees that finalized the standardization work. Subsequently, the standard agreed on was an ETSI standard (ETS 300-133). At the end of 1994 the first Ermes services were offered in France.

11.2.3.2 Ermes Architecture and Radio Interface

The Ermes standard defines all aspects of a paging system: services, networks, and radio interface (Figure 11.11). The part related to networks specifies the interfaces between the different network elements, the interface between different networks (to allow subscriber roaming), as well as access methods to the service for message transmission.

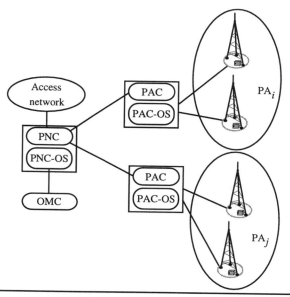

PNC: paging network controller	OMC: operating and maintenance center
PAC: paging area controller	PA: paging area
PNC-OS: PNC-operation system	PAC-OS: PAC-operation system

Figure 11.11 General structure of an ERMES network.

The Ermes standard presents several advantages compared with POCSAG. The 6.25-Kbps bit rate increases the mean capacity by a factor of about 3, compared with other existing networks (by 1.5 for the tone-only service, by 2 for the numeric service, and by more than 4 for the alphanumeric service). The spectral efficiency depends largely on the service type. Ermes spectral efficiency is better for long messages. Also the coding (error correction) and addressing increases the success of long message transmissions (including transparent mode transmission of files).

Receiver frequency agility allows tuning on different frequencies and eases the use of multiple frequencies by the operators (and thus the regulation work) by reducing the requirements for coordinating frequencies between different operators at country borders. And further, terminal numbering offers wide possibilities for users to roam between different networks.

The inclusion of all these functions has disadvantages, however, in the complexity of the standard with consequences on terminal prices, power consumption, and the difficulty of engineering the networks. The general structure of the radio interface [10] is represented in Figure 11.12.

The basic structure of the radio interface consists of a sequence of 60 min divided into 60 cycles each of one minute. Each cycle is broken down into five subcycles of 12 sec, which are then subdivided into 16 batches (denoted A to P). The receiver population is divided into 16 groups (Figure 11.13). Each group is allocated to a batch that is determined by the four least significant bits of the RIC address (Figure 11.14).

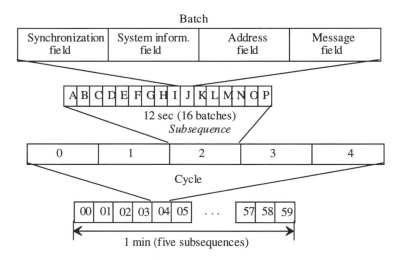

Figure 11.12 Channel structure at the Ermes radio interface.

Channel

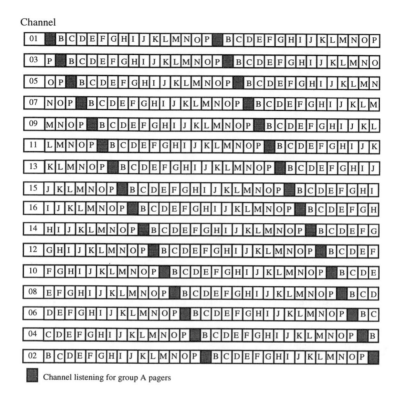

Channel listening for group A pagers

Figure 11.13 Channels synchronization and scanning procedure [10].

Operator identity	Zone code	3 bits
	Country code	7 bits
	Operator code	3 bits
Local address	Initial address	18 bits
	Batch number	4 bits

Figure 11.14 RIC format.

Transmitted messages are composed of several blocks. One block consists of nine codewords interleaved such that each codeword constitutes a matrix of 9 lines and 30 columns (Figure 11.15).

Codeword	Most significant bits			Least significant bits	
1	$C_{1,29}$	$C_{1,28}$	...	$C_{1,1}$	$C_{1,0}$
2	$C_{2,29}$	$C_{2,28}$	...	$C_{2,1}$	$C_{2,0}$
...	...	...	...	...	...
9	$C_{9,29}$	$C_{9,28}$	...	$C_{9,1}$	$C_{9,0}$

Figure 11.15 Codeword reading at transmission.

The bits are transmitted one column after the other; that is, in the order indicated in Figure 11.16.

Each batch starts with a synchronization block including two words each of 30 bits (preamble and synchronization word).

The system information part of a message includes two parts that follow each other:

- SI for *system information*, which includes the country code, the operator code, and the paging area code. It is reproduced in Figure 11.17.

- SSI for *supplementary system information*, which includes the local time and the date.

$C_{1,29}$	$C_{2,29}$	$C_{3,29}$	...	$C_{9,29}$	first,
$C_{1,28}$	$C_{2,28}$	$C_{3,28}$	...	$C_{9,27}$	then,
...	...	...	...	...	then, and
$C_{1,0}$	$C_{2,0}$	$C_{3,0}$	...	$C_{9,0}$	finally.

Figure 11.16 Interleaved codeword transmission.

Country code
Operator code
Paging Area (PA) code
External traffic indicator (ETI)
if = 0: external traffic will not be transmitted in this group
if = 1: external traffic will be transmitted in this group
Border area indicator (BAI)
Frequency subset indicator (FSI)
Cycle number
Subsequence number
Batch number

Figure 11.17 System information part.

The *message part* consists of an integer number of codeblocks. It includes a header of 36 bits, which can be followed by the external operator identity, an additional information part, and message data. The address field of a particular batch must include the initial addresses of the receivers to be called and which belong to this particular type of batch. It has a variable length that can be up to a maximum of 140 codewords. It includes an integer number of initial addresses. In Figures 11.18 and 11.19, the message formats for individual and group calls are reproduced.

The message characters (7 bits for alphanumeric and 4 bits for numeric messages) must be ordered sequentially in the information fields of the message codewords.

The transmitted data is formatted for the radio interface in four steps or levels:

- *Level L4*: information formatting. The coordination between system signaling and user messages is achieved at this level. The data transmitted is ordered according to a pre-defined format recognizable by the pagers.
- *Level L3*: error correction coding.
- *Level L2*: codeword interleaving.
- *Level L1*: modulation.

Ermes can operate in different modes: either in *time division, frequency division* or in *combined time and frequency division modes*. All modes use the same modulation. The frequencies used are between 169.4125 MHz and 169.8125 MHz with a channel separation of 25 kHz. The modulation is 4-PAM/FM.

Local address	22 bits
Message number	5 bits
External bit	1 bit
Additional information indicator (AII) if = 0: no additional information, if = 1: additional information	1 bit
Variable information field (VIF) Additional information type (AIT) Additional information number (AIN)	7 bits = 3 bits + 4 bits

Figure 11.18 Ermes message header format (individual call).

Second step					
Individual local address	Individual message number	External bit (0 or 1)	AII (1)	AIT (111)	AIN (common temporary address number)
Third Step					
Common temporary address	Message number (00000)	External bit (0)	AII (0 or 1)	VIF	

Figure 11.19 Ermes message header formats (group calls).

11.2.3.3 Network Synchronization

Network synchronization must satisfy the three following conditions:

- The radio receivers and transmitters must be tuned to the same channel.
- The transmission of identical signals coming from different transmitters within a unique paging area must be synchronized.
- The transmissions on different channels (e.g., managed by different operators) must be synchronized.

Base station synchronization. The network operates in a quasi-synchronous mode. Several transmitters in the paging area transmit the same information at the same time to improve system coverage. The maximum time difference between transmission of the same signals from different transmitters in the same paging area must not exceed 50 μs but preferably kept below 30 μs (10% of the symbol duration).

Network coordination. For the receivers to operate on different Ermes networks and to easily switch from two different networks it is necessary to coordinate the transmission times on the radio channels. The transmissions from the different networks must be coordinated so that the beginning of transmission for each of their cycles does not vary by more than ± 2 ms compared with the *common time reference* (UTC). That is, the beginning of a network cycle must be within 4 ms for all Ermes networks.

The *border area indicator* (BAI) only has significance for a receiver that is in its home network. When BAI = 1, the receivers operating in their home network must not tune to this channel. When BAI = 0, valid home network traffic is carried on this channel.

11.2.3.4 Identities and Area Splitting

The different identities defined in Ermes are:

- The RIC used by the system on the radio link to identify the receivers for which the message is intended. Its length is 35 bits. It consists of two main parts (Figure 11.14):
  - The operator identity, including the area code (3 bits), the country code (7 bits), and the operator code (3 bits),
  - The local address (22 bits), including the initial address (18 bits) and the batch number (4 bits) [11].

In a time division network transmission on the same frequency for each type of batch must occur at least once per minute in the coverage areas. The transmitters belonging to adjacent paging areas transmit in different subsequences to avoid interference between neighboring areas using the same frequency.

In a frequency division network, adjacent paging areas use different frequencies (Figure 11.20). No time division between the paging areas is implemented. Thus, in its home network a receiver must use different frequencies in different paging areas. When the paging area is adjacent to a network border, one bit of the area border indication is used in one of the system information (SI) partitions. An example of combining frequency and time division is shown on Figure 11.21.

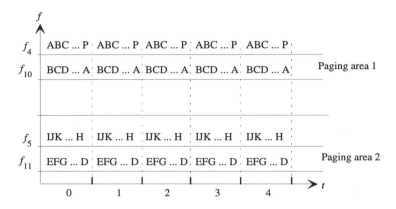

Figure 11.20 Frequency division mode example.

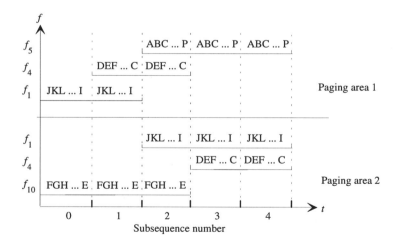

Figure 11.21 Combined frequency-/time-division mode example.

In a multifrequency network the system indicates to the receivers the channel number on which they must synchronize to receive their messages (Figure 11.22). The channel address is obtained by combining the *frequency subset indicator* (FSI) transmitted in the SI and the *frequency subset number* (FSN) stored permanently in the receiver. The FSI is transmitted to all receivers that have an FSN between 0 and 15. Each FSN defines a unique subset of 5 FSIs and each FSI defines a unique subset of FSNs to which messages for transmission must be sent. The FSIs are sent on the downlink. The mobile listens to a channel and detects the FSI included in the SI transmitted in each batch. If the network only uses one frequency, the FSI will be equal to 30; if it uses two frequencies, they will take the values 28 and 29, and so on. By matching the FSI to the FSN stored in its internal memory, the receiver can identify the batch (between *A* and *P*) to listen to and within which it will be informed if a message is sent. FSI/FSN combining allows the use of a dynamic number of channels in each Ermes network.

When the FSI is broadcast on the paging channel it indicates that only messages for pagers whose FSN is included in the FSI subset are transmitted.

When the FSI of the channel is equal to 27, only messages intended for receivers whose FSNs are equal to 12, 13, 14, or 15 are transmitted. Similarly, a receiver whose FSN is equal to 12 will have to obtain its messages from channels that broadcast the values of FSI equal to 12, 22, 27, 29, or 30. Note that the FSN does not directly correspond to the channel frequency.

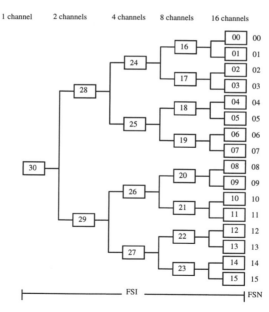

Figure 11.22 FSI subset structure.

Example 3: Receiving a long message by a pager (Figure 11.23)

Let the pager be a member of group *F*. By recognizing its initial address in this group the pager continues to listen to the following batches until it receives a message that is intended for it. This message can be in the batches immediately following the pager's group or in batches separated from *F* by a delay indicated by the system. This latter case occurs when following batches contain messages addressed to their relevant receivers. Part of the message was already included in the message part of the transmitted packets.

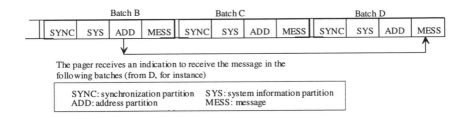

Figure 11.23 Example of long message reception.

11.2.3.5 Call Procedures

The Ermes standard can deal with two types of call. Individual calls, which are transmitted to a single mobile, and group calls, which are transmitted to several mobiles (e.g., in case of pagers belonging to the same company).

The messages can be transmitted to an Ermes pager in several different ways. The choice and the cost of which may vary from one operator to another and from one country to another. Nevertheless, the following access methods are usually available: access via a telephone terminal through a human operator, from a telephone terminal equipped with DTMF, from a packet switched data network (PSDN), from the PSTN, from analog or digital cellular mobiles, from a modem linked to a computer, or finally from an integrated services digital network (ISDN) terminal.

Individual call procedure. Messages transmitted to a receiver are transmitted in its batch and in any batch of the same subsequence or in the batches of following subsequences. Individual calls to mobile subscribers are processed in two steps:

- *First step.* A pager receives an indication of the address partition for its batch. It then starts searching for its address in the current batch and in subsequent batches. If messages are transmitted to pagers with identical initial addresses, or if several messages are transmitted toward a single pager, then the initial address is transmitted only once per message.

- *Second step.* The pager receives an individual message using its local address in the information field of the same batch or in subsequent batches.

Group call procedure. Calls to groups of pagers having individual RICs are processed using the principle of common temporary address (CTA). This method allows a message to be transmitted only once to all receivers in the group. Therefore, the message (the length of which may be large) is transmitted only once to a temporary address previously sent individually to the receivers. This procedure is performed in the three following steps:

First step. Each member of the group is addressed in its own batch and starts searching for its local address in subsequent batches.

Second step. Each member of the group receives an individual message, at its local address or with a full RIC if it is in a visited network, containing an AIN corresponding to one of sixteen common temporary address pointers. Each receiver of the group combines the four bits of the AIN with its initial address to form the CTA.

Third step. The message is transmitted only once with the CTA and with a message number equal to 0000 in the message header. All other parameters in the message must take the form of individual messages.

Transmission of long messages is accomplished by splitting them into submessages. The submessages are transmitted in order and contain the same message number. Each submessage is considered as one entire message.

An example with two receivers belonging respectively to batches *A* and *B* and becoming part of a temporary group is represented in Figure 11.24.

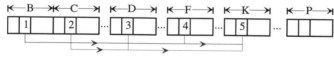

First step:
1: Initial address of the first pager transmitted in the address part of batch B.
2: Initial address of the second pager transmitted in the address part of batch C.

Second step:
3: Complete address of the second pager transmitted in the message part of batch D.
4: Complete address of the first pager transmitted in the message part of batch F.

Third step:
5: Message including CTA transmitted in the message part of batch K.

Figure 11.24 Batch call example.

Battery-saving techniques and radio network planning. These techniques can be implemented at different levels. At the batch level, by allocating a RIC at the high end of the addressable RIC population allows the pager to receive its message immediately after its initial address and then sleep. This is because addressing is always realized in descending order with the initial address transmitted first. A receiver can therefore switch off if it detects an address that is lower than its own. At the subsequence level, battery saving is an implicit function of the protocol because each receiver is addressed only during one of the 16 batches (*A* to *P*) of the subsequence and does not need to switch on at other times.

At the cycle level, battery saving is obtained by making the pager receive signals only for a subset of the five subsequences in a cycle. At the sequence level, it can be implemented so the pager receives only a subset of the 16 cycles of the sequence. Active cycles are determined by comparing the least significant bits of the cycle number with the least significant bits of the initial address of the receiver. The receiver can therefore be active during all cycles, every other cycle, once every 4 cycles, once every 8 cycles, or once every 32 cycles.

Planning constraints of an Ermes network require that:

- The call success rate within and at the boundary of a coverage area must be more than 95%.
- Limits on the differences in propagation times of two signals to the pagers is 30 µs, and thus, the maximum path length difference of two radio signals with amplitudes within 6 dB is equal to 9 km.

The message calling rates for a single-area system are as follows:

- 47.41 calls per sec for tone messages,
- 23.08 calls per sec for 15-digit numeric messages,
- 9.50 calls per sec for 30-character alphanumeric messages,
- 5.33 calls per sec for 80-character alphanumeric messages,
- 3,504 bps for transparent data transmissions.

11.3 CONCLUSIONS

Paging systems have witnessed a significant growth rate over the last few years, particularly with the general public where previously, up to the end of the 1980s, they were more popular among professionals. At the beginning of 1996 there were a little more than 120 million subscribers to one-way paging systems worldwide. Estimates have suggested that the paging industry will grow to 220 million subscribers by the year 2001 and have about an 8% penetration rate by 2008 [12]. The low cost of receivers and subscriptions has encouraged many young people, with limited financial resources, to adopt these systems. The most significant growth rates have been noted in Asian countries where there are double-digit percentage penetration levels. In 1998 the penetration level was more than 30% in Hong Kong and in Malaysia [13]. Competition from cellular systems is likely to swallow the professional user segment of the market. The GSM short message service (SMS), which is a two-way paging service available with cellular, looks a likely candidate to absorb this customer base.

One of the highest speed paging standards in the late 1990s is Ermes. This system offers, just like GSM, international roaming. The first Ermes commercial network was inaugurated in France in October 1994 by Infomobile. Then, the KOBBY service was launched on the Ile-de-France around Paris. Since then six new operators have launched networks, two of which are in France, two in Hungary, one in the Netherlands, and another in Saudi Arabia. By mid-1998, over 48 signatories had joined the MoU in about 30 countries (Europe, Middle East, Asian Pacific Coast) and about 11 countries have opened Ermes networks in Europe and the Middle East.

One of the Ermes development tasks has been to define a set of Chinese, Russian, and Arabic characters. Work is being directed to extend the Ermes standard to allow for message acknowledgment and two-way paging. More recently, a technique has been developed to allow a substantial increase in the number of tone messages (200%) and numeric messages (65%) going through the Ermes network without any impact on the optimization of traffic for alphanumeric messages.

The Ermes main competitor at the end of the 1990s is the FLEX standard developed by Motorola and supported by 50 other companies. The ITU approved FLEX as an international standard at the end of 1997. FLEX is able to cope with several bit rates (1,600 / 3,200 / 6,400 bps) which can be adapted to the traffic intensity and allows lower cost infrastructure in low traffic areas. It can operate on existing POCSAG networks, as can APOC developed by Philips, the other main competitor to Ermes. FLEX has been successful, mainly outside Europe with, at the end of 1997, about 20 million users.

Trends in paging services are towards the development of two-way paging (e.g., for message acknowledgment and greater user confidence) which can reduce the national spectrum requirements as messages need only be broadcast in areas where the pager is located. The REFLEX proprietary standard, an improvement on the FLEX standard, has been defined with this aim.

REFERENCES

[1]   Cox, D., "Wireless Personal Communications: What Is It," *IEEE Personal Communications*, pp. 20–35, Apr. 1995.

[2]   Tridgell, R. H., "Radiopaging and Messaging," *Mobile Information Systems*, Artech House, 1990.

[3]   Parsons, J. D., and J. G. Gardiner, *Mobile Communication Systems*, Halsted Press, 1989.

[4]   Diacre, J. L., and M. de Villepin, "La Radiomessagerie Unilatérale Publique," *Commutation et Transmission*, Numéro Spécial sur les Communications avec les Mobiles, 1993, pp. 85–92.

[5]   Sharpe, A. K., "Paging Systems," *Personal and Mobile Radio Systems*, edited by R.C.V. Macario, P. Peregrinus, 1992.

[6]   Sharpe, A. K., "Paging Systems," *Land Mobile Radio Systems*, P. Peregrinus, Ltd, 1985.

[7]   Remy, J. G., J. Cueugniet, and C. Siben, *Systèmes de Radiocommunications Avec les Mobiles*, Collection CNET-ENST, Eyrolles, 1992.

[8]   Berthelot, P., A. Carrie, J. Matamoros, and M. Pierrugues, "Les Réseaux Partagés: de l'Analogique au Numérique," *Commutation et Transmission* N° 2, pp. 59–69, 1989.

[9]   Tassan, L., "More power to the Paging Elbow," *Mobile Europe*, pp. 31–34, Mar. 1995.

[10]   Ermes Standard, "Radio Equipment and Systems (RES); European Radio Message System (Ermes) Receiver Requirements," DTBR/RES-04001, Nov. 1994.

[11]   Ermes Standard, "Paging Systems (PS); European Radio Message System (ERMES) Part 4: Air Interface Specification," DE/PS-2001-4, July 1992.

[12]   Quigley, P., "Paging: A Big Change in Tone," *Mobile Communications International*, pp. 77–88, Feb. 1998.

[13]   *Mobile Communications International*, July/Aug. 1998.

CHAPTER 12

CELLULAR NETWORKS

Since the early 1990s, the boom in mobile telecommunications has been seen in nearly all countries across the globe. Services are offered by many different types of systems, mainly cordless telephony, paging, private and professional mobile radio, and cellular radiocommunication. Among these, cellular is the one in which the share in this growth process is by far the most important, especially since the advent of international digital standards such as GSM and IS-95. At the end of 1998, Europe had about 70 million cellular subscribers (both analog and digital).

The cellular systems in operation at present are divided into two categories: the first-generation *analog* systems and the second-generation *digital* systems, (Figure 12.1). The next generation of mobile systems, called third-generation systems, relies on cellular techniques (see Chapters 6 and 7) and reuses the basic concepts of architecture, functionality, and services of these systems.

The main differences between the first-generation and the second-generation systems are as follows:

- First, the use of digital modulation in second-generation systems, and hence the origin of their name. The advantages offered by this kind of modulation are robustness of the signal (they can accept a C/I ratio lower than that of analog frequency modulation, while ensuring the same quality of service), a higher capacity (by means of speech compression that increases the number of logical channels for a given bandwidth), and better compatibility with modern digital fixed infrastructure.

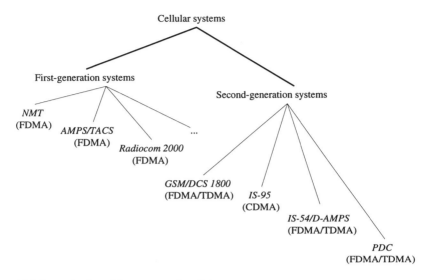

Figure 12.1 Categorization of the current main cellular systems.

- In general, the frequency of operation has increased (800 and 900 MHz, 1.8 and 1.9 GHz) and this has helped implement high-density systems. This is because as the frequency increases the propagation distance decreases. With second-generation systems the size of the cells can be reduced to a few hundred meters, which also allows higher traffic capacity per unit area.
- Particularly in GSM, system interfaces have been standardized and are non-proprietary, which allows an operator to purchase equipment from several manufacturers for various elements of the network.
- Network control is more decentralized. Distribution of the control functions is between the switches, base station controllers, base stations, and mobiles. This is illustrated by mobile-assisted handover (see Chapter 8).
- Management of roaming and handover (e.g., inter-MSC handover) is greatly enhanced thanks to the standardized signaling systems for communication between switches and the databases.
- The use of time-division multiple access (TDMA) (GSM or IS-54/IS-136) or code-division multiple access (CDMA) (IS-95) schemes.

In the first part of this chapter the main first-generation analog systems, which are NMT, AMPS, and Radiocom 2000, are discussed. In the second part, second-generation digital systems, such as GSM, D-AMPS, IS-95, and PDC, are introduced. The main characteristics of these systems, such as mobility management, operational aspects, and security were analyzed in Chapters 5 through 8. More details of the radio interfaces, the system architectures, and some important procedures relating to cellular systems are included here.

12.1 FIRST-GENERATION SYSTEMS

The aim of the first cellular mobile communications systems was to provide a mobile telephony service to a limited number of users moving over a wide area. The low-density traffic defined cells with large radii of several kilometers to several tens of kilometers using a limited number of radio frequency channels.

The first-generation cellular systems are based on a highly centralized architecture that relies heavily on the fixed telephone network (PSTN) infrastructure. The central element in these systems is the mobile services switching center (MSC), which gathers together all the functions of switching and processing of the radio resources (Figure 12.2).

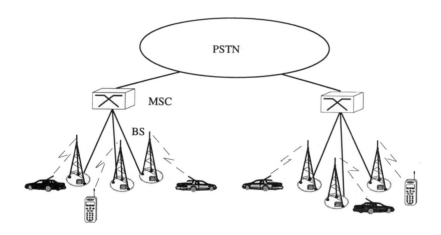

Figure 12.2 First-generation system general architecture.

Radiocom 2000 systems, NMT, and AMPS are typical examples of these types of networks. One of the first cellular systems was AMPS in the United States, opened in the 1970s. It is considered a *de facto* standard in North America. The second most widespread analog cellular standard in the world is the Nordic Mobile Telephony (NMT) system, designed in Scandinavia and implemented in several other countries.

12.1.1 Radiocom 2000 System

Radiocom 2000 is a French system developed in the 1980s by the MATRA Communication company. It is a system that is exclusively operated in France by France Télécom. It was first implemented in 1986.

Radiocom 2000 offers two types of service: the full-duplex cellular service and an enterprise network service where communication takes place in a half-duplex mode as in PAMR systems (See Chapter 9). With the full-duplex telephony service the subscribers are allowed to transmit and receive calls on all or part of the Radiocom 2000 network coverage. Regional or national subscriptions with call restrictions are also offered to users.

With the PAMR service, closed user groups can roam over the entire network. A fleet (or enterprise network) may include a maximum of 118 mobiles organized in subfleets (a maximum of 8 subfleets but with a maximum of 32 mobiles in any subfleet) and with up to 3 dispatcher terminals.

12.1.1.1 Architecture

The Radiocom 2000 network includes two basic entities (Figure 12.3):

- Base station management units (called a UGR, *unité de gestion du relais*), equivalent to a base station controller, manage the resources (e.g., radio channels), control the link protocols with the mobiles, and are capable of controlling several transmitters/receivers via a transmission/reception line interface unit (called a UER, *unité d'émission/réception*). They also manage subscriber location and billing information. A UGR is able to control up to six radio units.
- Radio units (called UR, *unités radio*), equivalent to base stations are connected to local switches (called CAA, *commutateur à autonomie d'acheminement*). A radio unit may be installed locally or remotely. Each local radio unit contains a minimum of 4 and up to 24 duplex channels and looks after a single cell. A remote radio unit can support four to eight radio paths and is connected by five to nine connections respectively to a UGR.

The UGRs are considered by the PSTN as private switches. A UGR is connected directly to a PSTN exchange and simulates a PBX. This connection with the PSTN ensures the establishment of call setups with PSTN subscribers and signaling exchange with other UGRs.

The user subscription data (subscription characteristics and location) are managed by a database located in a particular base station: this one is called the *nominal (home) base station* of the subscriber (it is similar to the HLR database in the GSM). This base station is a specific one selected from all the network base stations. The base stations where the subscriber's data are stored while roaming in their adjacent areas are called *visited base stations* (the same as the VLR databases in GSM). These generate billing information (e.g., date, hour, duration, number dialed) after each call made by the mobile in that area.

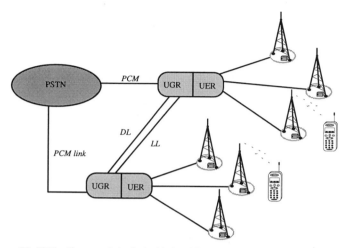

DL: X25 line (data transmission line)   LL: leased line

Figure 12.3 General architecture of a Radiocom 2000 network.

A Radiocom 2000 mobile is identified by a radio number that is only used and coded inside the Radiocom 2000 network and a directory number used by customers to call that mobile. These two numbers may be different.

In Radiocom 2000 high-density (RHD) networks, the organization is slightly different from the one presented above (Figure 12.4). More specifically, the UGRs do not manage subscribers. The structure of the RHD has the following characteristics:

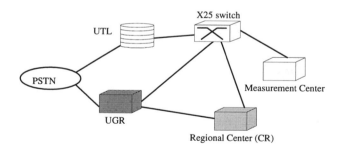

UTL: Unité de traitement de la localisation (location management unit)
UGR: Unité de gestion de relais (base station management unit)

Figure 12.4 RHD network architecture.

- First, at the UGRs, a measurement center (called CM, *centre de mesures*) is installed to analyze the mobiles' signal level.
- Second, a location management unit (called UTL, *unité de traitement de la localisation*) is implemented to manage the home and visitor subscribers' location files for all the UGRs belonging to the same geographical area. The UTL is connected to the PSTN as well as to the UGRs.
- Third, a regional center (called CR, *centre régional*) collects the measurement data about the calls in progress on the base stations in the same area. Its role is to manage handovers. If it detects deterioration in a communication session it triggers a handover, chooses the new cell, and carries out the handover procedure. The regional center is connected to the UGRs and the measurement centers by means of data circuits and leased lines. Finally, an operations and management center centralizes the control of subscriber information and supplementary services (e.g., authorization of new subscribers, barring).

12.1.1.2 Radio Interface

Various networks use the Radiocom 2000 system:

- The high-density network (RHD) opened in 1991, uses the frequency bands (called BDF, *bande de fréquences*) 929 to 933 MHz on the downlink and 884 to 888 MHz on the uplink with a duplex separation of 45 MHz. It is implemented in several large cities.
- Another network covering large metropolitan areas (Paris, Lyon, and Marseille) uses the 150-MHz band.
- Finally, a network, covering the rest of France uses the 414.8 to 418-MHz frequency bands on the uplink and 424.8 to 428-MHz bands for the downlink with a 10-MHz duplex separation. The channel separation is 12.5 kHz.

The FDMA scheme used in Radiocom 2000 splits the frequency bands into a maximum of 256 channels each separated by 12.5 kHz. Each base station has a coverage radius ranging from 20 to 30 km. Note that handover between the various systems is not possible.

The control channel uses a 1,200-bps FFSK (fast frequency shift keying) modulation. During communication, supervision of the connection is ensured by subaudio signaling using 50-bps FFSK modulation (see also Figure 12.5 for R2000 signaling).

Mobile access to the control channel is via slotted-ALOHA. Each mobile synchronizes to the control channel and when required, sends an access request message in one of the defined timeslots. All active mobiles when not engaged in a call remain synchronized to this channel. The call setup time is approximately 1.5 sec.

The signaling messages differ according to whether they are transmitted on a control channel or a traffic channel:

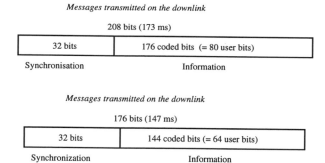

Figure 12.5 Signaling message structure in Radiocom 2000.

- On a control channel, signaling is transmitted in-band. On the downlink, transmission is synchronous and a message contains 10 information bytes. On the uplink, a message contains 8 information bytes. Data integrity is ensured by use of a Hagelbarger code in both directions.
- On a traffic channel, signaling is transmitted sub-band, that is, below 300 Hz. Transmission is synchronous with a message of two parts, each of 10 bits, transmitted in 100 ms.

12.1.1.3 Call Procedures and Location Management

During a full-duplex communication, control of the call is managed by the transmission of signaling frames on the voice channel (in-band transmission).

In the PAMR service, call control can only be performed by the receiver. For that reason, two mechanisms are defined:

- First, one called "anti-talk," which forces the mobile into receive mode after 40 seconds of continuous transmission,
- Then another called "dead man," which causes the mobile to go into transmit mode if it hasn't transmitted anything for 40 seconds.

Call procedures. For an incoming call the local switch (CAA) acts as the access point for the PSTN subscriber and routes the call to the home base station of the paged mobile. The home base station, which knows the location of the subscriber (i.e., the base station to which he has roamed), routes the call to the visited base station by using the Radiocom 2000 internal number of that subscriber.

For an outgoing call, a mobile transmits a message to its current base station, which answers with an acknowledgment (Figure 12.6). In order not to occupy a traffic channel during the call establishment phase, *off air call setup* (OACSU) is used. The network establishes the traffic channel with the calling mobile at the last moment only when the called subscriber has come "*off-hook.*" For this reason the called party may have to wait for a few hundred milliseconds before starting to communicate with the calling mobile.

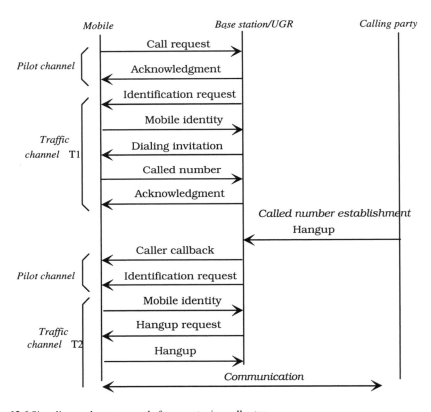

Figure 12.6 Signaling exchange example for an outgoing call setup.

Location management and cell selection. Radiocom 2000 uses a location updating method based on zone changes. A location area is typically equal to the coverage of a base station. The location updating procedure typically takes 10 to 30 sec (Figure 12.7).

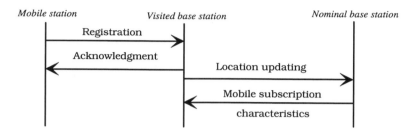

Figure 12.7 Message exchange for a location updating.

During the selection phase, the mobile station, in an idle state, searches for the control channel. It selects the cell where the control channel received signal strength is greater than 20 dBμV. If no control channel satisfies this criterion, the mobile starts looking for a channel where the signal strength is greater than 5 dBμV. Registration is achieved by transmission of a message from the mobile to the network. The corresponding base station acknowledges this message.

The selection of a new cell is triggered if the mobile loses the link with its current base station (i.e., if it has received fewer than n correct frames during the last 10-sec time window).

Handover procedure. In Radiocom 2000, handover can be processed only if the two cells belong to the same regional center. The procedure is centralized at the level of the UGR with the help of a CCC (*centre de changement de cellule* or cell transfer center). During a call, the mobile's current base station periodically measures the radio channel and transmits the measurements up to the CCC. As soon as the quality of a connection falls below a given threshold the CCC sends a command to the neighboring cells for measurement of the mobile signal. According to the results of these measurements and the traffic loads of the various candidate cells, the CCC selects a target cell. It then transmits a handover command to the mobile by indicating the target cell.

12.1.2 AMPS

Bell Laboratories developed AMPS in the early 1970s. In 1978, the first AMPS network was opened in Chicago and full commercial service of this cellular system started a few years later.

The pioneer of AMPS is the Improved Mobile Telephone Service (IMTS). The IMTS is a system that operates in the 150 and 450-MHz frequency bands. The duplex bands are separated by 5 MHz and the channel separation is 25 kHz.

The AMPS radio interface was defined by the American National Standard Institute (ANSI), Electronic Industries Association (EIA), and Telecommunications Industry Association (TIA). In 1982 the Federal Communications Commission (FCC) made specification standardization of the EIA standards. Canada and the United Kingdom adopted the same specifications (with slight differences to accommodate 25-kHz channel separations) in 1983 under the name of Total Access Cellular System (TACS). They were followed by Ireland, Spain, Italy, Malta, Austria, Russia, and Kazakhstan [1] in implementing TACS. AMPS/TACS is characterized by transmission of analog speech and the use of digital signaling. In the United States, the AMPS network includes 700 sites of various coverage. Roaming is possible between the U.S. and Canadian networks.

12.1.2.1 Radio Interface

AMPS uses the 800 to 900-MHz frequency band and contains 666 channels with a channel separation of 30 kHz. When two competing operators cover the same geographical area, the 20-MHz frequency allocation is split into two parts. Each network then has 312 speech channels and 21 control channels.

The duplex frequency split (i.e., the difference between transmit and receive frequency) is 45 MHz. A base station is capable of processing a unique transmission channel (control channel) and usually up to 29 traffic channels. The signaling uses FSK direct digital modulation at either 10 Kbps for AMPS or 8 Kbps for TACS together with Manchester encoding, that is, a transition from zero to one represents an NRZ "1" and a one to zero transition represents an NRZ "0." The signaling messages have from 40 (forward channel) to 48 bits (reverse channel) and use a BCH (40,28 and 48,36 respectively) error correcting code. The signaling message is transmitted five times after the initial bit sync, word sync, and digital color code.

The supervision of a traffic channel (TCH) is achieved by a Supervisory Audio Tone (SAT) signal summed with the speech signal before the modulation of the carrier. Three SATs are used: 5,970, 6,000, and 6,030 Hz. This is a high-pitched tone that acts like a marker. When a mobile tunes to its assigned channel it looks for the right supervisory audio tone. Upon hearing it, the mobile returns the tone to its base station on the reverse voice channel with minimal phase shift. Should it not hear the expected tone then the traffic channel is rejected. This system assists in the prevention of co-channel interference should a mobile be able to hear the same RF channel from a nearby cell due to propagation anomalies. The SAT also has another use and that is of assistance in handovers. The returned phase of the SAT allows calculation of the mobile to base station distance. This distance can be evaluated with an accuracy approaching 100m and can be used by the handover procedure. Mobile power control is performed over a range of 32 dB in 4 dB steps. The mobile power is reduced when close to its current base station to minimize interference caused by intermodulation products at its own base site and also reduces the probability of co-channel interference at distant cells on the same reuse pattern.

Additional signaling between the base station and the mobile is performed by means of a signaling tone (ST) which is equivalent to the data modem sending continuous "1s" or "0s" (10 kHz in AMPS and 8 kHz in TACS). For instance, the ST is used by mobiles for disconnection (1.8 sec of continuous tone), a hookflash (400 ms of continuous tone), the confirmation of a handover request (50 ms of continuous tone), and an alarm (continuous tone).

The identification of the mobiles requires two numbers: the mobile identification number (MIN) constituted of 34 bits representing the ten numbers of the directory and the electronic serial number (ESN) made of 32 equipment identification bits (registered at the manufacturing stage). Moreover, the station class mark (SCM) of the mobile indicates its maximum transmission power.

AMPS discontinuous transmission (DTX) is defined by using two power levels:

- A high DTX mode, where the mobile transmits at a normal power level,
- A low DTX mode, where the mobile transmits with a power level 8 dB below the high DTX power level.

In the first versions of AMPS, DTX was implemented by complete interruption of transmission during the speech blanks and capture problems with co-channel transmitters were found to occur.

Three groups of mobiles are defined in AMPS. Group 1 mobiles are provided with a maximum power of 4W, group 2 mobiles with a maximum power of 1.6W, and group 3 mobiles with a maximum power of 0.6W. In TACS there are 4 classes of mobile specified starting at 10W down through 4W, 1.6W to 0.6W.

12.1.2.2 Architecture and Services

AMPS is based on a centralized structure with mobile service switches called mobile telephone switching office (MTSO) connected to the base stations (Figure 12.8). The MTSO is considered to be a PBX by the PSTN.

The MTSO supplies the switching functions, the control of the base stations, and the interface with the fixed network. The base stations operate the traffic channels and control channels used for call establishment (usually one control channel per cell).

The specification allows a cellular geographical service area (CGSA), to be covered by two operators as indicated in Figure 12.8.

12.1.2.3 Mobility and Call Processing

For the establishment of an outgoing call, the mobile scans the 21 control channel frequencies assigned to its network. It selects the most powerful channel and if it has already registered sends an FSK digital message on the uplink by sending the dialed number as well as its MINs and ESNs.

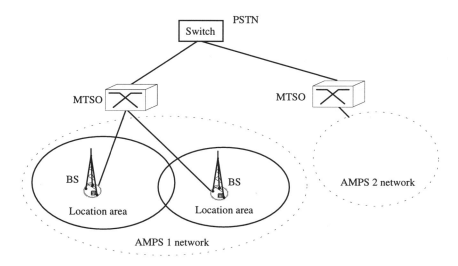

Figure 12.8 AMPS structure.

The exchange of messages during handover in AMPS is represented in Figure 12.9. The principle of the handover in AMPS is described in Chapter 8. Throughout the communication, the network evaluates the distance between the mobile and its base station and the quality and strength of signal. During the handover, the communication is interrupted for 300–400 ms (Figure 12.10).

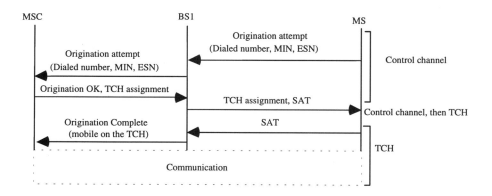

Figure 12.9 Message exchange during the establishment of an outgoing call.

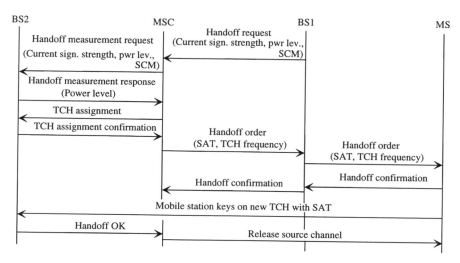

Figure 12.10 Message exchange during a handover in the AMPS system.

Mobility management requires the transmission, by each base station, of system identity (SID) transmitted on the broadcast channel.

12.1.3 NMT System

The NMT system is the result of a common specification developed by the telecommunications departments in the Nordic countries (Norway, Sweden, Finland, and Denmark) at the end of the 1970s. The first NMT systems date back to 1981. Since that time many countries across the world have adopted it.

12.1.3.1 Radio Interface

The NMT system initially used a 2×5 MHz allocation located in the 450-MHz band with the following frequencies for the uplink channel: 453–457.5 MHz; and for the downlink: 463–467.5 MHz. This system is called NMT 450.

The channel separation is 25 kHz with a duplex band separation of 10 MHz in NMT 450 and 45 MHz in NMT 900. Thus, 180 channels are used in NMT 450 and 1,000 channels in NMT 900.

The signaling is binary and uses subcarrier FFSK modulation at 1,200-bps. Similar to Radiocom 2000 a Hagelbarger error-correcting code is used against signal fading. The channel throughput is about 460 bps. The call capacity on the control channel is six calls per second.

Like AMPS, the supervision of communication is carried out by the transmission of an infraband signal SAT that can have four values in the 4-kHz frequency band (3,955, 3,955, 4,015, and 4,045 Hz).

To cope with the increase of traffic in the high-density areas, and like the Radiocom 2000 system, an NMT 900 system using the 900-MHz band has been defined. NMT 900 is an evolution of the 450 standard. More specifically, it includes handover functionality and international roaming. It also uses FFSK modulation. The forward channel is in the 935–960-MHz frequency band with the corresponding uplink 45 MHz below. The bandwidth of each channel meets the specification usually associated with channel separations of 25 kHz. However, by use of cellular planning adjacent channels never appear in the same cell, channels are allocated in accordance with a frequency plan using a 12.5-kHz raster.

12.1.3.2 Network Architecture

NMT is based on a star structure, the center of which is a switching center, called a Mobile Telephone eXchange (MTX) where the apexes are the base stations (Figure 12.11).

The NMT has a two-level hierarchical structure. One part represents the interface with the public network while the other ensures interconnection with base stations through a mobile two-way trunk circuit (MBTC).

The base stations themselves supervise the quality of the radio communication with the help of the SAT pilot tone. This tone is transmitted by the base station to the mobile, which returns it on the uplink. By evaluating the SNR, the base station is able to evaluate transmission quality.

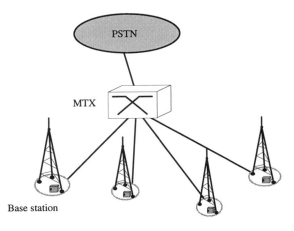

Figure 12.11 General structure of an NMT network.

As is the case with all first-generation cellular systems, the MTX is the heart of the NMT system. It manages the following functions:

- Signaling exchange between the base and mobile stations,
- Processing and exchange of data to and from other MTXs,
- Control of signal level measurements to the base stations and evaluation of the results (handover),
- Switching of the incoming and outgoing calls on the best available base station,
- Control of the calls when the signal level falls below a certain threshold,
- Disconnection of the call when the SNR falls below a predetermined level.

An MTX manages two functions: the *home MTX* for the permanent registration of the profiles of tens of thousands of subscribers (e.g., 50,000) and the *visitor MTX* for the temporary registration of profiles for about 10,000 subscribers located in the area served by the base stations connected to that MTX.

Calling channels and *traffic channels* are defined in the standard. A calling channel is normally used to page mobiles but can be used to carry traffic if all other traffic channels are busy. A calling channel is allocated to each cell and can support about 50 traffic channels. Base stations connected to the same MTX cover a *service area* that can be subdivided into several areas called *traffic areas* (Figure 12.12).

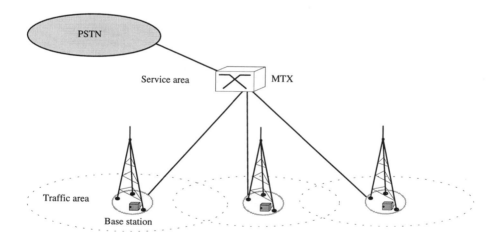

Figure 12.12 Service area and traffic area.

Signal-level measurement carried out by the base station during each communication allows the MTX to control the power of the mobile and process handovers. Four power levels are defined, which allows adjustment of the cell size.

In NMT 900 identification of a subscriber is achieved using a subscriber identity security (SIS) number. Security is strengthened by the use of a three-digit password. This function authenticates the mobile (and therefore the subscriber) and gives protection to the subscriber. Authentication between the MTX and the mobile station is based on the challenge-response method. A secret authentication key (SAK) is implemented in the mobile station and also at the MTX level in the authentication register (AR). In Figure 12.13, we reproduce an example of call setup message flow.

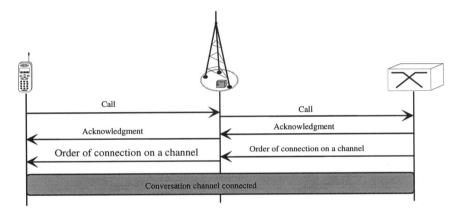

Figure 12.13 Incoming call message flow.

12.1.3.3 Handover

To process a handover, the MTX must measure the following parameters:

- The SNR (estimated by the SAT evaluation mechanism),
- The mobile signal level (allowing the network to determine the location of the mobile according to the various base stations that can receive it).

When the SNR falls below a certain level (30 dB, typically) the base station sends an alarm message to the MTX. The MTX requests the neighboring base stations (from 5 to 16 typically) to measure the signal level of the mobile. These measurements are carried out at 140-ms intervals and reported to the MTX by each of the neighboring base stations. After processing these measurements the MTX selects the best base station to which the mobile should be connected. A command is sent to the mobile station to go to a traffic channel on this new base station. After access and identification of the mobile is confirmed, the call is reestablished. The total handover duration is about 2 sec, with a total disruption of the communication channel of between 300 and 1,700 ms. A handover can only occur between two cells belonging to the same MTX. If the mobile moves out of range of its current MTX, the quality of communication worsens progressively with no chance of being transferred to a new cell.

12.2 SECOND-GENERATION SYSTEMS

In Table 12.1 the main characteristics of second-generation system radio interfaces are reproduced.

Table 12.1

Radio interface characteristics of the second-generation cellular systems

Systems	IS-54 (D-AMPS)	IS-95	GSM	DCS 1800
Multiple access	TDMA/FDMA	CDMA/FDMA	TDMA/FDMA	TDMA/FDMA
Frequency band (MHz):				
Downlink (MHz)	869-894	869-894	935-960	1,710-1,785
Uplink (MHz)	824-849	824-849	890-915	1,805-1,880
Channel spacing:				
Downlink (kHz)	30	1,250	200	200
Uplink (kHz)	30	1,250	200	200
Modulation	π/4 DQPSK	BPSK/QPSK	GMSK	GMSK
Speech coder bit rate (Kbps)	7.95	8 (variable)	13	13
Frame (ms)	40	20	4.615	4.615

12.2.1 The GSM System

The GSM standard initially developed by ETSI within a European context has developed rapidly to become a world standard. A majority of countries over the world have adopted GSM, DCS 1800, or PCS 1900 systems, which are all based on the same original GSM specifications. By the end of 1997 there were 203 commercial networks across the world with about 60 million subscribers using GSM systems.

Initiated at the end of the 1980s, GSM standardization work has been planned over three phases. This allowed the start up of the first GSM networks in the early 1990s while still improving the standard and progressively introducing new enhanced services.

In the following section the main characteristics of the GSM system (Figure 12.14) are briefly described. The basic principles and details of the different interfaces in the GSM standard are dealt with in other documents [2, 3].

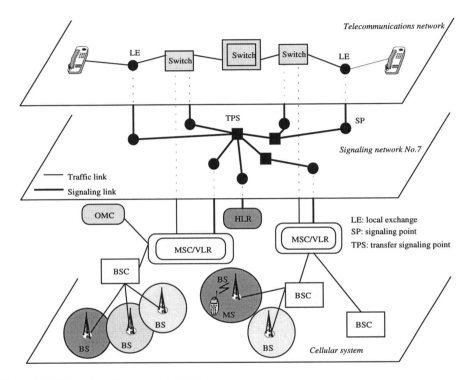

Figure 12.14 General architecture of the GSM.

12.2.1.1 GSM Radio Interface

The frequency bands used by GSM are 890-915 MHz for the uplink and 935-960 MHz for the downlink channels. The frequency carriers are separated on a 200-kHz channel raster to minimize the impact of intersymbol interference. There are 124 carriers available in each direction, in other words, 124 bidirectional physical channels. The bit rate per carrier is about 271 Kbps. The GSM system uses the Gaussian minimum shift keying (GMSK) method of modulation.

The different logical channels (Table 12.2) defined for signaling and traffic are combined into the eight slots on each GSM carrier.

Speech transmission uses a 13-Kbps regular pulse excitation–long-term prediction (RPE-LTP) coder. A speech block is sampled every 20 ms and is coded as 260 bits. With the addition of error protection, the sample increases to 456 bits. To the 50 most significant bits (group 1A) are added three parity bits. These together with the group 1B bits are convolution coded. Four tail bits are then added. The 78 remaining bits (group 2) are left unprotected (Figure 12.15).

Frequency hopping and power control procedures are used optionally. A voice activity detector (VAD) minimizes the power consumed by the mobiles and, as a consequence, the global interference level.

Table 12.2

GSM logical channels

Type of channel	Name	Direction	Use
Traffic (several types):	TCH	↙↗	Speech or data
• Full rate	TCH/F	↙↗	
• Half rate	TCH/H	↙↗	
• 1/8 rate	TCH/8	↙↗	
Signaling:			
• Broadcast channel	BCCH	↙	General information broadcast
• Synchronization channel	SCH	↙	Mobile station time synchronization
• Frequency correction channel	FCH	↙	Mobile station frequency synchronization
Common control channels:	CCCH		
- Access grant channel	AGCH	↙	Resource (channel) allocation
- Paging channel	PCH	↙	Mobile station paging
- Random access channel	RACH	↗	Channel request
Dedicated signaling channels:	DCCH	↙↗	User-network signaling
• Nonassociated channels	SDCCH	↙↗	Location updating, call setup
• Associated channels:	ACCH		
- Slow channel	SACCH	↙↗	Radio measurements report
- Fast channel	FACCH	↙↗	Obtained by TCH frame stealing

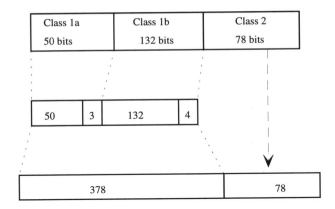

Figure 12.15 Data encoding in GSM.

A TDMA frame is composed of eight slots each of length 577 μs (total 4.615 ms). These frames are assembled into *multiframes* (schematically, 26 frames for a traffic multiframe and 51 for signaling and low-throughput traffic multiframes), then into *superframes* and finally into *hyperframes* (see Chapter 3).

The bursts transmitted on the radio path are subdivided into several groups:

- *Normal bursts* (Figure 12.16): (= two blocks of 57 information bits separated by a training sequence used in the equalization procedure, some head bits and some guard time),
- *Frequency correction bursts* (= 142 bits of all "1s"; i.e., a pure frequency used to adjust the synthesizers of the receivers),

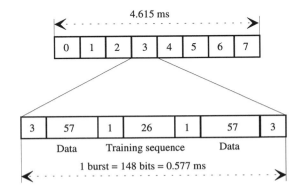

Figure 12.16 Normal burst structure.

- *Synchronization bursts* (= synchronization sequence of 64 bits of data identitifying the base station and containing information about the slots used in the frame),
- *Access bursts* (transmitted by a mobile to access the network).

12.2.1.2 Architecture

A GSM network is composed of several entities: the switches (MSC), the databases (HLR, VLR), the base station controllers (BSC), the base stations transceiver system (BTS), the operating and maintenance centers for the radio and networks subsystems (OMC-R and OMC-N), and the mobiles (MS).

Mobile stations. Three kinds of mobile stations were originally defined. Class 1 has now been deleted from the specification as networks have been rolled out to cater to handportable use and no manufacturer has put such equipment into series production. These are:

- Installed mobile stations (class 1), with a 20W output power (deleted),
- Transportable or vehicle mounted mobiles (class 2), 8W,
- Handheld mobiles (class 3, 5W; class 4, 2W; and class 5 with a 0.8W output).

2W handheld mobiles are currently the most widespread in GSM networks. Their ubiquitous use has compelled operators to achieve high coverage density in areas that have been traditionally reserved for more powerful mobiles (e.g., rural areas, and highways).

Base stations. The BTS provides the radio path between the mobiles and the network, with each BTS covering a cell and managing:

- Radio transmission/reception (e.g., modulation/demodulation, equalization, and interleaving),
- The physical layer (e.g., transmission in TDMA, slow frequency hopping, coding, and ciphering),
- The link access layer (LAPDm),
- The quality and RSSI measurements of the received signals.

Base station controller. The BSC supervises one or several BTSs. The BSCs and BTSs grouped together are the radio subsystem (*base station subsystem* or BSS). This manages the area radio resources; in other words, the allocation of channels, power control of the BTS and mobile station, and handover. It also provides the interfacing functions between the MSC and the BTSs.

A BSC can manage several tens to several hundreds of BTSs, according to the equipment providers.

Mobile services switching center. The MSC supervises one or several BSCs and processes the mobile calls. It is connected to the PSTN by means of a signaling network. It processes calls between mobiles and the fixed network, handovers, interconnection with the fixed network (switching functions). It also looks after visiting mobile stations by means of its VLR. An MSC can include a gateway MSC function (G-MSC) to provide the interface between the GSM network and any other network, generally the PSTN.

Home database. The home database or *Home Location Register* (HLR) contains the subscription characteristics of the subscribers of the GSM network to which they belong. The HLR updates the VLRs roamed to by these mobiles and answers requests from these VLRs using SS7 signaling exchange procedures. A GSM network can contain, according to the number of subscribers, one or several physical HLRs.

The HLR contains user subscription information along with identities, international mobile subscriber identity (IMSI), mobile subscriber ISDN number (MSISDN) (see Section 12.2.1.4), the type of subscription (e.g., restrictions, supplementary services), and location information (identity of the VLR onto which the mobile has roamed). An HLR can store the registration details of several hundreds of thousands of subscribers.

Visitor database. The visitor database or Visitor Location Register (VLR) contains all the necessary characteristics for the management of roaming mobiles that have registered in its service area. It contains, more specifically, data such as their IMSI, MSISDN, TMSI (see Section 12.2.1.4), the kind of subscription, and the location area. It allocates the mobile station roaming number (MSRN) used to direct all incoming calls. Typically, the VLR can store the registrations of several tens of thousands of subscribers.

Operating and maintenance center. The operating and maintenance center (OMC) is used to update, consult, and maintain the NSS/BSS. The OMC is connected to the MSCs and BSCs through an X25 network.

The SIM card. One of the significant innovations in the GSM system, compared with first-generation systems, is the introduction of the SIM card. It allows any GSM subscriber to use any GSM terminal once a SIM card has been inserted. From a network point of view this card allows the user to personalize the terminal (e.g., access to services, routing of calls).

The SIM card is a chip-based smart card which is necessary for a GSM subscriber to have access to the services for which a subscription has been paid. It is inserted in the mobile terminal. The main data stored in the SIM card includes the following: identity of the subscriber, personal password, subscription data (e.g., networks authorized, roaming area), authentication and ciphering algorithms, abbreviated numbers, last number received, and last visited location area. At the operators discretion part of the SIM card may be set aside for storing telephone numbers. This allows the subscriber to transport his personal phone book between terminals.

Only the SIM card allows access to GSM services. A GSM terminal without a SIM card does not allow the user access to the subscribed-to services. Without a SIM card, a GSM terminal could, if the network operators allow it, have access to the emergency services.

In summary, the subscriber will only have access to GSM services with a GSM terminal if he or she has already subscribed to these services, inserted the SIM card in the terminal, and entered a valid password or PIN.

12.2.1.3 GSM Services

GSM includes strengthened security procedures such as authentication, ciphering, and the use of temporary identities (see Chapter 5).

One of the objectives of GSM is to offer services that are compatible with the ISDN; in other words, to offer comparable speech quality, data transmission service, and supplementary services (see Appendix 12A for more details).

With the introduction of ISDN, telecommunications services are split up into two main categories: *teleservices* and *bearer services*. Figure 12.17 summarizes these two kinds of services.

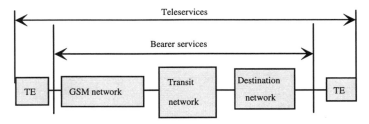

TE: terminal equipment

Figure 12.17 GSM teleservices.

Teleservices. The teleservices are those services that relate to the terminal equipment (e.g., telephone, videotex, and mail). They include radiotelephony, emergency calls, teletex, videotex, access to an X400 mail service, and group 3 fax. Also included is the Short Message Service (SMS). This is point-to-point transmission of data with a maximum of 160 characters to (mobile originated) or from (mobile terminated) a service center and point-to-multipoint (cell broadcast) to all mobiles in a cell.

GSM offers several supplementary services that are applicable to some basic services, and more specifically to the speech service. These include calling line identification presentation (CLIP), call holding, closed user group (CUG); call forwarding with its variants such as call forwarding unconditional (CFU); call forwarding not reachable (CFNR); on busy (CFB); multiparty service; call restriction with several variants, barring of all outgoing calls, barring of outgoing international calls, barring of outgoing international calls except those directed to the home public land mobile network [PLMN] country, and barring of all incoming calls.

GSM bearer services. The bearer services offer transmission capacity between access points in the network. They offer a low-level connection to GSM and offer basic data services. They cannot be used as they are and require upper layer protocols.

The bearer services offered include data transmission at 1,200 bps, 9,600 bps in synchronous or asynchronous modes from the mobile network into the PSTN/PSDN/ISDN, asynchronous circuit switched data from 300 to 9,600 bps, synchronous circuit switched services at 1,200 to 9,600 bps, and PAD access services that allow asynchronous and synchronous access to a packet-switched data networks.

High-speed circuit switched data (HSCSD) and general packet radio service (GPRS) are being offered in GSM Phase 2+. GPRS provides a virtual connection that can exist indefinitely, the radio spectrum being used only for short transmission periods or during data reception. This service offers bit rates that can reach from 70 to 100 Kbps. The main reason for this service is to improve GSM data services, which in phase 2 are unable to offer throughputs higher than 9,600 Kbps. Some applications for this service are reproduced in Table 12.3 [4].

Table 12.3

GPRS applications

Service	Application	Type
World Wide Web	Mobile computing	Transactions
File Transfer Protocol	Mobile computing	Transactions
Electronic mail	Mobile computing	Mail
Telnet	Mobile computing	Bidirectional
Video	Mobile computing	Conference
Financial transactions	Telematics	Tele-action
Emergency system	Telematics	Tele-action
Automatic toll	RTTI (road traffic and transport informatics)	Distribution
Road navigation	RTTI	Distribution
Fleet management	RTTI	Distribution

The HSCSD protocol is optimized for applications that require a high bit rate and where time constraints are important. Initially four concatenated slots can be allocated to offer bit rates up to 38.4 Kbps.

GPRS allows services like videoconferencing, low bit rate video and CD-quality sound. In GSM Phase 2+ the basic new data service is circuit-switched data at 14.4 Kbps.

12.2.1.4 Identities Related to Mobility

The GSM standard defines several equipment identities for the management of subscriber mobility (Figure 12.18).

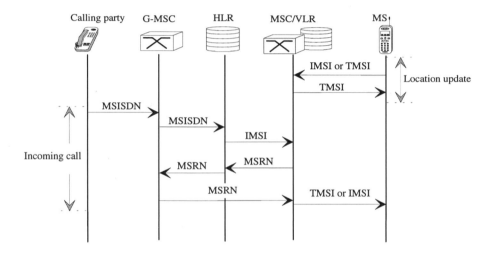

Figure 12.18 Use of the various numbers in GSM.

International mobile subscriber identity (IMSI) is a number that identifies each subscriber, and allows him to be paged (when the TMSI is not available). The IMSI is international and does not vary with time. In order to keep the location of a subscriber confidential, it is rarely used across the radio air interface.

The temporary mobile subscriber identity (TMSI) is a number that varies with movement of the mobile. It is allocated locally (by the VLR) and is not used outside the area managed by that VLR. The TMSI is used in the establishment of a communication to identify the caller or called mobile.

The mobile subscriber ISDN number, also called mobile station ISDN number (MSISDN) is a number that identifies the subscriber to the "external world." In other words, this is the number that is dialed by a person who would like to communicate with a GSM subscriber.

The mobile subscriber roaming number (MSRN) is a number used to route the calls from the home MSC towards the MSC in contact with the mobile station. It is allocated temporarily to the MS by the visited network VLR for the duration of a call.

Figure 12.18 shows the use of these various identities during the location updating and incoming call procedures.

12.2.1.5 GSM Handover

Figure 12.19 shows an example of message flows in the case of an intra-MSC handover.

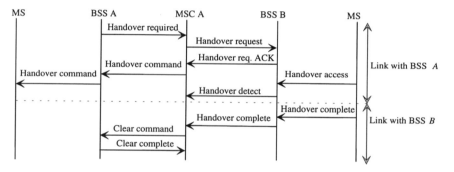

Figure 12.19 Intra-MSC handover procedure in GSM.

12.2.2 The D-AMPS System

The IS-54 North American digital cellular (NADC) or digital AMPS (D-AMPS) as it is also known, TDMA radio interface standard was developed by the EIA/TIA subcommittee TR 45.3 to replace analog AMPS. As is the case with GSM, the multiple-access method combines FDMA and TDMA techniques and defines control and traffic channels. The frequencies used are the same as those in analog AMPS. This allows the migration of analog channels to digital according to demand and in a way which progressively increases network capacity [5]. During initial implementation, dual mode mobiles can use AMPS analog or D-AMPS digital channels.

IS-54 uses DQPSK π/4 modulation with a 48.6 Kbps channel bit rate. The channel spacing is 30 kHz, which gives a good spectral efficiency of 1.62 bps/Hz. The speech coding is VSELP. Speech data are interleaved over two timeslots. For full-rate working it means that 3 users may be supported per RF carrier.

Six 8.1-Kbps channels per carrier are defined. On the downlink channel, the duration of the frame is 40 ms and each slot (6.67-ms duration) carries 312 bits, which are distributed as follows: 260 user bits, 12 signaling bits in a SACCH, 28 synchronization bits, and 12 digital verification color code (DVCC) bits used to identify the frequency of the channel on which the mobile is synchronized. Each frame thus carries 1,944 bits (or 972 symbols).

On the uplink channel, a slot includes a guard space with a six-bit duration, during which time no signal is transmitted (G in Figure 12.20) and six bits of ramp up time (R in Figure 12.20), to allow the transmitter to reach maximum power.

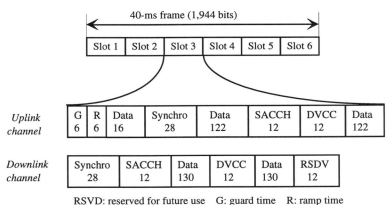

Figure 12.20 Radio interface IS-54 structure.

Full-rate and half-rate channels are defined:

- A full-rate channel consists of two slots per user per frame (slots 1 and 4, 2 and 5, or 3 and 6), which offers a transmission bit rate of 13 Kbps in each direction,
- A half-rate channel, which consists of one slot per user per frame.

The architecture of a D-AMPS network is similar to that of GSM. It contains an HLR, VLRs, MSCs, and base stations. The only standardized interfaces are the radio interface (IS-54) and MSC to MSC network interface (IS-41).

A second phase standard has been developed by the TIA under the name of IS-136. This standard implements digital control channels (instead of sharing the analog network ones as in IS-54) to enable stand-alone digital TDMA handsets. IS-136 has effectively replaced IS-54.

12.2.3 The IS-95 System

In March 1992, the TIA established the TR-45.5 subcommittee with a charter to develop a spread-spectrum digital cellular standard. In July 1993, the TIA approved the CDMA IS-95 standard.

IS-95 is, in the main, a radio interface standard as is IS-54. A cellular network based on the IS-95 radio interface standard can use network standards such as IS-41 and this is commonly used in IS-95 networks. Yet what should be underlined is that IS-95 could equally be implemented on a GSM-like network, where only the radio interface would be different from the one defined in the GSM standard.

12.2.3.1 IS-95 Radio Interface

The channels used are in the 824- to 849-MHz frequency band for the uplink and the 869 to 894-MHz frequency band for the downlink. The channel bandwidth is 1.2288 MHz (which corresponds to about 41 AMPS channels). The uplink and downlink channels are separated by 45 MHz. Accurate power control is essential, which makes this system complex together with the various algorithms that allow for system optimization [5]. The system uses an error-correcting code combining interleaving, voice activity detection, speech coding with variable bit rate, and a RAKE receiver technique to maximize system capacity.

The data are transmitted at 9.6 Kbps with an 8.55 Kbps CELP (IS-96A) speech coder and are protected by an error correction code (Figure 12.21). The data bit rate is segmented into interleaved blocks each of 20 msec and coded with a convolutional code $K = 9$ with a code rate of 1/2 on the uplink (i.e., a gross bit rate of 19.2 Kbps), and at a code rate of 1/3 on the downlink (i.e., a 28.8 Kbps bit rate). The CDMA development group and TIA have also specified a new vocoder called the *enhanced variable rate vocoder* (EVRC) or IS-127 [6].

The data, transmitted at 19.2 Kbps, are modified by means of a long code for ciphering. The modified bits are then coded before transmission using a Walsh orthogonal code of dimension 64 (i.e., producing a bitrate of 1.2288 Mbps).

A code is allocated to the mobile for the duration of the call. After Walsh coding the data are separated into I and Q paths, each one coded by means of a short code. The bit streams generated then feed a QPSK I/Q modulator.

The cells use the same frequency band and the same Walsh codes, but with a time shift of the bit sequence in the two short codes (which are the same, used at all the sites and by all mobile stations). This allows each CDMA channel (i.e., each base station) to be identified in a unique way.

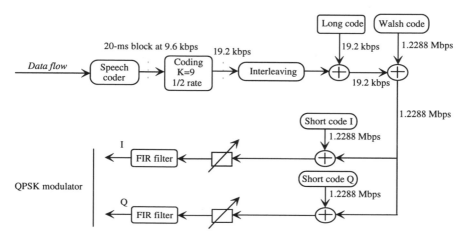

Figure 12.21 Transmission chain (the uplink or forward channel).

Each base station transmits a PN sequence (pilot PN sequence) on a channel (the Pilot channel) and is identified by the offset (or shift) of this sequence. More precisely, in the final stages of the encoding of the radio link from the base station to the mobile, CDMA adds a special pseudorandom code to the signal that repeats itself after a finite time. Base stations in the system identify themselves one from another by transmitting different portions of the code at any given time. In other words, the base stations transmit time-offset versions of the same pseudorandom code. The signal strength of the pilot channel is typically 4 to 6 dB higher than that of the other channels. The general structure of the uplink and downlink channels is shown below.

12.2.3.2 IS-95 Downlink Channel

The IS-95 system defines a forward link combining FDMA and CDMA multiple-access techniques. The PN codes allow a mobile to identify the signals transmitted from various base stations. All the signals transmitted by a base station on a specific CDMA channel use the same PN code. The code allocated to a mobile station by the base station is determined by the terminal number and the subscriber identity (there are 2^{41} different codes).

Four kinds of channel are defined: the pilot channel, the sync channel, the paging channel, and the traffic channel. These channels are further explained in Section 12.2.3.4.

In the example of Figure 12.22, seven paging channels (the maximum number allowed) and 55 traffic channels are shown. It is, nevertheless, possible to replace the paging channels with traffic channels (similarly, a traffic channel could be replaced by a paging channel), up to a maximum number of 63 traffic channels and 0 paging channels.

Interleaving is uniquely implemented at bit level and introduces a 20-ms time delay between the first and last bits of each transmitted frame. This is the only interleaving used in this system.

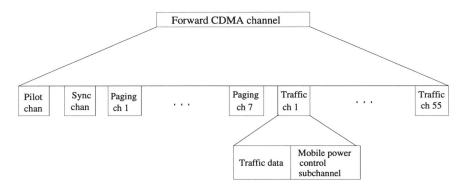

Figure 12.22 Structure of the IS-95 downlink channel.

12.2.3.3 IS-95 Uplink Channel

The uplink channel uses the same PN modulation as the downlink channel. Information is interleaved over 20-ms periods called *frames*. The frames are similar (i.e., have the same length) on the access, paging, and traffic channels. On the sync channel, they have a duration of 26.666 ... ms. Two kinds of channels are defined on the reverse link: access channels and traffic channels.

In Figure 12.23, an example shows the channels received by a base station on a particular sector antenna. Each reverse channel can include up to 62 traffic channels and up to 32 access channels per paging channel.

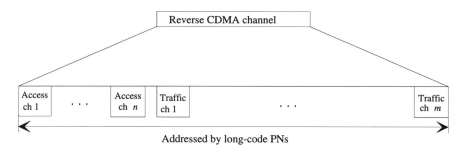

Figure 12.23 Structure of the IS-95 uplink channel.

12.2.3.4 IS-95 Channels

Pilot channel. The pilot channel is a broadcast channel and is transmitted exclusively by base stations. As stressed before, the PN pilot sequence is the same for all base stations; each base station uses a time offset of the PN pilot sequence to identify its forward channel. The pilot channel helps the mobiles recover this offset without the necessity of first decoding the identity of the base station. After acquisition of the code offset by the mobile, this channel is used as the modulation reference for all other signals transmitted by this base station.

Synchronization channel. This channel is used during the system acquisition phase by mobiles attempting to be synchronized to the base station. After acquisition, the mobile does not reuse the synchronization channel until power down. The only message transmitted by the synchronization channel is the *sync channel message*.

Paging channel. After the synchronization phase, the mobile determines its paging channel and starts listening to that channel. The paging channel is selected from all the available paging channels by use of a "hash-coding" technique. A maximum of seven paging channels can be defined per carrier.

The data transmitted on a paging channel are of four types:

- *Overhead.* System parameters (e.g., paging channel configuration, registration parameters), access parameters (access channel configuration and control parameters), a list of neighboring stations, and a list of CDMA channels.
- *Paging.* Paging towards one or several mobile stations.
- *Channel assignment.* Traffic channel allocation message or indication of change of paging channel or an indication of transfer of the mobile station to an analog system.
- *Order.* Other commands such as registration acknowledgment through to transmit restriction orders to individual mobile stations (e.g., *service option request, service option response, registration request, base station challenge confirmation, lock, lock until power-cycled, maintenance order, release, status request*).

The paging channel uses a method involving the use of slots (note: *slots* and *frames* are different from those in TDMA mode). A slot lasts 200 ms and a frame 20 ms (a frame corresponds to the information unit transmitted on the CDMA channel). These slots are able to be used at a rate of 1 every 128 sec up to 1 every 2 sec (the slots are numbered 0 to 639). Messages transmitted to a particular mobile station are only transmitted in its predetermined slot(s). During registration, the mobile indicates the slot or slots that it will listen to from the base station.

The paging channel can have three different bit rates: 2.4, 4.8, or 9.6 Kbps. The paging channels in the same system all have the same bit rate.

Access channel. On this channel mobiles transmit call requests, responses to a page, an order, or a registration message. One or several access channels (32 maximum) are associated with a paging channel. The access channel is slotted.

The base station answers these messages with a message transmitted on the corresponding paging channel. In a similar way, a mobile answers a message on the paging channel by transmitting on one of the corresponding access channels.

A mobile randomly chooses an access channel, from among the free access channels, and simultaneously a PN time alignment, from among the free PN time alignments. It then sends its access packet. If no acknowledgment is received, the mobile retransmits its packet at a slightly higher power level.

Frames and signaling on the traffic channel. Both the forward and reverse traffic channels use the same structure, which consists of 20-ms frames. When a mobile has accessed a traffic channel, the signaling is transmitted on that channel. The frames are transmitted with a bit rate of 9.6, 4.8, 2.4, or 1.2 Kbps. This bit rate can vary from one frame to another: the receiver detects the frame bit rate and processes it accordingly. For example, when a speech burst is to be sent the system immediately transmits at a high bit rate. When there is nothing to be transmitted, the bit rate is reduced. This technique helps minimize levels of interference in the system.

Three kinds of traffic are statistically multiplexed on the traffic channel:

- Primary traffic (speech or data),
- Secondary traffic (one or several data flows),
- Signaling traffic.

Secondary and signaling traffic have a higher priority than primary traffic.
The four kinds of messages transmitted on the traffic channel are:

- Call control messages,
- Handover control messages,
- Power control messages on the *forward channel*,
- Security and authentication messages.

12.2.3.5 IS-95 Network and Control Architecture

The signaling protocols are synchronized at the bit level. On all channels, messages have the same structure. The structure of a message at the highest layer has the following format: *message + stuffing bits.*

The signaling system architecture includes the physical layer, the logical link layer, and process control at level 3 (Figure 12.24). The basic functionality of the mobile station and more particularly call establishment, power control, handover, management, authentication, and registration are controlled through these layers.

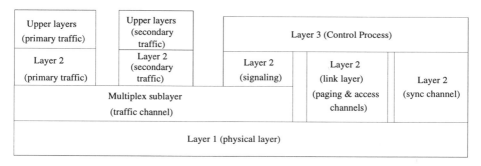

Figure 12.24 IS-95 layer structure.

When a mobile is not in a call the signaling functions are carried by the sync channel, the paging channel, and access channel (on the forward link).

Layer 1 (the physical layer) manages the transmission and reception of frames. It is the same for all system channels. The multiplex sublayer is used by the traffic channel to multiplex and demultiplex the signaling traffic and user traffic. This layer also determines the frame category. There are 14 frame types, differentiated by parameters such as transmission bit rate, the kind of information—signaling or data.

Layer 2 (link layer) handles the transmission and reception of signaling messages. The layer 2 protocols vary slightly on the sync, paging, and access channels, as well as on the traffic channel. Layer 3 (process control layer) processes the messages and looks after control of the mobile station.

12.2.3.6 Some IS-95 Procedures

Registration or location updating. This is the process through which the mobile informs the network of its location. In IS-95, eight kinds of registration are defined:

- Registration when the mobile is powered up or when it enters the system (i.e., comes from another system),
- Power-down de-registration,
- Time based registration,
- Distance based registration,
- Area based registration,

- Parameter change registration,
- Base station ordered registration,
- Implicit registration: when the mobile makes access to the system, the base station deduces its location.

Switching center functions. When macrodiversity techniques are used (as is in soft and softer handovers; see Section 12.2.3.5), signals containing the same information can be received from one or several cells and are forwarded to the switch. The switch directs the mobile traffic to selector equipment. The selector compares the signal quality indicators (the estimated signal quality is the average of the SNR measurements taken during a 20-ms time interval) of the messages received by two or more cells and selects the corresponding bits with the best quality. The same choice is carried out for each incoming frame.

Outgoing call. This procedure is shown in Figure 12.25. Most of the steps are common to all message exchanges. Once the origination message is transmitted on the access channel and the channel assignment message is received from the base station on the associated paging channel traffic can be received on the forward traffic channel. The mobile may now start to transmit on its traffic channel.

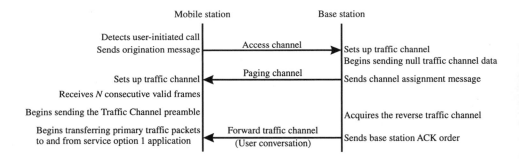

Figure 12.25 Signaling exchange for an outgoing call establishment.

Incoming call (Figure 12.26). Upon receiving a page message on its paging channel, the mobile transmits a page response on the access channel. The base station transmits a channel allocation message on the paging channel together with the traffic on the assigned forward traffic channel. When the subscriber answers (the terminal rings on receipt of the alert message), the reverse traffic channel starts transporting user information.

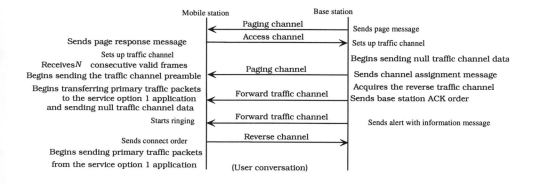

Figure 12.26 Signaling exchange for an incoming call establishment.

Handover procedure. During a call each mobile scans its neighboring cells. If the level of a neighboring pilot channel signal approaches that in the current cell, the mobile station sends a message to the network (Figure 12.27). This message contains the identity of the new cell and indicates the power of the corresponding received signal.

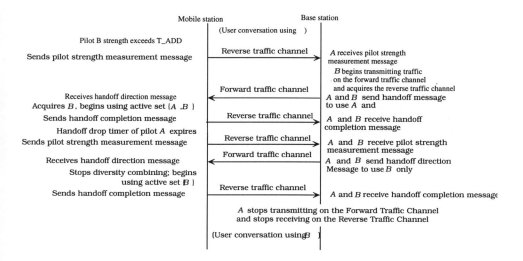

Figure 12.27 Example of a handover signal exchange.

Three types of handover are defined in IS-95:

- *Soft handover* occurs whenever the mobile enters a new cell. The switch triggers the handover by establishing a link to the mobile through the new cell, while maintaining the former link. As long as the mobile station is located in the overlapping area between the two cells, the call transits via both cells, resulting in the elimination of the ping-pong effect. The original cell ceases communication as soon as the mobile station has a good connection in the new cell. What should be highlighted is that soft handover differs from the *softer* handover. This is a handover that occurs between two sectors belonging to the same cell. It is transparent to the switch.

- *Hard handover* occurs when the mobile enters a new group of base stations not related to the previous one; in other words, it has a different frequency allocation plan or frame offsets.

- *Analog handover* occurs when the call is transferred to an analog voice channel (in the case where terminals are operating in dual mode: CDMA and analog).

12.2.4 Personal Digital Cellular System

The personal digital cellular (PDC) system is used in Japan. Development was started by the Japanese MPT in 1989. The specifications were issued in June 1990 and the standard was published in April 1991. The first commercial network using the full-rate system [7] started two years later.

A PDC network consists of the following entities (Figure 12.28):

- Mobile gateway switching centers (MGCs) in control of call routing and interconnection between the PDC network and other networks,
- Mobile communication control centers (MCC), in charge of base station control, including handover and mobile location registration,
- Base stations,
- Home location register (HLR), which includes terminal equipment numbers as well as their current locations,
- Gate location register (GLR), which records transient roaming calls.

12.2.4.1 PDC Radio Interface

In the first phase, an 11.2-Kbps full-rate codec was used, but in 1995 a half-rate codec with a 5.6-Kbps rate and a quality equivalent to that of the full-rate codec was introduced (see Table 12.4 for PDC radio interface characteristics).

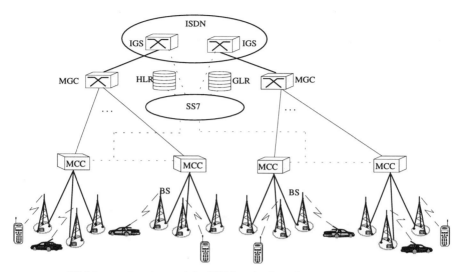

IGS: interconnection gateway switch   HLR: home location register
MGC: mobile gateway switching center   GLR: gate location register
MCC: mobile communication control center

Figure 12.28 PDC network architecture.

In Figure 12.29 the logical structure of the PDC system is shown and in Figure 12.30, its frame structure. As in GSM, the logical channels can be classified into two main groups: the TCH (used to carry user traffic) and the CCH (used to carry system control traffic). The different logical channels defined in the PDC are quite similar to those defined in GSM. These are detailed in the following sections.

The CCH can be further classified into:

- The common access channel (CAC), used to carry signaling information for access management,
- The user-specific channel (USC), dedicated to the control signals of each user during a call. These CCHs are subject to layered signaling protocols. The BCCH, SCCH, and PCH are multiplexed over a superframe structure.

The BCCH is a point-to-multipoint unidirectional control channel that carries operator identification number, restriction information, physical structure of the control channel (e.g., frequency, slot number) for each basestation and maximum transmission power, authorized for the terminals in the cell.

Table 12.4

PDC radio interface characteristics

Specifications	*PDC*
Frequency band (downlink)	810–826 MHz
	1,429–1,453 MHz
Frequency band (uplink)	940–956 MHz
	1,477–1,501 MHz
Symbol rate	21 ksymbol/s (42 Kbps)
Frequency band separation	130 MHz
	48 MHz
Carrier spacing	50 kHz
Modulation	π/4-QPSK
Access	TDMA
Channel/carrier	3 (full rate)
	6 (half rate)
Voice codec	VSELP

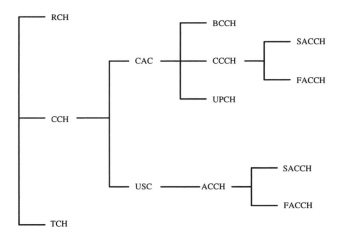

Figure 12.29 PDC logical channel structure.

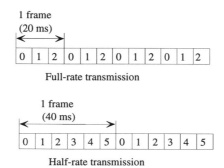

Figure 12.30 PDC frame format.

The PCH is a point-to-multipoint unidirectional control channel used to page terminals.

The SCCH is a point-to-point bidirectional channel used for call establishment requests and responses to these requests. It carries signaling information other than paging messages. Accesses on this channel follow a random access protocol.

The UPCH is a point-to-multipoint bidirectional channel used to transmit user packet data.

The ACCH is a point-to-multipoint bidirectional channel used to carry signaling information during a call and is associated with a TCH. A distinction is made between the SACCH (slow ACCH) and the FACCH (fast ACCH). The SACCH is used to transmit signaling information at a slow bit rate and uses the RCH slot, whereas the FACCH is implemented by stealing frames from the TCH to transmit information at a higher bit rate.

In Figures 12.31 and 12.32 the PDC burst formats are shown:

The RCH (Radio Channel) is a point-to-point nonlayered channel because real-time responses are required on this channel. It is used to transmit the received signal parameters (RSS, BER), power control, and time alignment information.

12.2.4.2 PDC Procedures

In a PDC system each mobile station monitors the RSS level and the BER which are averaged over one superframe period. Each terminal also monitors the RSS level of the neighboring cells during idle timeslots. In Figures 12.33 and 12.34, the message flows for outgoing and incoming calls are shown. Mobile-assisted handover in PDC is carried out using these measurements. The received signal level information is transmitted over the uplink RCH and the base station transmits a power control signal back to each terminal.

G	R	P		SW	No.1	No.2	No.3	Q	G
54	4	48		32	21	21	21	1	78

Uplink

Synchronization burst

R	P	TCH (FACCH)	SW	CC	SF	SACCH (RCH)	TCH (FACCH)	G
4	2	112	20	8	1	15	112	6

TCH, FACCH, SACCH, and RCH bursts

R	P		SW	No.1	No.2	No.3	Q	Post
4	102		32	21	21	21	1	78

Downlink

Synchronization burst

R	P	TCH (FACCH)	SW	CC	SF	SACCH (RCH)	TCH (FACCH)
4	2	112	20	8	1	21	112

TCH, FACCH, SACCH, and RCH slot formats

G: guard time   R: ramp time   Post: postamble   P: preamble   SF: steal flag

TCH: TCH bits   SACCH: SACCH bits   FACCH: FACCH bits   CC: color code

SW: synchronization word   Q: tail bit   RCH: radio-channel housekeeping bits

Figure 12.31 PDC burst format (1).

R	P	CAC	SW	CC	CAC	G
4	48	66	20	8	116	18

Uplink
(first burst)

R	P	CAC	SW	CC	CAC	G
4	2	112	20	8	116	18

Uplink
(second and
latter bursts)

R	P	CAC	SW	CC	CAC	E
4	2	112	20	8	112	22

Downlink

CAC (BCCH, SCCH, and PCH) slot formats

G: guard time   R: ramp time   Post: postamble   P: preamble

SW: synchronization word   CC: color code   E: collision control bits

Figure 12.32 PDC burst format (2).

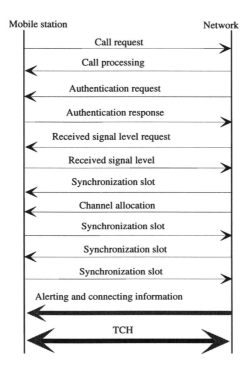

Figure 12.33 Outgoing call in PDC.

Location Registration

When a terminal wants to update its location, it sends a location registration message to the base station. When the base station receives this message, it sends back an authentication request to the terminal. According to the nature of this request, the terminal sends the necessary information back to the base station. When the authentication process has been successful, the BS sends a location registration acknowledgment message to the terminal to indicate that registration is completed.

When location registration is carried out in standby mode, these messages are transmitted on the SCCH (Figure 12.35). During a call these messages are transmitted over the SACCH or the FACCH.

12.2.4.3 PDC Services

The PDC system supports the following services:

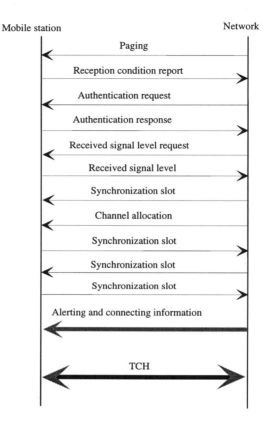

Figure 12.34 Incoming call in PDC.

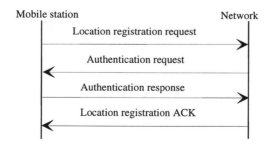

Figure 12.35 Location registration in PDC.

- Voice (full-rate and half-rate codec),
- Supplementary services (e.g., call waiting, voice mail, three-party call, call forwarding),
- Nonvoice data (up to 9.6 Kbps),
- Packet-switched data (PDC-P).

12.3 CONCLUSIONS

In addition to the use of the digital techniques at the radio interface, progress in the development of cellular systems has shown that between the first and second generations, there has been an important decentralization of functionality and a growing independence of the cellular networks from the fixed networks. Effectively, where analog systems were considered as extensions of the fixed network, second-generation cellular systems have been designed to be fully independent networks. This trend will continue with third-generation systems (that are dealt with in the general conclusions of this book). The adoption of more and more sophisticated techniques will help improve cellular network capacity (e.g., a decrease in speech transmission bit rates, use of intelligent antennas). As services become increasingly rich in features and are introduced onto cellular systems (notably, in GSM 2+, where operators are themselves able to develop their own services), they will complicate already complicated systems still further, in particular in the area of increased international roaming functionality.

Thus, two trends characterize the evolution of these networks. First is the radio part, where processing of radio aspects is increasingly sophisticated together with the advent of microcells caused by the need to provide high capacity and to offer a good quality of service; second, a network part, which is becoming more and more complex in its management of logical control (e.g., services and interfunctionality with the intelligent network).

APPENDIX 12A

GSM FUNCTIONALITY IN ITS DIFFERENT PHASES

Phase 1

Phase 1 defines a limited number of services and functionality. The most important are international roaming, SIM card, ciphering, authentication, 13-Kbps-rate speech transmission, short message service, and data services. The main advantages of GSM are its ability to integrate future services and functionalities, for instance, conformity with intelligent networks (INs) and the ISDN as well as terminal innovation.

Phase 2

Phase 2 brought several new services to the user, as well as functional improvements such as the 900-MHz band extension (50 new channels situated just under the GSM frequency band), the introduction of half-rate channels for voice and data (which could allow traffic capacity to double), functions allowing the implementation of microcells (e.g., distinction between slow and fast mobiles), the use of various types of ciphering algorithms (A5-X), improvements at the signaling level, and optimization of phase 1 procedures.

In phase 2, two categories of services are implemented in the network:

- Services with evolution of phase 1:
  - Call forwarding services (CFU, CFB, CFNRy, CFNRc),
  - Call restriction services (BAOC, POIC, POIC-exHC, PAIC, BIC-Roam).
- New services:
  - Call identification services such as:
    Calling line identification presentation (CLIP), which allows the called party to know the identity of the calling line. The identity of the calling user is not presented if the CLIR service is implemented or if the network does not allow

this service.

Calling line identification restriction (CLIR), which allows the calling user to protect his or her identity by deciding not to present it to the called user. It applies to all the services except for short messages.

Connected line identification presentation (COLP).

Connected line identification restriction (COLR).

Call waiting (CW).

Call hold (HOLD), which temporarily disconnects and reconnects a call.

Closed user group (CuG), which defines user groups with different call restriction levels.

Multiparty (MPTY), which allows the calling of up to six subscribers during the same communication.

Advice of charge (AoC), which provides online call charging information with the possibility of association with public telephone applications.

- Phase 2 SIM card
- Phase 2 supplementary services facilitate call management from the user point of view, for example:

Ability to choose the terminal language with menus and short messages displayed on the screen,

Storage of the AoC, control, and security mechanisms,

Introduction of a second PIN code allowing another level of protection and control (e.g., allows a user to be protected from a fraudulent usage of certain services or data),

Storage of supplementary services control information (e.g., procedures for activation, registration),

Reinforced control for short message service parameters.

The SIM card also adds the benefit of extended memory capacity.

Phase 2+

Although phase 2 services are more than enough for a large number of subscribers and network operators, phase 2+ has been planned. This phase deals with items such as:

- Professional-oriented functions: virtual networks (e.g., private numbering plans),
- General improvements for the public networks: new data services, optimized routing, new access methods (e.g., UPT),
- Definition of GSM variants for specific applications (e.g., user groups, geographic areas),
- GSM-specific improvements such as roaming between GSM and DCS networks, new power classes, high speed mobiles and packet radio,
- DECT access to the GSM infrastructure,

- GSM for wireless local loop applications,
- Multiple MSISDNs for the same SIM card (e.g., professional number and private number) to allow different services and charging.

GSM 2+ includes the following supplementary functions [8]:

- Advanced speech call items (ASCI): group and broadcast calls with priorities and preemptions,
- General packet radio service (GPRS): packet access mode for data transmissions,
- High-speed circuit-switched data (HSCSD): allows rates up to 64 Kbps with the allocation of several timeslots to one user,
- Customized applications for mobile network enhanced logic (CAMEL): intelligent network capabilities for specific network operator service provision during roaming,
- Enhanced full-rate (EFR) codec: allows improved voice quality,
- Multiband GSM terminals: 900 MHz, 1.8 GHz, and 1.9 GHz,
- Dual-mode terminals: DECT/GSM.
- Enhanced Data rate for GSM Evolution (EDGE) which envisages higher channel data rates by use of multilevel modulation schemes when close to a base station.

Beside the standardization of these functions there are also the following:

- The development of dual-mode terminals: satellite-GSM proposed by certain satellite PCN operators, such as IRIDIUM,
- Introduction of several radio interfaces, such as the CDMA IS-95 interface.

REFERENCES

[1]   Cellmer, J., "Réseaux cellulaires de radiocommunications mobiles," *Techniques de l'Ingénieur*, E 7 360, June 1994.

[2]   Mouly, M., and M. B., Pautet, "The GSM System for Mobile Communications," published by the authors, 1992.

[3]   Lagrange, X., P., Godlewski, and S., Tabbane, *Réseaux GSM-DCS*, Hermès, France, Sept. 1996.

[4]   Brasche, G., and B., Walke, "Concepts, Services and Protocols of the New GSM Phase 2+ General Packet Radio Service," *IEEE Communications Magazine*, pp. 94–104, Aug. 1997.

[5]   Pahlavan, K., and A. H., Levesque, *Wireless Information Networks*, Wiley and Sons, NYC, 1995.

[6]   Mera, N., and A., McDermott, "The Next Step," *CDMA Spectrum*, pp. 10–12, Mar. 1998.

[7]   Sampei, S., "Applications of Digital Wireless Technologies to Global Wireless Communications," Feher/Prentice Hall, 1997.

[8]   Webb, W., "GSM, UMTS, and the Third-generation," *GSM Quarterly*, pp. 12–16, July 1996.

CHAPTER 13

WIRELESS DATA NETWORKS

Two-way data transmission services in mobile networks rely on a wide variety of systems and technologies. They can be provided by two systems, one that is primarily designed and developed for voice transmission services, and another that is developed and optimized specifically for data transmission services. First- and second-generation cellular and cordless telephony systems may be regarded as among the first group with Ardis and Mobitex in the second. Other systems, such as TETRA and third-generation cellular (UMTS), have always been regarded as carriers of both voice and data.

With GSM, for instance, data transmission mechanisms defined in phases 1 and 2 require communication setup delays of several seconds, delays that do not meet the constraints of bidirectional data exchanges modes. The Short Message Service (SMS) enables the transfer of messages with a maximum length of only 160 characters. This was one of the reasons that the General Packet Radio Service (GPRS) data transmission has been introduced into GSM phase 2+ (see Section 13.2.5).

In this chapter, networks that are specifically designed for data transmission are discussed. These networks differ from those for which data services came as second to voice services. Before dealing with these systems, the characteristics that differentiate data traffic transmission from voice traffic transmission are discussed.

Data traffic is quite different from voice traffic. The characteristics of voice traffic are well understood (e.g., Poisson arrival rate, exponentially distributed duration of bursts and silence) and are, therefore, predictable. Also, voice communications generally last for several tens of seconds, so call setups of several seconds have little impact on the quality of service (with the exception of PMR).

Data transmission traffic characteristics are much more random. This kind of traffic consists of short-duration messages, in the case of transactional modes, that do not require the establishment of a circuit as is the case with voice communications (e.g., reservation of a connection complete with a call set-up phase). Packet transmission is more appropriate to this kind of service than the circuit connected mode as used for mobile voice services. However, circuit switching is more efficient for large file transfers. Packet switching is more efficient and consequently less costly for bursty applications that transmit small amounts of data at every transmission. This fundamental difference is the main reason that has led to the definition and design of systems dedicated to data transmission.

First, applications of data transmission (present or future) are addressed. These can be split into two main groups: applications that predominantly address the professional user and those that are likely to affect the general public. The spectrum of applications addressing professionals is as varied as the number of professional activities. The most common are mobile computing, fleet management, after-sales service, maintenance, measurement, electricity metering, alarm signaling, credit-card checking, security (e.g., police, army), car rental, remote vending machine control, train control, and store inventories [1].

Among the applications which affect the general public are automatic driving assistance systems (e.g., automatic toll collection, road-traffic information, automatic guidance), electronic mail, word processing, database access, and personal digital assistants (PDA) [2].

Two major types can be identified from among dedicated wireless networks offering these kinds of services: the *wireless local area networks* (wireless LAN) covering a few hundred square meters and *wide area networks* which can cover the country.

The wireless LAN is mainly dedicated to applications used on wired local area networks. They are characterized by a larger variety of applications and standards. There are a wide variety of products that are designed to proprietary or standardized specification using both infrared or radio frequency connections. These have witnessed significant development since the early 1990s. These are examined in the first part of this chapter and in particular, the HIPERLAN and IEEE 802.11 standards.

Wide area data transmission networks have similarities with cellular systems, even though they are based on a much simpler structure. In the second part of this chapter the ARDIS system (the oldest), MOBITEX (the most extensive system at the moment), CDPD, and GPRS networks are introduced.

13.1 WIRELESS LOCAL AREA NETWORKS

The *wireless LAN* concept has been in existence since the early 1980s. This kind of network has seen significant development in the United States since the release of the industrial, scientific, and medical (ISM) bands in the mid-1980s. Early products emerged at the beginning of the 1990s.

These systems are characterized by a very large diversity of products and technologies in quite a similar way to first generation residential cordless systems. It is for this reason that standardization processes have been initiated to prompt better harmony between products. Thus, definition of the IEEE 802.11 standard in the United States and HIPERLAN in Europe were started.

If the major objective of wireless LANs is to allow high bit rate communication for low-mobility users over a limited coverage area (building or campus) [3], there are no accurate definitions of the needs and objectives of the wireless LANs concept. As proof of this there is a multitude of diverse products that offer bit rates ranging from hundreds of bits per second to more than 10 Mbps.

The important characteristics for a wireless LAN are, according to [4]:

- Bit rate (main characteristic in the LAN field),
- Protocols (to allow interoperability among networks),
- Coverage,
- Power consumption,
- Safety related aspects (the importance of this being similar to that of other mobile systems).

The most well-known wireless LANs manufacturers are: AT&T, Aironet Wireless Communications, DEC, InfraLAN Technologies, Motorola, Proxim, and Windata [5].

13.1.1 Types of Wireless LAN Systems

From an architectural point of view, two types of wireless LANs can be highlighted (Figure 13.1): a *bus architecture* (similar to an Ethernet-like LAN architecture) and a *star architecture* (where a central station gathers and controls all network traffic).

For the transmission medium the wireless LANs can use either infrared (IR) or radio. The IR wireless LANs are generally dedicated to communication between fixed terminals. In the case of the radio wireless LAN, mobility can be offered in areas such as offices and on campuses. The three main radio frequency bands currently in use are:

- 18 to 19 GHz,
- ISM bands (902–908 MHz, 2.4–2.5 GHz, and 5.8–5.96 GHz),
- 5 GHz.

13.1.2 Classification According to the Technique Used

The air interface can be either IR or radio (Figure 13.2) and the different characteristics of each technology are discussed in the following sections.

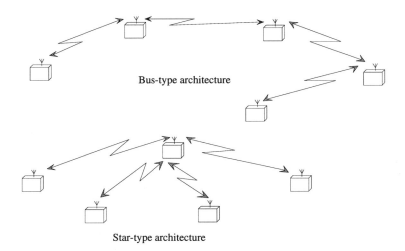

Bus-type architecture

Star-type architecture

Figure 13.1 Different wireless LANs architectures.

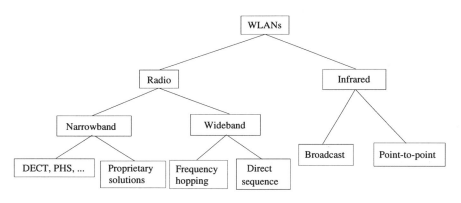

Figure 13.2 Wireless LAN categorization.

13.1.2.1 Infrared

Infrared technology is well suited to data transmission applications where the transmitters and receivers are fixed. The transmission technique is the same as that used by the general public for remote control of electronic equipment (e.g., televisions and videos), resulting in low-cost designs. An IR wireless LAN is also easy to install because it does not interfere with radio systems. Moreover, they only interfere with each other if they are close together.

The main drawbacks are low communication range and their vulnerability to masking from obstacles. In addition, IR signals can be affected by the sun or other sources of IR light (in this case filters can be of use).

Two kinds of systems using IR have been developed: the *directional* and the *broadcast* IR wireless LANs. The directional IR wireless LAN requires the use of directional sources and receivers that have to be lined up. This technique obtains a higher SNR than when the energy is transmitted in all directions. Multipath propagation scattering is lower, bit rates and possible ranges are consequently higher. They are well suited to file transfer applications between central sites and servers. An IR wireless LAN can reach bit rates close to that of a wired Ethernet LAN (in other words, likely to reach 10 Mbps) over distances ranging from 100 to 200m in line-of-sight conditions as long as the transmitter or receiver is not moving.

The broadcast IR wireless LAN has the advantage of being simple to install, as it does not require a direct transmitter/receiver path. The receivers can recover signals coming from multipaths. Yet, they can only offer limited bit rates and coverage. They are well suited to medium-sized networks and short-range communications.

13.1.2.2 Microwaves

The microwave wireless LAN is currently the most common. At these high frequencies signal strengths are significantly attenuated by dense structures (e.g., concrete walls). This means that frequency reuse can be very high.

13.1.2.3 Spread Spectrum

Wireless LANs based on the spread-spectrum technique (see Chapter 3) have been mainly developed in the United States since the release of their ISM bands. These systems are currently the ones that are most commercialized. Direct sequence and fast-frequency-hopping spread-spectrum techniques have both been used and equipment has been implemented on PC (PCMCIA) cards.

Note: A PC card allows the connection between a microcomputer with a PCMCIA port and a mobile terminal (which should be capable of supporting data transmission functions). This card, similar in size to a bankcard, operates like a modem that fits in the PCMCIA port of the laptop.

The advantage of spread-spectrum in wireless LAN applications comes from two interesting features of these systems: first, the possibility for several systems coexisting in the same environment by using a multiple access method such as CDMA, and second, multipath propagation resistance that ensures reliable high bit rate communication. Once the spreading codes have been randomly chosen there is no longer a requirement, as with TDMA and FDMA, to have explicit coordination between the various terminals. Coordination is even less desirable in heterogeneous environments where several pieces of equipment can use different technologies. Nevertheless, the use of power control, which is a necessary mechanism for optimum operation of CDMA, cannot be achieved in the absence of the coordination between systems. This is why the first spread-spectrum systems in the ISM band did not use CDMA even though some products have used several codes simultaneously to increase transmission throughput. When the terminals transmit their bursts quickly, they use all the available spectrum so CDMA band splitting is not adopted. The majority of current systems use the frequency hopping spread-spectrum method. These products fit applications ably designed for small areas, including the case where several terminals are distributed on several floors of a building and served by only one system. Wireless LANs based on spread-spectrum technologies, due to their inherent coding, are clearly more secure from the point of view of confidentiality than their wired network counterparts.

13.1.3 Wireless LANs Applications

The first wireless LAN application is the *extension of the existing cabled LANs*: here, the task consists of extending the coverage capacity of the LAN to reach areas where cabling is difficult to install. This is typical of buildings with large production areas, historic buildings where cabling works are prohibited, and small-sized offices where the maintenance of cabled networks would be too expensive.

The second application is the *interconnection of LANs* located in different buildings. Most of the manufacturers have designed their products to meet this application (adding directional antennas and optimizing the transmission power) to allow them to operate as inter-network bridges. Finally, the third most important application is of supplying data transmission services for mobile users provided with laptops in campus-like areas.

Some wireless LANs applications are listed below [6]:

- LAN in the office environment,
- Medical applications inside hospitals or clinics allowing the transfer of patient data in real time from PC laptops or PDAs,
- Shops and malls for use at point of sale terminals in conjunction with a bar code reader,
- Maintenance in airports or ports helping detect failures quickly and the automation of technical failure diagnoses by expert systems stored in central computers.

In Figure 13.3, the three main groups of services offered by wireless LANs are shown [7].

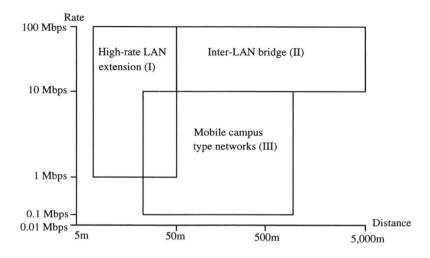

Figure 13.3 Categorization of data transmission networks.

The first category of services is offered for coverage of less than 50m radius. These services have been the center of activity for the standardization bodies and are considered as the most commonplace form of wireless LAN. This category includes the first-generation high bit rate wireless LANs as well as the PC cards used for laptops.

In Table 13.1, some quality-of-service requirements for wireless LANs are reproduced.

13.1.4 The HIPERLAN Standard

Initiated by an ETSI committee as early as 1991, the final approval of the *High Performance Radio Local Area Network* (HIPERLAN) standard [8, 9] occurred in 1996. The high bit rate and low transmission power constraints for short-range communications (10–100m) have prevailed during the definition process of this standard. The major characteristics of the HIPERLAN standard are defined in Table 13.2.

Table 13.1

QoS for Wireless LANs

Application	Bandwidth (Kbps)	Transmit delay (ms)	BER
Voice telephony	13–64	< 140	$< 3.10^{-5}$
High-quality audio	1,400	< 500	$< 3.10^{-5}$
Video telephony	32–2,000	< 100	$< 10^{-7}$
Telefax group 4	64	< 200	$< 10^{-5}$
Television	15–4,000		$< 10^{-10}$
File transfer	64–2,000	> 1,000	$< 10^{-8}$

Table 13.2

Characteristics of the HIPERLAN standard

Transmission frequency/power	5,150–5,300 MHz / 10 mW ... 1W
Receiver sensitivity	50, 60, 70 dBm
Channels	5 (FDMA)
Channel bandwidth	23.5294 MHz
Mobile maximum speed	1.4 m/s (5 km/h)
Functions	- Acknowledgment, CRC
	- Point-to-point link, connectionless
	- Variable packet size
	- Time-bounded transmission services
Modulation	HBR: GMSK
	LBR: FSK
Throughput	HBR: 23.5294 Mbps
	LBR: 1.47060 Mbps
Maximum duration of a burst	1 ms
Error correction (FEC)	(31, 26, 3) BCH-type coding
Error detection	32 bit CRC

HBR: *high bit rate*   LBR: *low bit rate*

The HIPERLAN standard defines a virtual radio subnetwork that could be interconnected to the computer systems in an indoor environment. The terminals can be mobiles with speeds that do not exceed 36 km/h. A handover mechanism is not specified in the standard.

Some possible applications of HIPERLAN, besides the transmission of computer or office data, include the following: teleconferencing; video; medical data transmission (e.g., scanners, cardiograms); maintenance services at train stations, airports, and garages; remote-controlled robots in contaminated sites; and temporary data transmission.

The HIPERLAN standard is based on the OSI layered architecture, part of which is reproduced in Figure 13.4.

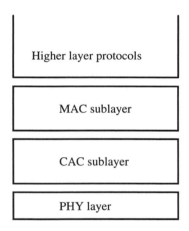

Figure 13.4 HIPERLAN reference model.

The physical layer (PHY) processes the following functions:

- FSK or GMSK modulation and demodulation,
- Bit and burst synchronization,
- Burst transmission and reception,
- Training sequence insertion,
- Error correction coding (FEC),
- Indication of radio signal strength (RSSI).

The channel access control (CAC) sublayer controls communication on the shared channel and the determination of the channel access priorities. It therefore ensures the following functions:

- Random access according to the elimination/yield non-preemptive multiple access (EY-NMPA) procedure (see also Section 13.1.4.2),
- Acknowledgment generation,
- Detection of errors (CRC).

The medium access control (MAC) sublayer main function is to specify the timing of the various packets transmitted and to detect the presence of the HIPERLAN to control access. It therefore ensures the following functions:

- Transmission of Hiperlan MAC service data unit (HMSDU) data blocks,
- Collection of information about neighboring HIPERLAN networks,
- Establishment and maintenance of routing tables,
- Transmission in multihop mode (see Section 13.1.4.3),
- Control of mobile status.

The terminals are identified in a unique manner. To every HIPERLAN network is allocated an identification number (between 0 and 2^{31}) and a 32 character name. These identities are used at the MAC level to identify different HIPERLAN networks.

13.1.4.1 HIPERLAN Radio Interface

The HIPERLAN bandwidth is 150 MHz. It can be located in one of two different frequency bands: 5.15 to 5.30 GHz, the only one considered at the moment by CEPT and 17.1 to 17.2 GHz. Five channels are defined in each band. In the 5-GHz band, the central frequency is located at 5.176468 GHz. Frequencies are separated from one another by 23.5294 MHz referenced to this central frequency. The modulation used is GMSK, which reduces adjacent channel interference and obtains a good efficiency from the amplifiers. The packet error rate (PER) threshold is located below 10^{3}.

The channel coding is of BCH form with an interleaving distance over 16 codewords. The blocks of 416 user bits are coded and result in 496-bit blocks. An error protection code (block error correction code) capable of correcting at least two random errors and grouped errors of fewer than 32 bits is also used. The data packets to be transmitted are made up of, at most, 47 blocks. The equalization mechanism used to combat intersymbol interference uses an equalization sequence of 450 bits in each packet.

The exchanged messages are small (a few hundreds of bytes) and their frequency is at a few messages per second. This allows the supply of asynchronous services with gross bit rates of between 1 to 20 Mbps with limited delays (e.g., for the voice and video services) with user bit rates ranging from 64 Kbps to 2,048 Kbps. The range of coverage of a HIPERLAN station is up to 50m if the bit rate is 20 Mbps and up to 800m if the bit rate is 1 Mbps. The service area is determined by the range of the direct path. For greater coverage, terminal groups can be interconnected by a wired system.

The general format of a HIPERLAN packet is reproduced in Figure 13.5.

Content	Octet
Size indicator (= n)	2
	1
Packet type	3
Data	4 - n

Figure 13.5 HIPERLAN packet format.

The data field contains information such as destination, source, and residual time (see definition in Section 13.1.5.2).

The standard defines two kinds of burst: *access bursts* and *data bursts*. The access burst consists of a predetermined sequence of 32 bits. The data bursts can be subdivided into two groups (Figure 13.6):

- Low bit rate-high bit rate (LBR-HBR) data bursts, which include a low bit rate part, and a high bit rate part,
- LBR data bursts with only a low bit rate.

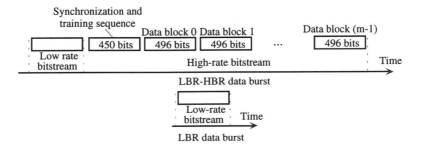

Figure 13.6 Different kinds of bursts in the HIPERLAN standard.

13.1.4.2 Channel Access Mechanism

At the MAC level, there are two kinds of address (one address consists of 48 bits):

- Individual addresses with only one service access point (SAP) and used as a source address and as a destination address,
- Group address identifying a SAP group, used as a destination address only.

The MAC protocol is based on a channel-sensing mechanism. If the channel is detected as free for a given period (1,700 bits) the terminal may transmit immediately. If not, channel access follows three phases: *priority determination, elimination,* and *yield* (Figure 13.7).

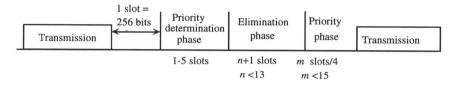

Figure 13.7 Access phases in the HIPERLAN standard.

During the priority phase only the highest priority stations can be in contention: a station with priority p transmits a burst in the slot $(p+1)$ if it hasn't heard a higher priority burst (the highest priority is 0). As soon as a burst is transmitted, the priority phase is suspended and the elimination phase starts. During this phase, the stations that have transmitted a burst during the first phase attempt to gain access to the channel by each transmitting a burst. The duration of this burst is such that the probability of obtaining a burst of duration equal to n_e slots is given by the formula:

$$P_e(n_e) = p_e^n e.(1 \quad p_e) \text{ if } 0 \leq n_e < n$$

and

$$P_e(n_e) = p_e^n \text{ if } n_e = n.$$

After the transmission of the burst the station listens to the channel. If it detects another burst it stops its access process. Thus, only the station that has transmitted the longest burst is able to resume the access procedure to the channel. There is a possibility that more than one station may transmit bursts of the same length. At the end of this phase (the duration of which is equal to the longest burst), at least one terminal, and possibly more, is authorized to resume the access procedure.

The remaining competing stations enter a yield phase. During this phase, each contending station reports, while listening to the channel, its presence by transmitting a burst for a random time. This random time is calculated in accordance with the following formula:

$$P_y(n_a) = 1/(m+1) \qquad \text{for } 0 \le n_a \le m.$$

If a station detects a transmission other than its own, it withdraws from contention. Only the last station (i.e. the one which transmitted the longest) is finally authorized to transmit. Because of this decentralized mechanism, the collision rate decreases to less than 3%.

Example: Channel access contention
Suppose there are four stations A, B, C, and D as shown in Figure 13.8. Stations A, B, and C have priority 2 and station D has priority 4. During the priority phase, A, B, and C transmit a packet in slot 3. Station D may not transmit before slot 5. Because station D has heard packets from the other stations on the channel it suspends its access.

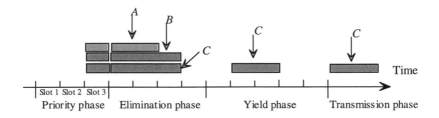

Figure 13.8 Example of channel access contention.

During the elimination phase, station A transmits a burst of 3 slots duration and stations B and C transmit a burst that lasts 4 slots. After hearing signals on the channel from stations B and C, A immediately stops its access attempt. During the yield phase B waits for a random time of 4 slots and C waits only 1 slot. Station B, when it senses C's transmission, suspends its access attempt and leaves the channel free for C.

HIPERLAN defines a quality of service (QoS) for packet transmission. The QoS is qualified by the use of three parameters, the initial values of which are defined by the application. The quality of service is measured by the three following parameters:

- *User priority* (range between 0, the highest priority, and 3),
- *Message lifetime* (range between 0 and 1,600 ms) specifies the maximum time between the message transmission request at the source and message receipt at the receiving entity,

- *Residual lifetime* (range between 0 and 1,600 ms) specifies how much time is left after transmission between two MAC points.

As a multihop technique is used (see definition in Section 13.1.4.3), the values of the parameters for message and packet residual lifetimes are included in the transmitted packet. Packets that cannot be delivered in time are deleted.

13.1.4.3 Message Transmission

The MAC protocol uses a multihop technique that extends the range of communication beyond the maximum range of a single station. Multihop transmission means that a packet is transmitted between two remote stations A and B by being relayed through stations located between A and B. For this, each station entity is declared as a *forwarder* or *nonforwarder* (Figure 13.9). A nonforwarder never retransmits a message that it receives. A forwarder retransmits the messages received if necessary.

Each message contains routing information under the name of $\{R_{Dest}, R_{Next}, R_{Dist}\}$, where R_{Dest} is the destination address located at R_{Dist} hops from the current terminal and can be reached via a terminal with the address R_{Next}. For this reason, each terminal is required to store information about each connection with its neighbors in its neighbor information database $\{N_{Nbor}, N_{Status}\}$, where N_{Nbor} is the address (or identity) of the next node and N_{Status} is the state of the node N_{Nbor} at the N_{Status} state. The N_{Status} parameter can take three possible values, which indicate:

- Either the considered terminal has an asymmetric connection with its neighboring terminal (it can only receive *or* transmit messages addressed to its neighbor), or
- The connection is symmetrical (the two stations can exchange messages in a bidirectional manner), or
- The connection is symmetrical and the neighboring terminal has been used as a multipoint relay.

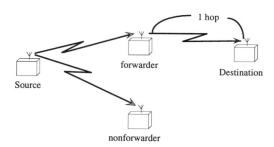

Figure 13.9 Forwarder and nonforwarder concept.

A terminal also processes a *hello information base* provided by its neighboring entities {H_{Dest}, H_{Status}, H_{Next}} specifying that the terminal with the address H_{Dest} and is in state H_{Status} can be connected with an H_{Next} neighboring address terminal.

13.1.4.4 Power-Saving Mechanism

Some functions and protocols are defined to allow terminals to save power when in idle mode. For example, a station called *p-saver* announces that it will move to a listening mode from standby periodically and only for a short period of time. Other stations that would like to send it a message, called *p-supporters*, therefore, can only send their messages to this station in that time interval.

Stations that relay transmissions can also announce the time that they will transmit, which allows the other stations to go into standby for the rest of the time.

An extra mechanism defines the two-speed transmission mode. Packets include a short header transmitted at the low bit rate of 1.4706 Mbps, which contains sufficient information to enable the station to decide whether it should listen to the rest of the packet.

13.1.5 The 802.11 Standard

The work of the 802.11 committee of the Institute of Electrical and Electronics Engineers (IEEE) started in 1990 with the objective of defining wireless standards to allow inter-operability between products from different manufacturers. Three projects were introduced: one based on infrared technology and the remaining two looking into direct sequence and frequency-hopping spread-spectrum techniques. Radio technology was used in the first studies.

The standard was finalized in early 1997. The specifications describe basic configuration, transmission procedures, bit rate constraints, and transmission range characteristics. The IEEE 802.11 standard is primarily aimed at PDAs and laptop computer applications.

13.1.5.1 Radio Interface

The three groups of physical radio interfaces upon which the 802.11 protocol are implemented are the following:

- 2.45-GHz ISM band frequency-hopping spread spectrum (with hopping rates greater than 2.5 hops per second),
- 2.4-GHz ISM band direct sequence spread spectrum,
- Infrared.

Where the radio interface is used, it is characterized by the following major parameters:

- Bandwidth: 2,400 to 2,482 MHz with channels separated by 1 MHz,
- Two level GFSK modulation for the 1-Mbps bit rate (99% of the energy concentrated within the 1-MHz band),
- Four level GFSK modulation for a 2-Mbps bit rate,
- Receiver sensitivity better than -80 dBm at 1 Mbps and 75 dBm at 2 Mbps, for a BER lower than 10^5.

In the case of direct sequence spread-spectrum, 11 central frequencies are defined, but only three of these channels can be used.

Because fast frequency hopping spread-spectrum systems offer significantly more channels (i.e., frequency hopping patterns), than a system with direct sequence (i.e., using PN codes), they are more useful in dense environments where cells overlap and with each cell having several adjacent cells. Both interfaces require transmitter powers of at least 100 mW in order to reach ranges that can be as great as 100m indoors. These ranges vary according to the desired bit rates as well as the layout and design of the building.

13.1.5.2 802.11 Topologies and Link Sublayers

Two basic topologies have been defined (Figure 13.10):

- The *bus topology*, where stations have access to the backbone network (called the distribution system) via access points (denoted AP),

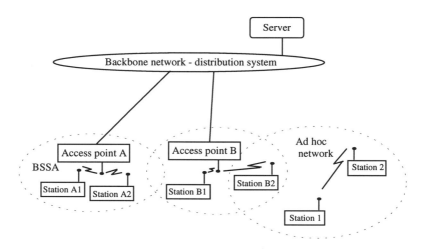

Figure 13.10 Possible topologies with the 802.11 standard.

- The *meshed network topology*, where stations communicate between themselves over to an ad hoc network, regardless of any infrastructure or base station.

Bus topology. This kind of topology is used to cover buildings or a campus. It consists of deploying several APs, the coverage areas of which overlap each other to provide complete coverage. Stations are associated with a given access point, a basic service set (BSS), or other members of the AP.

Meshed network topology. An ad hoc network is a multihop radio network in which the mobiles communicate between themselves directly or by one or several intermediate mobiles. This is necessary because of the absence of a backbone network or any other form of centralized network management system. Thus, as paths change quickly because of the movement of mobile terminals and intermediate mobiles, the routing algorithms of ad hoc networks also have to react quickly to the topology changes [11]. This network structure presents another advantage which is the ability to set up a temporary network very easily (e.g., at a conference hall or an exhibition room).

Level 2 protocols. The MAC protocol allows these two topologies to coexist and be interconnected on the same site.

The data link control (DLC) layer or layer 2 is divided into three sublayers (Figure 13.11):

- 802.2, control layer,
- 802.10, security and integrity layer,
- 802.11, medium (wireless) access control layer.

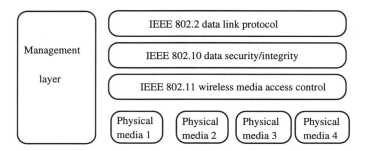

Figure 13.11 Protocol layers in the 802.11 standard.

13.1.5.3 Random Access

The MAC layer, as defined in the IEEE 802.11 specification, includes some basic functions and several optional functions. Its main characteristics are support of access point oriented topologies or ad hoc networks, synchronous or real-time traffic (for time-limited services), and power control.

The first access method defined is the distributed coordination function (DCF) derived from the CSMA/CA family (Figures 13.12 and 13.13). Because radio access does not allow the direct use of the CSMA/CD protocol, one of its variants has been adapted to the mobile radio environment.

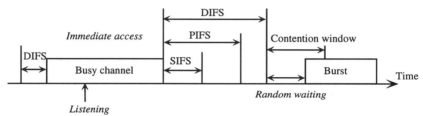

Figure 13.12 Access mechanism in the 802.11 standard.

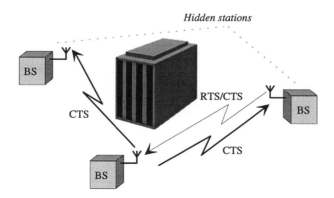

Figure 13.13 CSMA protocol adapted to wireless LANs.

The CSMA/CA protocol is designed in such a way as to minimize the probability of collisions. To facilitate this, the waiting time between channel release and transmission by a contending station can take one of three different values. This duration, called the InterFrame Space (IFS), depends on the priority of the contending station.

- If a station has a high priority, it waits for a short period: a short IFS (SIFS), which is also the case for acknowledgments.
- If a station has a medium priority, it waits for a point coordination function IFS (PIFS) period. This priority level allows a station with short lifetime bursts to transmit them ahead of asynchronous bursts.
- If the station has a low priority, it will wait for a distributed coordination function IFS (DIFS) period.

In addition the CSMA/CA protocol uses a random waiting period, after the contention period, to reduce the collision probability. This waiting time follows a uniform distribution (in the number of slots) and its maximum duration is the contention window (CW). The station doubles the CW value after each unsuccessful access (up to a maximum limit).
The access protocol operates in the following manner for each station likely to want channel access:

- It listens before to the channel state to determine if it is idle or busy.
- If the channel is idle for a period longer than DIFS, it starts transmitting.
- If not, the station attempts its transmission after a time that is equal to DIFS plus a random waiting period.

The duration of this last period is calculated using the following formula:

int(CW[ ]*rand[ ])*slot time

Where *CW[ ]* is an integer between $CW_{min} = 7$ and $CW_{max} = 255$, *rand[ ]* is a random number between 0 and 1 and the *slot time* is equal to the minimum duration for determination of the channel state. This duration is equal to the duration required by the modem to move from the receive mode to transmit, or conversely, plus the radio propagation duration. There are seven *CW* values between (CW_{min}), 15, 31, 63, 127, and 255 (CW_{max}).
To reduce the collision probability after each unsuccessful transmission the *CW* takes the next value in the series until it reaches CW_{max}. The *CW* remains equal to CW_{max} for following attempts. Immediate positive acknowledgments are used to indicate the successful receipt of each message. The receiver transmits an acknowledgment after a time SIFS, which is shorter than a DIFS, immediately upon receipt of the message. The acknowledgment is transmitted without the receiver needing to listen to the channel beforehand. If an acknowledgment is not received the message is supposed lost and a re-

transmission is scheduled. To overcome the problem of hidden stations (stations which cannot be heard by all other stations) the 802.11 protocol incorporates the use of control bursts:

- Request to send (RTS). A potential transmitter sends this to a receiver.
- Clear to send (CTS). It is sent by a potential receiver to authorize the transmission in response to an RTS message. Because of the additional signaling overhead, this mechanism is not used for short-burst transmissions.

The access procedure using control frames is as follows:

- A station (*A*) wishing to transmit sends a short RTS message.
- The receiver (*B*) sends a CTS message only if the channel is not busy.
- Station (*A*) starts transmitting only after it has received the CTS message.

The CTS message is assumed to have been heard by all other stations likely to interfere with reception, which solves the hidden station's problem.

13.1.5.4 Power Saving

Most mobile stations have limited battery capacity. For this reason, an optional mechanism for power saving is included in the MAC protocol. When a station is in the power-saving mode it can neither receive nor transmit.

Some different procedures are used in the two groups of topology. For example, with a backbone based system a station listens periodically to receive messages transmitted by the access point. If it receives a control burst indicating that the access point has data addressed for it, it acknowledges by sending a burst indicating that it is ready to receive data.

13.1.6 Conclusion

Although the spectrum of wireless LANs applications is very wide, wireless LAN products have seen the majority of their growth in vertical markets such as retail stores, warehouses, and manufacturing plants. This is where salespeople and other workers roam with portables and handheld devices and enter inventory or sales data, or call up data from central databases. It should be noted that the current situation shows that wireless LANs, in the majority of cases, have not replaced wired LANs. Usually, they are being used to create wireless extensions to wired LANs.

13.2 WIDE AREA WIRELESS DATA NETWORKS

This section introduces wide area wireless data networks, as opposed to other radio mobile systems such as PMR/PAMR, cellular, and cordless that were presented in previous chapters and which are capable of transmitting both voice and data.

As with PMR networks (see Chapter 9), which are mainly dedicated to voice services, some wide area wireless data networks are based on dedicated channel techniques, such as the *Advanced Radio Data Information System* (ARDIS), and others are based on trunking techniques. Wide area wireless data networks serve fast moving mobiles, and offer significant communication ranges but with a relatively low transmission bit rate [3]. Their growth has been far less significant than bidirectional voice systems even though they have been deployed in several countries over the last few years.

13.2.1 Types of Systems and Evolution

As wide area wireless data networks are optimized for the transmission of data over a radio interface their performance, from the point of view of capacity, is better than other types of paging networks (e.g., cellular systems). In addition they offer better transmission quality. The datagram method of operating liberates the user from connection establishment procedures, as required in circuit mode connections, and allows the radio terminal to access the radio channel only when data needs to be transmitted.

Wide area wireless data networks use specific protocols to minimize connection setup time and guarantee data transport. For example, in the case when a receiver cannot be found a mailbox stores the received messages either at the center or at the base station on which the mobile is registered. Generally, the terminals consist of a laptop computer equipped with a radio modem and communications software. Connection from the mobile network to servers via fixed data networks enables remote data access, updating, or processing, for example.

First-generation systems offer limited transmission bit rates. For example, they are 4.8 Kbps for ARDIS and of 1,200 bps for Mobitex. Base station transmission power levels are significant (several tens of watts), which can cover considerable areas with relatively few base stations. The low bit rates and limited capacity of these first networks led the operators to adopt high tariffs, which have been for a long time one of the reasons why these networks have not witnessed growth similar to other kinds of mobile radio systems.

The ARDIS and CDPD systems have improved to reach bit rates of 19.2 Kbps. A reduction in cell sizes with the implementation of microcells has improved the capacity of these systems. Microcellular networks have been deployed to serve those users who move very little, or are fixed.

In the main, wide area wireless data networks can be considered as extensions of fixed data networks (e.g., X25 PSDN) outside the walls of a company. For this reason, the service provided should be similar to the one offered by the fixed networks. The most well-known systems at the moment are ARDIS, Mobitex, CDPD, and GPRS. These are introduced in the remainder of this chapter.

13.2.2 ARDIS

ARDIS is the forerunner of mobile data transmission networks in the United States. It offers a bidirectional data transmission service. Its origin goes back to the early 1980s when IBM decided to supply its traveling engineers and technicians (about 20,000) with a means of communication allowing them, for example, to order spare parts, and to transmit service reports at a distance. About ten central computers distributed over the country could be accessed with the help of this system. At that time, the network was called digital communication system (DCS).

Motorola, which was the supplier to IBM of the radio network equipment, decided to deploy its own network and to offer services to other companies. In 1990, IBM and Motorola merged their networks and set up ARDIS, which was opened, as a commercial venture, to all companies interested in data transmission business services. In 1995, Motorola acquired 100% ownership of ARDIS and in 1998, the American Mobile Satellite Corporation bought ARDIS.

This network amounted, at the end of the 1990s, to around 1,400 base stations in the United States which covered around 400 metropolitan areas and served around 40,000 users. The first external users of ARDIS were the police—for vehicle license plate checking—and taxicab companies. Today, there are many ARDIS applications: emergency use (e.g., the earthquake of Los Angeles in 1994), insurance reporting, financial services (e.g., home-delivered personal bank loan), and supply of other services (e.g., distribution). In 1993, three new services were added to the ARDIS network: national roaming, a fourfold increase in data transmission rate (which went from 4.8 Kbps to 19.2 Kbps in several cities), and the setting up of satellite bridges and connections for interoperability with other networks. In 1994, ARDIS introduced a real time, national bidirectional mail service called *personal messaging*. This service allows any PSTN user equipped with a DTMF phone to send a numeric or alphanumeric message to an ARDIS subscriber.

13.2.2.1 ARDIS Radio Interface

The ARDIS network is located in the 800-MHz frequency band. Access is made in FDMA mode and uses Gaussian Frequency Shift Keying (GFSK) modulation. The channels have a separation of 25 kHz and the duplex channel spacing is 45 MHz. ARDIS is based on two proprietary air-interface protocols: the MDC-4800 which works at 4,800 bps and is implemented over the whole ARDIS network, and the RD-LAP protocol which is at 19,200 bps, for a useful bit rate of 8 Kbps, and was implemented in 1999 in a few service areas.

The power of a base station is 40W, which covers 15 to 25 km. The mobile has a 4W transmitter. The RD-LAP version of ARDIS uses a 4-ary FSK modulation. The packets have a maximum size of 256 bytes. Access to the channel is achieved by means of a DSMA based protocol.

Only packet data transmission is offered. The network is designed to give indoor coverage, which puts a high constraint on the frequency reuse mechanism. To ensure indoor coverage, ARDIS transmits the same message from several base stations, as against the RAM network (see Section 13.2.3.4), where a transmission is only carried out from a single base station [5].

13.2.2.2 ARDIS Architecture

The ARDIS network includes leased lines for communications between network controllers and routers connected to the TYMNET and TELENET value-added networks (Figure 13.14). Leased lines are used for the connection of the ARDIS network to users' host computers. The four network control centers provide access to the users' central sites over various types of connection (e.g., asynchronous, X25).

The Radio Frequency Network Control Processor (RF/NCP) is a network component that manages the radio resources. This entity chooses the best BS to serve each communicating mobile terminal. In order to do this it evaluates the signal strength received from a mobile terminal at all base stations that can hear it. It then selects the best BS to communicate with that mobile terminal and sends the next message via that BS.

The RF/NCP is connected to a Message Switch (MS) which represents the heart of the ARDIS network. The Message Switch routes messages to their destination, stores subscriber information, and initiates accounting and billing. It can also be an access point for host computers and thus has to be capable of protocol conversion. ARDIS supports several standard communications protocols (e.g., SNA 3270, X25, synchronous, asynchronous) which allows direct connection of customers' computers to the network.

13.2.3 The Mobitex System

The Mobitex system was developed by Ericsson and Swedish Telecom [4]. The first Mobitex network was opened in 1986 in Sweden. Norway, Finland, Canada, Great Britain, and the United States (the latter with RAM Mobile Data, which opened in 1990) have all installed Mobitex networks. At the end of the 1990s there were Mobitex networks in about 20 countries in the world over the 5 continents.

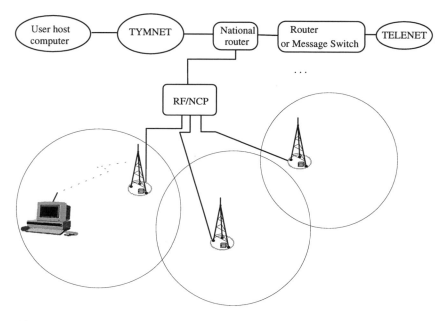

Figure 13.14 Structure of the ARDIS network.

Initially designed for carrying voice and data at 1,200 bps most of the Mobitex networks are now only used for data transmission services operating at 8 Kbps. Data are transmitted in a packet transmission mode (a packet contains a maximum of 512 user bytes). Optimization of radio channels is obtained by using a trunking method [12]. The Mobitex specifications are in the public domain and are based on the OSI model.

12.2.3.1 Mobitex Radio Interface

In the United States Mobitex uses the 900-MHz band. The reverse (or uplink) channel uses frequencies between 896 and 901 MHz, and the forward (or downlink) channel uses frequencies between 935 and 940 MHz. In the rest of the world most systems are in the 410–450-MHz band. The channel separation is 12.5 kHz and the bit rate is 8 Kbps. The modulation used is GMSK. Both ARQ and FEC mechanisms protect transmission.

Each base station manages several FDMA channels (16 maximum) each operating in TDMA mode. Communication is semi-duplex. A mobile attempts access after scanning several base stations and selecting the one which has the best error rate and is the most powerful.

Power control is used for both high power (i.e., 100 mW to 10W) and low-power mobiles (i.e., 100 mW to 4W).

The basic data packet is known as an MPAK (Mobitex PAcKet) which consists of up to 512 octets of user data (Figure 13.15). Three different kinds of data packet are defined: *text*, *data*, or *status*. A status MPAK has just one octet of data, which allows up to 256 possible different status codes. For transmission over the air interface the MPAK is mapped into blocks which have error correction and detection properties. The blocks are preceded by a frame header, which only has a parity check. The first or *primary block* contains the first 12 bytes of the MPAK with subsequent blocks (*following blocks*) containing the next 18 bytes and so on up to a maximum of 31 *following blocks*. This is greater than required for transmission of the MPAK but allows for additional address blocks when the data are destined for several users. Block interleaving is used to spread the errors within the block to assist the (12,8) Hamming code in the correction of burst errors.

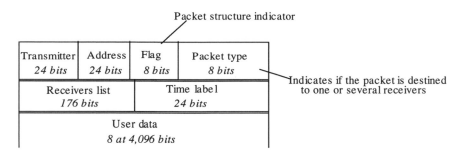

Figure 13.15 Structure of MPAK packets (545 bytes) in RAM.

The packet transmission mode achieves handover without loss of information. Packets that may be lost (i.e., those for which no acknowledgment has been received by the network) are retransmitted automatically by the network. Handover is instigated by the mobiles and not by the network (as is the case with current cellular networks).

13.2.3.2 Mobitex Architecture

The structure of a Mobitex network contains three functional levels:

- A radio layer for the transmission of information between the mobile users and the infrastructure. This layer supervises the terminals and manages the registration and automatic location procedures.
- An interconnection layer ensures the transport of data from the radio layer between various mobile terminals or between mobile terminals and fixed terminals. This

layer also supports connection over different wired transmission transports, such as a packet switched data network.
- A management and supervision layer controls the operation and maintenance of the network as well as management of users.

Message switching is achieved at the lowest possible level. This allows:

- Data exchanges between two mobile users inside the same cell only involving the local base station,
- A base station to operate in autonomous mode even when the link between it and its local switch is lost.

13.2.3.3 Mobitex Access

Upon arrival in a cell a mobile must first acquire frame synchronization before attempting to access the network. It waits for an <FRI> frame in which network and access parameters are transmitted, particularly the positions and lengths of the access slots (Figure 13.16). A multiframe contains a cycle of slots limited by two successive <FRI> frames [13]. Channel access uses a dynamic S-ALOHA access procedure.

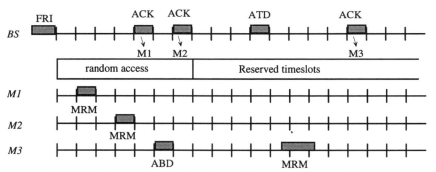

FRI: Defines the length of the timeslots and defines the slots reserved for random access
MRM: Transports the packets at the network level
ACK: Frame acknowledgment
ABD: Long frame transmission request
ATD: Answer to ADB: Authorization of MRM frame transmission

Figure 13.16 Transmission on Mobitex.

The load on a base station channel can be reduced by migrating the mobiles on to other channels in the same cell or to neighboring cells.

13.2.3.4 Example of Mobitex Networks: RAM and MOBIPAC

The U.S. RAM Mobile Data Network was installed during 1989 and expanded over several years. It contains more than 900 base stations with 40 switches covering the continental United States and Hawaii. It covers more than 90% of the professional U.S. population and offers roaming between the different service areas. Figure 13.17 represents its architecture.

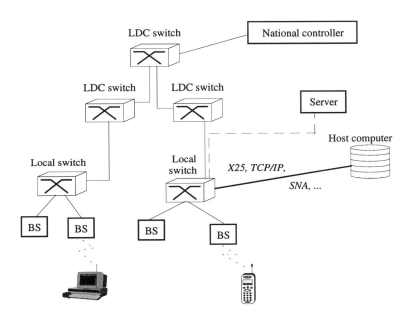

Figure 13.17 Structure of the RAM network.

Shown below is another example of a Mobitex system implementation, the Mobipac network of France Telecom (Figure 13.18). This network was opened in 1993 to commercial traffic and closed in 1996 due to an insufficient customer base. Very similar networks are running successfully in the United Kingdom, Netherlands, and Australia.

The radio layer is managed by base stations. Each base station transmits information about its service area. Besides their radio transmission function, the base stations also act as local exchanges. Thus, for two mobile terminals located in the same cell and likely to communicate with each other a connection can be made directly at the base site without the need for connection to the rest of the network. The base station reports internal activity to the local exchange (MOX) which passes it up to the Network Control Center (NCC) for billing purposes.

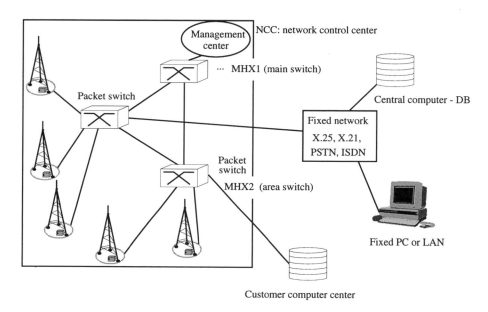

Figure 13.18 Example of architecture of a Mobitex network (e.g., the Mobipac network).

The interconnection layer is distributed on two levels. The first level includes local or regional switches (MOXs) and routes messages between the different base stations it supervises. It also has ports for connection to fixed terminals. For communication between terminals linked to base stations connected to different regional switches, a national switch (MHX) is used as a second level of interconnection. The national switch is a transit center gathering routing information on all mobiles in the network. It automatically ensures location management. The national switch is linked to the operations and maintenance center of the overall network, the NCC.

Each regional switch is able to manage various kinds of connection for directly wired terminals. These had access to the Mobipac network through either leased lines or an X25 network. Users had direct access both to their computer centers and databases and to their own company LANs.

13.2.4 Cellular Digital Packet Data

The cellular digital packet data (CDPD) technology was developed in 1992 by 10 U.S. cellular operators (Ameritech, Bell Atlantic, GTE, IBM, McCaw, Nynex, Pacific Bell, Sears Technology Services, Southwestern Bell, and US West) and IBM. It is a packet data transmission cellular network standard. At the end of the 1990s CDPD networks covered 195 markets in the U.S., and about 53% of the U.S. population.

The origin of this initiative comes from the determination of the cellular operators to offer data services to their customers (who were around 10 million in 1992 and more than 20 million by 1997). The main constraint was that this service had to be independent from the available wireline data transmission networks. The originality of CDPD technology is that it uses AMPS frequencies that are not carrying voice traffic. These channels are shared between the AMPS analog networks (see Chapter 12) and the CDPD networks, with a priority given to AMPS voice services.

CDPD enables transmission of data in a packet mode in the 800-MHz band at a bit rate of 19.2 Kbps. The main advantage of frequency sharing is to minimize infrastructure costs as the AMPS sites were already installed. The first networks were deployed by the end of 1993 and commercial service launched in 1994. By 1996 the CDPD service covered over fifty metropolitan areas in the United States. The applications targeted for CDPD networks services [4] include electronic mail, parcel delivery recording, inventory control, credit card checking, security, weather information services, and road traffic information.

13.2.4.1 CDPD Basic Principle

The basic idea consists of deploying the CDPD network by using AMPS network software and hardware. The CDPD system transmits data packets on the cellular networks voice channels when they are not used such as during call set up. This technology has several advantages, such as transmission at a bit rate four times higher than that of other similar services, a flexible architecture, and the possibility of carrying data and fax. A significant advantage with the CDPD concept is that, because this service is implemented within an available network, the installation of a CDPD network costs far less than any other similar network (e.g., MOBITEX). Compared with cellular systems CDPD services offer a large number of technical advantages, such as:

- They are fast (response time for a database query less than 5 sec),

- They require no call establishment time, whereas a cellular system generally requires a communication establishment time of up to 30 sec,

- They are based on TCP/IP, simplifying communication with fixed networks (the Internet, in particular).

Classic security mechanisms such as authentication and ciphering are used. Moreover, because the data are transmitted on packet mode on different channels, eavesdropping is virtually impossible and almost perfect confidentiality is ensured for all communications.

13.2.4.2 CDPD Radio Interface

The channels used by the CDPD networks are the same as those of the AMPS networks (i.e., their separation is 30 kHz). GMSK modulation is used.

The transmission gross bit rate is 19.2 Kbps which corresponds, once the message overheads and the correcting codes have been eliminated, to a user bit rate of between 9 and 13 Kbps (Figure 13.19). The transmitted blocks have 378 bits (or 63 symbols) of coded data using a Reed-Solomon code to which 42 control bits are added making a total of 420 bits transmitted every 21.875 ms (producing a bit rate of 19.2 Kbps).

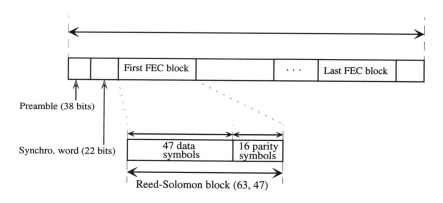

Figure 13.19 Structure of a CDPD data block.

The 42 control bits are distributed as seven flags of six bits each in the packet (Figure 13.20). Frames of various sizes can be carried in several blocks if their size is greater than 35 bytes. If not, they are gathered into one block and transmitted. A color code is included at the beginning of each packet. Color codes have the same function as the AMPS SATs. Yet, in addition to their co-channel interference detection function, the CDPD color codes are used by mobiles when crossing areas. Cells belonging to the same cluster have the same color code called a *group color code*. The cells belonging to the same location area also use the same area color code.

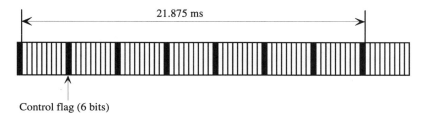

Control flag (6 bits)

Figure 13.20 Structure of a CDPD block on the forward link.

13.2.4.3 CDPD Network Architecture

A CDPD network consists of several network entities (Figures 13.21 and 13.22). A mobile end-system (M-ES), which is a computer equipped with a radio modem, has access to the CDPD network over the radio interface (*A* interface).

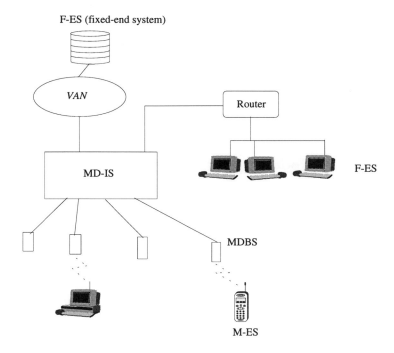

Figure 13.21 Network architecture of a CDPD system.

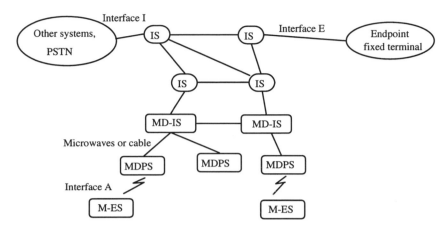

Figure 13.22 CDPD system architecture.

Each M-ES has one or several network entity identifiers (NEIs) which can be a connectionless network protocol (OSI CNLP) address or an Internet protocol (IP) address used for routing of the packets. The CDPD specification also defines the equipment (host computers mainly) that is connected to the CDPD network through wired connections. These are called fixed end systems (F-ES). The network entity that manages the routing is the mobile data intermediate system (MD-IS). An MD-IS ensures the following functions:

- Mobile home function (MHF), provided by the service operator in the user's home domain. It is similar to the HLR in GSM.
- Mobile serving function (MSF), provided by the MD-IS where the mobile station is operating. It is similar to the VLR in GSM [14].

The M-ES identifies itself to the network by sending its network equipment identifier (NEI). When an M-ES moves between cells, it registers at the new MD-IS using the mobile network registration protocol (MNRP). The visited MD-IS informs the home MD-IS via the mobile network location protocol (MNLP). Prior registration and authentication is required before a session can be started.

A CDPD network can be composed of a single network run by a single service provider or as a network joining several service provider domains (Figures 13.23 and 13.24).

When CDPD networks are interconnected, some routers and intermediate networks are used [15]. The CDPD standard defines a random access protocol called digital sense multiple access (DSMA), similar to the CSMA/CD. The choice of channel is made by the processing entity of the radio resource. Hence, each Mobile Data Base Station (MDBS) maintains a list of free channels (i.e., those not in use for voice traffic).

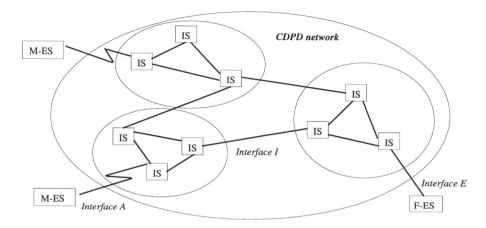

Figure 13.23 CDPD network made with one service provider only.

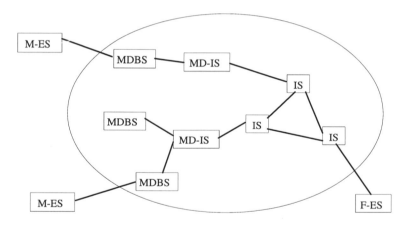

Figure 13.24 CDPD network made of several service providers.

13.2.4.4 Sharing the Frequency Groups Between AMPS and CDPD

The deployment of a CDPD network is realized in areas where an AMPS network is already in operation. AMPS networks have established frequency plans that could, nevertheless, change as the network changes. Frequency allocation in the CDPD network is based on the host AMPS network frequencies and can be achieved in two different ways:

- By frequency sharing, where the frequencies are used by both systems (AMPS and CDPD). The CDPD system uses mechanisms that are able to detect the use of these frequencies by the AMPS system. In the case when the CDPD system is using a radio channel needed by the AMPS host network, the CDPD network has to release it within 40 ms.
- By dedicating frequencies, where if one or more frequencies are available and not in use by the AMPS network in a given area, they can be allocated to the CDPD network. These frequencies will be used exclusively by the CDPD network.

When sharing frequencies with AMPS, CDPD must free the occupied channel whenever a voice communication is set up. The following procedure is used. Each MDBS monitors the voice channels by sensing the power entering its antenna. If a power ramp is detected, the channel relinquishment procedure is triggered. The MDBS sends a special signal to close down the channel within 40 ms after power ramp up detection. This signal either indicates the new CDPD channel, if it knows it, or the MDBS finds a new idle voice channel and starts transmitting an identification signal on it. The mobile terminal hops to the new CDPD channel if told by the MDBS or in the second case, searches for the new one.

Some effort has been made to ensure that the cell borders of AMPS and CDPD networks coincide with one another to avoid co-channel interference in the different cell groups (see Section 6.2.4). The AMPS and CDPD transmission levels are adjusted initially so that the borders correspond with each other.

13.2.5 General Packet Radio Service

The first General Packet Radio Service (GPRS) specifications were published at the end of 1997 after work in ETSI, which had started in 1994. The main objective of GPRS is to provide users with a mobile packet data interface that can interconnect with other data networks. This interface is normally embedded in a GSM network that was originally designed for circuit-switched services. The important aspect of GPRS is its total compatibility with GSM architecture. GPRS can be considered, similar to CDPD and the AMPS networks, as a data network that shares the GSM radio access infrastructure. GSM circuit-switched and GPRS services have to coexist in the same environment with minimum change to the system. Radio resources are allocated dependent on the requirements of the GPRS subscriber. In this section are presented the main principles of GPRS. Protocol layering and procedural details are not presented here. For more details, the reader can refer to [16–20].

Note: In the remainder of this section, the term "GSM services," or simply "GSM," relate to the services which are independent of GPRS (e.g., voice calls, SMS, circuit-switched data transfer), and "GPRS" relates to all packet data services.

13.2.5.1 GPRS Services and Mobile Stations

GPRS bit rates are sufficient to support applications such as Web surfing, compressed video transmission, and File Transfer Protocol (FTP).

GPRS services are classified into 2 main groups:

- Point-To-Point (PTP) services which can be either connection oriented (PTP-CONS, PTP Connection Oriented Network Service) for connection to an IP network for instance, or connectionless oriented (PTP-CLNS, PTP Connectionless Oriented Network Service) for connection to a packet data network (PDN) such as X.25. PTP services are available in the first phases of GPRS rollout.
- Point-To-Multipoint (PTM) services include multicast services (PTM-M, Point-To-Multipoint Multicast) or group services (PTM-G, Point-To-Multipoint Group). PTM services will be available as a second step.

GPRS terminals are able to support various network protocols. A GPRS mobile station can operate in one of the three following modes [16]:

- *Class A*: a mobile station that can use both GPRS and GSM services and can support GPRS and GSM services simultaneously.
- *Class B*: a mobile station which can use both GPRS and GSM services but can only operate one type of services at a time, either GPRS or GSM.
- *Class C*: a mobile station that has access to GPRS services only.

A GPRS user can open several simultaneous network sessions with his or her terminal. In order to do this the *Packet Data Protocol* (PDP) has been defined. It is supported by a set of information stored in the mobile, the SGSN (*Service GPRS Support Node,* see Section 13.2.5.3) and the GGSN (*Gateway GPRS Support Node,* see Section 13.2.5.3) which enables data exchange with a packet data network. The corresponding PDP context is characterized by the following information:

- PDP network type (e.g., IP, X.25).
- Terminal PDP address (e.g., IP address). This address can be allocated dynamically during the PDP session.
- Current SGSN IP address.
- Network Service Access Point Identifier (NSAPI).
- Quality of Service.

13.2.5.2 GPRS Radio Interface

GPRS channels are mapped on to the same physical channels as GSM. The logical channels, which are defined for GPRS, are very close to those in GSM. Three main types of channels are defined (see classification on Figure 13.25):

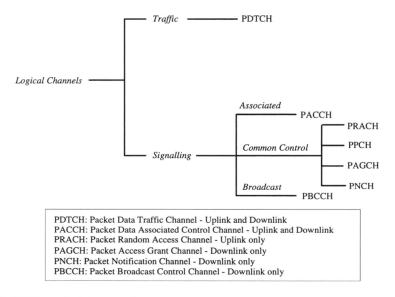

PDTCH: Packet Data Traffic Channel - Uplink and Downlink
PACCH: Packet Data Associated Control Channel - Uplink and Downlink
PRACH: Packet Random Access Channel - Uplink only
PAGCH: Packet Access Grant Channel - Downlink only
PNCH: Packet Notification Channel - Downlink only
PBCCH: Packet Broadcast Control Channel - Downlink only

Figure 13.25 GPRS logical channel definitions.

- Packet Common Control Channel (PCCCH),
- Packet Broadcast Control Channel (PBCCH),
- Packet Traffic Channel (PTCH).

PRACH is an access channel used by GPRS mobiles to access the network and request a channel.

PPCH is the paging channel to send messages to mobiles. A mobile involved in a packet transfer and which receives a circuit-switched call (e.g., a GSM voice call) is paged on the PACCH (see hereafter).

PAGCH is a channel on which channel allocation messages are transmitted to mobiles that have sent a channel request on the PRACH. When a mobile is already involved in a packet transfer the channel allocation message is sent on the PACCH.

PNCH is used to inform a group of mobiles of a PTMM (Packet Transfer Mode Message) transmission.

PBCCH is a broadcast channel that conveys *System Information* messages. In the case when the PBCCH is not present, the GPRS network specific information is transmitted on the GSM BCCH (see Chapter 12).

PDTCH is a traffic channel. Several Packet Data Traffic Channels can be allocated to a single mobile for multislot operation, which allows instantaneous bit rates of up to about 170 Kbps. PACCH is an associated channel that is used to transmit signaling (e.g., paging, channel assignment, power control commands).

The physical channels dedicated to GPRS traffic are called Packet Data Channels (PDCH). At least one PDCH acts as a master and carries the control signaling (through Packet Common Control Channels) and traffic with dedicated signaling (through Packet Traffic Channels). This type of channel is called the Master Packet Data Channel (MPDCH). Other PDCHs, acting as slaves, only carry PTCHs. These are called the Slave Packet Data Channels (SPDCH).

Four coding schemes are defined for transmission over the radio path [17]. In Table 13.3, the four types of coding scheme are detailed.

Table 13.3
GPRS coding schemes (extract from [17])

Coding scheme	Code rate	Data rate (Kbps)
CS-1	1/2	9.05
CS-2	2/3	13.4
CS-3	3/4	15.6
CS-4	1	21.4

The mapping of GPRS logical channels on to the physical channels (i.e., time slots in the multiframe GSM structure) can alter dynamically, as opposed to circuit-switched GSM where the mapping can only be changed by the operator and remains static during operation. A cell that supports GPRS can instantaneously have GPRS radio resources allocated. A physical channel can either be allocated to GPRS or to GSM services. This allocation may vary rapidly with network load. A slot allocated to a PTCH that consists of a PDTCH and a PACCH, may, at the end of the communication, be allocated to a TCH/SACCH for a GSM voice call.

13.2.5.3 GPRS Network Architecture

The GPRS network architecture is based on GSM system architecture. It is presented on Figure 13.26.

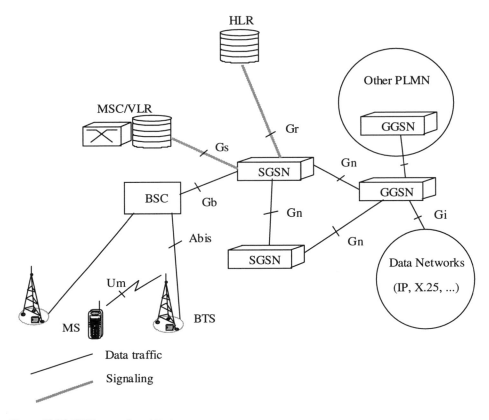

Figure 13.26 GPRS network architecture.

The *Service GPRS Support Node* (SGSN) is a packet router, which includes additional mobility management functions. It is connected to one or several BSSs and is responsible for the delivery of packets to mobile stations that are located in its service area. The *Gateway GPRS Support Node* (GGSN) is a node that acts as a logical interface to external data networks. Location management is controlled by the HLR as in GSM. Additional information is stored in the HLR, which is the relationship between the PDP address and the IMSI.

GPRS uses specific identities that differ from the GSM identities. A GPRS mobile station can be allocated a *Packet Temporary Mobile Station Identity* (P-TMSI) similar to the TMSI (see Chapters 5 and 12). Class A terminals can be allocated a TMSI and a P-TMSI.

When accessing the network, a mobile with no P-TMSI will use a *Temporary Link Layer Identity* (TLLI). This TLLI is a random number that will be replaced by the P-TMSI once the mobile has been attached to the network (i.e., after having processed the *Attach* procedure).

13.2.5.4 GPRS Main Functions

Packet routing. Two different encapsulation schemes are defined in GPRS (Figure 13.27). The first is used for transmissions between GSNs (SGSN or GGSN). This type of encapsulation is by means of a common tunnel protocol allowing different packet data protocols to be used even if they are not supported by the SGSNs. The second type of encapsulation occurs between the mobile and the SGSN, and is used to decouple the logical link management from the network-layer protocols.

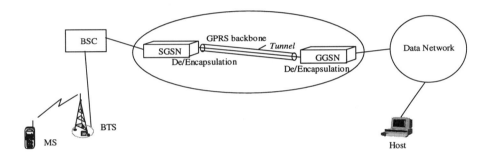

Figure 13.27 GPRS packet routing.

Mobility management (Figure 13.28). Mobility management in GPRS involves Routing Areas (RAs). RAs are defined as groups of cells, the same as location areas (see Chapter 8). When in a *Ready* state (see following paragraph), the mobile updates its SGSN at each cell change. When in the *Standby* state, the mobile updates its location at each RA change. When the mobile changes its SGSN, the new SGSN makes a request to the old SGSN to send the mobile mobility management information. The old SGSN then deletes the mobile's MM information. In addition, periodic RA updates can be implemented. Inter/Intra SGSN RA updates can either be combined with LA updates or performed alone. There are three MM defined states: *Idle, Standby,* and *Ready.* In Table 13.4, the services and procedures that the mobile can perform in each state are shown.

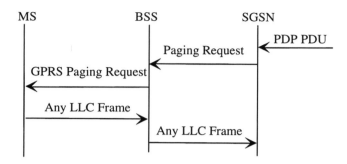

Figure 13.28 Paging procedure in GPRS (extract from Figure 32, [16]).

Table 13.4

Mobility management states

State	Allowed services or procedures
Idle	- PLMN selection
	- GPRS cell selection/re-selection
	- PTM-M
Standby	- PLMN selection
	- GPRS cell selection/re-selection
	- PTM-M
	- PTM-G
	- RA update
	- Paging for PTP or PTM-G
Ready	- PLMN selection
	- GPRS cell selection/re-selection
	- PTM-M
	- PTM-G
	- RA update
	- Paging for PTP or PTM-G
	- Cell updating
	- PTP transmission and reception

On Figure 13.29, the transitions between the three states are represented.

Mobility Management State Model of MS

Figure 13.29 Mobility management transition states (extract from Figure 12, [16]).

Security functions. GPRS includes some security functions, which are:

- *Subscriber authentication* which is implemented in the same way as in GSM. In GPRS, the SGSN acts in the role of the VLR.
- *User identity confidentiality* is ensured by means of the TLLI.
- *P-TMSI signature* is used when the MS is attached to the network and acts in a similar way to a GSM TMSI for user location confidentiality (see Chapter 5).
- *Ciphering* occurs from the MS to the SGSN and not just over the radio path as in GSM.

13.2.5.5 GPRS, a Path Between Second- and Third-Generation Systems

With throughputs of more than 100 Kbps, GPRS enhances applications such as messaging and Internet access. It is a good migration path towards the 2 Mbps (see Chapter 14) speed of third-generation wireless systems. First GPRS applications were implemented in GSM networks at the end of 1999.

13.3 CONCLUSIONS

Mobile data networks with local or wide area coverage are part of the mobile radio scene but have shown the lowest growth. Yet the boom in e-mail, data services, and Internet access together with significant growth in the population of cellular subscribers should accelerate the growth of data services. Besides which, with the emergence of new services such as information access, security, location, and remote banking, an impact on requests for wireless data services should be expected. On the other hand, programs such as the intelligent vehicle highway system (IVHS) or road transport informatics (RTI) with the aim of providing drivers with information and assistance should also enhance the growth of mobile data transmission systems.

In Table 13.5, the various systems examined in this chapter are shown with a comparison of their characteristics.

Table 13.5

Comparison of different systems dedicated to data transmission

	Wireless LANs	*Wireless wide area data networks*
Coverage	Indoor, campus	City, region, country
Offered mobility	Low (some tens to some hundreds of meters)	Extended (several hundred to several thousands of kilometers)
Throughput	High (less than 10 Mbps)	Low (less than 20 Kbps)
Typical applications	Data transmission among PCs	Remote connections to central sites (e.g., database queries)
Standards	HIPERLAN, IEEE 802.11, proprietary standards	ARDIS, MOBITEX, CDPD, GPRS

REFERENCES

[1]   LaMaire, R.O., A. Krishna, P. Bhagwat, and J. Panian, "Wireless LANs and Mobile Networking: Standards and Future Directions," *IEEE Communications Magazine*, pp. 86–94, Aug. 1996.

[2]   Newman, D., "Wireless LANs: How Far? How Fast?" *Data Communications*, pp.77–86, Mar. 21, 1995.

[3]   Cox, D., "Wireless Personal Communications: What Is It," *IEEE Personal Communications*, pp. 20–35, Apr. 1995.

[4]   Pahlavan, K., and A. H. Levesque, *Wireless Information Networks*, Wiley & Sons, New York, 1995.

[5]   Muller, N.J., *Wireless Data Networking*, Artech House, 1995.

[6]   Chelouche, M., S. Héthuin, and L. Ramel, "Digital Wireless Broadband Corporate and Private Networks: RNET Concepts and Applications," *IEEE Communications Magazine*, pp. 42–51, Jan. 1997.

[7]   Falsafi, A., K. Pahlavan, and G. Yang, "Transmission Techniques for Radio LANs - A Comparative Performance Evaluation Using Ray Tracing," *IEEE Journal on Selected Areas in Communications*, Vol. J-SAC 14, No. 3, pp. 477–491, Apr. 1996.

[8]   ETSI Radio Equipment and Systems (RES), "HIPERLAN - Services and Facilities," Vers. 1.0, Dec. 1992.

[9]   Johnson, I. R., T. A. Wilkinson, A. E. Jones, S.K. Barton, M. Li, A. Nix, I. Marvill, and M. Beach, "On Suitable Codes for Frame Synchronization in Packet Radio LANs," *IEEE Vehicular Technology Conference 1994*, Stockholm, Sweden, June 20–22, 1994.

[10] ETSI TC-RES/RES10, Radio Equipment and Systems (RES), "HIPERLAN Requirements and Architecture," ETSI-Server, work item RES 10-07, Jan. 1997.

[11] Das, B., and V. Bharghavan, "Routing in Ad Hoc Networks Using Minimum Connected Dominating Sets," *Proceedings of the International Conference on Communications*, Montréal, Canada, pp. 376–379, June 8-12, 1997.

[12] Juhen, P., and Y. Lossouarn, "MOBIPAC, Le 3RD de France Télécom Mobiles Data," *SIRCOM '93*, Dec. 1993.

[13] Stein, P., "La Communication de Données avec les Mobiles, Facteur d'Efficacité," *Ericsson Review No. 4*, pp. 104–110, 1991.

[14] Frankel, Y., A. Herzberg, P. A. Karger, H. Krawczyk, C. A. Kumzinger, and M. Yung, "Security Issues in a CDPD Wireless Network," *IEEE Personal Communications*, pp. 16–27, Aug. 1995.

[15] Sreetharan, M., and R. Kumar, *Cellular Digital Packet Data*, Artech House, 1996.

[16] ETSI, "General Packet Radio Service (GPRS), Service description, Stage 2," GSM 03.60, Version 6.3.1, Final draft 1999.

[17] ETSI, "General Packet Radio Service (GPRS), Overall description of the GPRS radio interface, Stage 2," GSM 03.64, Version 6.0.1, 1998.

SELECTED BIBLIOGRAPHY

[1]   ETSI, "General Packet Radio Service (GPRS), Service Description," GSM 02.60, 1997.

[2]   ETSI, "General Packet Radio Service (GPRS), GPRS Tunnelling Protocol (GTP) across the Gn and Gp interface," GSM 09.60, 1997.

[3]   ETSI, "General Packet Radio Service (GPRS), Serving GPRS Support Node (SGSN) - Visitors Location Register (VLR); Gs interface network service specification," GSM 09.64, 1997.

ABOUT THE AUTHOR

Sami Tabbane received a B.S. degree in Computer Science from the Université d'Orsay, Paris, in 1986 and graduated from the École des Mines de Paris engineering school in 1988. From 1988 to 1991, he worked on his Ph.D. in the field of networks and computer science at École Nationale Supérieure des Télécommunications (ENST) of Paris. During his Ph.D., he worked within the European PROMETHEUS project on the medium access control (MAC) protocols and on TDMA slot synchronization algorithms. He started his work on location management during his visit to WINLAB (Rutgers University, NJ) in 1992. When he joined the Centre National d'Etudes des Télécommunications (CNET) in Issy-les-Moulineaux, he worked on third-generation systems within the RACE European project focusing on mobility aspects (handover and location management). Since September 1994, he has worked with the École Supérieure des Communications de Tunis engineering school, where he researches and lectures on mobile systems. He is involved in several government commissions dealing with telecommunications. He is the co-author of the book *Les réseaux GSM/DCS* and the author of *Réseaux Mobiles*. He is a member of the IEEE.

INDEX

Wireless Technicians's Handbook, Andrew Miceli

For further information on these and other Artech House titles, including previously considered out-of-print books now available through our In-Print-Forever® (IPF®) program, contact:

Artech House
685 Canton Street
Norwood, MA 02062
Phone: 781-769-9750
Fax: 781-769-6334
e-mail: artech@artechhouse.com

Artech House
46 Gillingham Street
London SW1V 1AH UK
Phone: +44 (0)20 7596-8750
Fax: +44 (0)20 7630-0166
e-mail: artech-uk@artechhouse.com

Find us on the World Wide Web at:
www.artechhouse.com